Lecture Notes in Computer Science   14633

Founding Editors

Gerhard Goos
Juris Hartmanis

The series Lecture Notes in Computer Science (LNCS), including its subseries Lecture Notes in Artificial Intelligence (LNAI) and Lecture Notes in Bioinformatics (LNBI), has established itself as a medium for the publication of new developments in computer science and information technology research, teaching, and education.

LNCS enjoys close cooperation with the computer science R & D community, the series counts many renowned academics among its volume editors and paper authors, and collaborates with prestigious societies. Its mission is to serve this international community by providing an invaluable service, mainly focused on the publication of conference and workshop proceedings and postproceedings. LNCS commenced publication in 1973.

Colin Johnson · Sérgio M. Rebelo · Iria Santos
Editors

Artificial Intelligence in Music, Sound, Art and Design

13th International Conference, EvoMUSART 2024
Held as Part of EvoStar 2024
Aberystwyth, UK, April 3–5, 2024
Proceedings

Springer

Editors
Colin Johnson ⓘ
University of Nottingham
Nottingham, UK

Sérgio M. Rebelo ⓘ
University of Coimbra
Coimbra, Portugal

Iria Santos ⓘ
University of Coruña
Coruña, Spain

ISSN 0302-9743   ISSN 1611-3349 (electronic)
Lecture Notes in Computer Science
ISBN 978-3-031-56991-3   ISBN 978-3-031-56992-0 (eBook)
https://doi.org/10.1007/978-3-031-56992-0

This Springer imprint is published by the registered company Springer Nature Switzerland AG
The registered company address is: Gewerbestrasse 11, 6330 Cham, Switzerland

Paper in this product is recyclable.

Preface

This volume contains the proceedings of EvoMUSART 2024, the 13th International Conference on Artificial Intelligence in Music, Sound, Art and Design. The conference was part of Evo*, the leading event on bio-inspired computation in Europe, and was held in Aberystwyth, UK, as a hybrid event, between Wednesday, 3rd April, and Friday, 5th April 2024.

While the utilisation of Artificial Intelligence for artistic purposes can be traced back to the 1970s, the use of Artificial Intelligence for the development of artistic systems is a recent, trending, and significant area of research. There is a growing interest in the application of these techniques in fields such as visual art and music generation, analysis and interpretation; sound synthesis; architecture; video; poetry; design; and other creative tasks.

The main purpose of EvoMUSART 2024 was to bring together practitioners who are using Artificial Intelligence techniques for artistic tasks, providing the opportunity to promote, present, and discuss ongoing work in the area. As always, we strove to foster a fun, inclusive, welcoming, and constructive environment that encouraged networking and contribution.

EvoMUSART has grown steadily since its first edition in 2003 in Essex, UK, when it was one of the Applications of Evolutionary Computing workshops. Since 2012, it has been a full conference as part of the Evo* co-located events. At the same time, under the Evo* umbrella, EvoApplications targeted research on the application of bio-inspired computing, EuroGP focused on the technique of genetic programming, and EvoCOP was dedicated to evolutionary computation in combinatorial optimisation. The proceedings for these co-located events are available in the LNCS series.

EvoMUSART 2024 received 55 submissions. The peer-review process was rigorous and double-blind. The Programme Committee, listed below, was composed of 59 members from 20 countries. We selected 17 of these papers for long oral presentations (31% acceptance rate), while 8 works were presented in short oral presentations and as posters, meaning an overall acceptance rate of 45%.

As always, the EvoMUSART proceedings cover a wide range of topics and application areas, including generative approaches to music, visual art, and design. This volume of proceedings collects the accepted papers. As in previous years, the standard of submissions was high and good-quality papers had to be rejected. We thank all authors for submitting their work, including those whose work was not accepted for presentation on this occasion.

The work of reviewing is done voluntarily and generally with little official recognition from the institutions where reviewers are employed. Nevertheless, professional reviewing is essential to a healthy conference. Therefore, we particularly thank the members of the Programme Committee for their hard work and professionalism in providing constructive and fair reviews.

Evo* 2024, of which EvoMUSART 2024 was part, would not have been possible without the substantial contributions of a large group of people. We are grateful to the support provided by SPECIES, the Society for the Promotion of Evolutionary Computation in Europe and its Surroundings, for the coordination and financial administration.

We would also like to thank Nuno Lourenço (University of Coimbra, Portugal) for his dedicated work as Submission System Coordinator. We thanks the Evo Graphic Identity Team, Sérgio M. Rebelo, Jéssica Parente and João Correia (University of Coimbra, Portugal), for their dedication and excellence in graphic design. We are grateful to Zakaria Abdelmoiz (University of Malaga, Spain) and João Correia (University of Coimbra, Portugal) for their impressive work managing and maintaining the Evo* website and handling the publicity, respectively.

We credit the invited keynote speakers, Jon Timmis (Aberystwyth University, UK) and Sabine Hauert (University of Bristol, UK), for their fascinating and inspiring presentations. We would like to express our gratitude to the Steering Committee of EvoMUSART for their help, advice, support and supervision. In particular, we thank Juan Romero (University of A Coruña, Spain) and Penousal Machado (University of Coimbra, Portugal).

We extend special gratitude to Christine Zarges (Aberystwyth University, UK) for her role as a local organiser, to Aberystwyth University, UK to Sophie Bennett-Gillison, and to the Aberystwyth Arts Centre for organising and providing an enriching conference venue.

Finally, we would like to express our continued appreciation to Anna I. Esparcia-Alcázar, from SPECIES, Europe, whose considerable efforts in managing and coordinating Evo* helped build a unique, vibrant, and friendly atmosphere.

April 2024

Colin Johnson
Sérgio M. Rebelo
Iria Santos

Organization

Organising Committee

Conference Chairs

Colin Johnson   University of Nottingham, UK
Sérgio M. Rebelo   University of Coimbra, Portugal

Publication Chair

Iria Santos   University of A Coruña, Spain

Program Committee

Gilberto Bernardes   University of Porto, Portugal
Daniel Bisig   Zurich University of the Arts, Switzerland
Oliver Bown   University of New South Wales, Australia
Jean-Pierre Briot   LIP6, Sorbonne University – CNRS, France
Marcelo Caetano   CNRS-PRISM, France
F. Amílcar Cardoso   University of Coimbra, Portugal
Miguel Angel Casal Santiago   photoilike, Spain
Vic Ciesielski   RMIT University, Australia
João Correia   University of Coimbra, Portugal
Camilo Cruz Gambardella   Monash University, Australia
João Miguel Cunha   University of Coimbra, Portugal
Hans Dehlinger   University of Kassel, Germany
Georgios Diapoulis   Chalmers University of Technology, Sweden
Jonathan E. Rowe   University of Birmingham, UK
Edward Easton   Aston University, UK
José Fornari   UNICAMP, Brazil
Björn Gambäck   Norwegian University of Science and Technology, Norway
Pablo Gervás   Complutense University of Madrid, Spain
Carlos Grilo   Polytechnic Institute of Leiria, Portugal
Varvara Guljajeva   Hong Kong University of Science and Technology (Guangzhou), China

Contents

The Forest: Towards Emergent Collaborative Art Through Human Swarming

Razanne Abu-Aisheh, Khulud Alharthi, Tom Didiot-Cook, Henry Hickson,
Suet Lee$^{(\boxtimes)}$, Mickey Li, Avgi Stavrou, Georgios Tzoumas, and Sabine Hauert

Bristol Robotics Laboratory, University of Bristol, University of The West
of England, Bristol BS16 1QY, England
suet.lee@bristol.ac.uk

Abstract. The Forest is a modular system of pillars inviting an audience to create an emergent audio-visual landscape through collective interactions over time. The pillars are designed to invoke reaction from participants through function and visual aesthetic, and act as a mediator for human-human collaboration–whether direct or indirect–towards an evolving piece of art to be experienced as a whole. In this paper, we show two variations of the Forest each with two pillars, which we demonstrated at public outreach events. The first variation has no explicit interaction rules whereas in the second variation, we introduce "human swarm" rules towards emergent patterns in the installation. We show that the "human swarm" rules are able to produce diverse emergent outputs through simulation. Finally, we present anecdotal observations of the outcomes from outreach events, where interaction feedback empowers individuals in their creative exploration to produce consensus art.

Keywords: Interactive Art · Emergent Art · Human Swarming

1 Introduction

Art encompasses a diverse array of human creative expressions and interpretations, ranging from visual arts to performing arts, literature, and architecture. Its connection with human creativity is profound, acting as a transformative process in which humans re-shape their environment, thus becoming magicians of their surroundings [8]. This act of transformation highlights our fundamental role as shapers of the world and establishes a deep-rooted connection between artistic expressions and our innate capacity for creativity.

Interactive art adds another layer to this connection by engaging the audience directly in the creative process. An early example is Allan Kaprow's Eat

---

R. Abu-Aisheh, K. Alharthi, T. Didiot-Cook, H. Hickson, S. Lee, M. Li, A. Stavrou
and G. Tzoumas—All authors contributed equally.

C. Johnson et al. (Eds.): EvoMUSART 2024, LNCS 14633, pp. 1–16, 2024.
https://doi.org/10.1007/978-3-031-56992-0_1

from 1965. Eat was an interactive environment designed in a way where people had the freedom to engage with it as they wish, becoming participants in the creation of the artwork themselves through individual decisions and interactions with the environment. Advances in digital technology have expanded these interactive possibilities, enabling artworks to respond in real time to participant interactions, thereby shifting the artist's role from creator to facilitator of creativity.

Swarm robotics introduces a decentralized approach to collaboration, drawing inspiration from natural systems such as ant colonies and bee swarms [6,20]. This concept extends to human collectives, where large-scale cooperation occurs amidst a backdrop of heterogeneity and complexity [7,15,16]. These complex interactions give rise to emergent social phenomena, forming the basis of our interest in exploring how an interactive system, governed by swarm algorithms, can shape the actions and experiences of individuals within a "human swarm."

This paper introduces "The Forest", a modular system of pillars designed to create an emergent audio-visual landscape through collective human interactions. The pillars serve as a medium for participants to influence and transform the artwork, fostering both direct and indirect collaborations. We aim to explore the dual role of individuals as both creators and audience, examining how the design of swarm algorithms and the interactive system can guide the emergent art.

We present a method for a modular design that enables distributed swarm-like interaction, fostering human-swarm behaviour to create a collaborative, generative artwork experience. Through selected parameter choices, we showcase two variations of a two-pillar design and analyze the dynamics of collective interaction over time. The objective is to provide insight into how the principles of swarm robotics and human collectives can be applied to create a collaborative art-making process, resulting in a unique and evolving piece of art (Fig. 1).

Fig. 1. A conceptual implementation of the Forest: tree "pillars" have two points of interaction each, visualized as green circles. Participants are free to walk among the pillars. Each pillar will have audio output and may also produce light output through LEDs. (Figure produced using graphics from *vecteezy.com*) (Color figure online)

2 Related Work

Generative art is an expansive field, encompassing artistic creations from music to visual images, where artists set autonomous systems in motion to contribute to the final artwork [10]. The concept extends beyond digital means, as seen in the rule-driven approaches of Kenneth Martin's abstract paintings and the chance-based compositions in music [11,14]. In the realm of swarm-inspired generative art, swarm algorithms have been leveraged to synthesize existing artwork or create new ones, drawing on the beauty of natural processes. These implementations have varied from ant colony-inspired music generation to interactive dance and architecture, where human input ranges from passive reception to the active design of the systems [4,5,18,19].

The role of human interaction in the generative art process is evolving, with works exploring active collaboration between humans and systems. Examples include swarm robots serving as a canvas for interactive painting [2,3], and installations such as Improbotics [12] and SwarmArenaNTT [17], where both artificial and human actors engage in a continuous feedback loop of creation and response. However, these instances primarily focus on human-system interactions rather than facilitating human-human collaboration within the generative process.

Our study is particularly interested in systems that not only involve human-system interaction but also enhance human-human collaboration, influenced by generative inputs from the system. Projects like Reactable [13] and Sound Forest [9] illustrate this by enabling participants to collaboratively compose music, revealing that social dynamics significantly influence the level of cooperation and the resulting creation. Other initiatives aim to foster positive social encounters through interactive urban installations, demonstrating that the nature of human interaction - whether following implicit or explicit rules–shapes the collective experience and emergent art. Building upon these ideas, our work seeks to integrate explicit rules of swarm algorithms to promote efficient collaboration and enrich both the participatory experience and the artistic output. We show two variations of the Forest which were tested in reality in public events: one with unlimited participants and the other with 3–6 participants. The first variation was designed to have no explicit interaction rules, whereas for the second we added rules to constrain and guide user interaction. These rules were tested in simulation to ensure diversity of emergent experiences in the installation. Finally, we share anecdotal descriptions and observations from real-life demonstrations which go towards validating our design choices to facilitate human swarming for collaborative art.

3 The Forest

The Forest has been created to explore human-swarm generative art. The concept revolves around the idea of multiple pillars or 'trees', through which participants can interact and from which an evolving audio-visual art piece will emerge.

Table 1. Design parameters for Variation 1 and Variation 2 of The Forest

Parameter	Variation 1	Variation 2
No. pillars	2	2
No. participants	No limit	3–6
Topology of pillars	Independent nodes, 2 m apart	Independent nodes, 2 m apart
No. interactions	4 points per pillar	6 points per pillar
Pillar Features & Functionality	Placing an object on a sensor, triggers a sound, removing the object stops the sound	Interaction points are capacitive touch sensors, touching these sensors will cycle through the colour of the tube and colours are mapped to fixed notes Notes play in sequence
Inter-connectivity of pillars	No connection	No connection
Interaction rules	Free Interaction	Rules are given to each participant

The Forest is an ecosystem in which components are modular and distributed: in our design, we aim to create an environment in which participants will interact both with each other and the system itself and through their interactions, will impact the ecosystem and be impacted in turn as part of a meaningful experience. The Forest as a whole then becomes a living, adaptive, and dynamic system - a collective composed of the individual nodes (the trees/pillars) and those who inhabit the space (the participants). The overall audio-visual-sensory experience of the Forest is the resultant output of the collaborations of the participants - the human swarm.

We consider the following parameters in the design of the system: (1) Number of pillars, (2) Number of users, (3) Topology of pillars, (4) Number of interaction points, (5) Pillar features and functionality, (6) Inter-connectivity of pillars, and (7) Interaction rules.

By varying each of these parameters, we vary the space of possibility for collective art to emerge. In particular, we are interested in how the design of interaction rules, whether implicit (e.g. through physical constraints or design choices) or explicit, will impact and shape the way users engage with the Forest towards an emergent outcome, much like individuals in a swarm.

3.1   Variation 1: Implicit Interactions

The first variation is based around implicit interactions, where participants can interact with the installation in any way they choose and are only constrained

Fig. 2. Nature (left) and Urban (right) pillar exploring implicit interaction

by the implied 'rules' of interaction built into the design of each pillar. This installation is themed around the idea of natural and human-built environments and features two pillars, one representing nature and the other urban city-scapes as shown in Fig. 2. The parameters selected for this variation are listed in Table 1.

Pillar Design. Each pillar is 2 m tall with a central aluminium extrusion, around which the two designs are based. The nature pillar features a series of laser-cut plywood panels, curved and wrapped in branches and leaves. The urban pillar features a series of laser-cut transparent acrylic, straight edged with hard corners to represent urban architecture. Each pillar is fitted with a Raspberry Pi with audio hat and a set of four speakers, distributed around the pillars at different heights (Fig. 3).

Each pillar is fitted with four RFID readers, each covered with a 100 mm diameter 3D printed white disc, making them clearly visible to users. Alongside

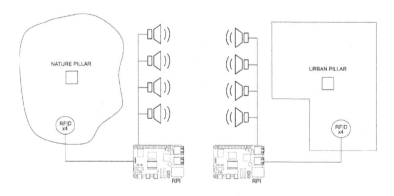

Fig. 3. Top down schematic of Variation 1

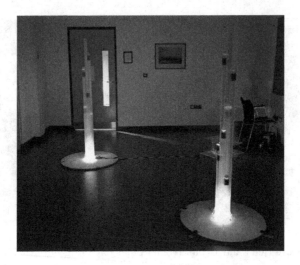

Fig. 4. The two pillar setup for variant 2 of The Forest

the pillars are a selection of objects, each of which has an RFID tag mounted to its base. The objects are designed to fit into the "nature versus urban" theme ranging from fire engine toys to pebbles. Each RFID tag has an associated audio sample that is related to the object itself. The RFID tags are processed by each pillar's Raspberry Pi and the relevant audio samples are selected and played through the pillar's speakers. The samples are played using the sonic-pi music generation tool [1].

Interaction Rules. Although we do not have explicit interaction rules or constraints in this variation, the choice of design parameters forms implicit constraints for the interactions available to participants. By adding obvious interaction points around the pillar, and designing the system such that there is an instant and audible response as soon as a participant places an object on a reader, rules of interaction with the pillars are implied and intuitive. This encourages participants to form an implicit human swarm, each making changes to their environment, and reacting to changes from others.

The users are invited to place objects on the pillars which will trigger sounds from either natural or man-made environments, thus producing a collective soundscape. The objects are assigned a sound based on their visual aesthetic giving the user an instant connection between different objects and the sounds they produce. There is then a natural set of rules for the participants in that they are able to complete only a finite number of actions and each action has a clear outcome.

3.2   Variation 2: Explicit Interactions

In this second variation, we have explicit rules to guide user interactions. The pillars do not have themes and the aesthetic design is the same for both pillars. The parameters selected for this variation are listed in Table 1 (Fig. 4).

Pillar Design. A single pillar is built out of six perspex tubes arranged in a hexagon pattern as shown in Fig. 5. Each pillar is wired up to a Teensy micro-controller which detects touch, and controls the Leds to light up each tube. A ·capacitive touch sensor, made using copper tape, is used for detecting touch. Each pillar's micro-controller is connected over serial to a central Raspberry Pi acting as the controller. The Raspberry Pi uses a sound hat to drive the 4 speakers for creating the emergent music.

Each tube has a different height and features a touch sensor at the top of the tube. When the user maintains contact with the touch sensor, the coloured light in that tube gently moves through the colour spectrum. By changing the colours of the individual tubes, a 'pattern' emerges for the pillar as whole. For example, the pillar could have alternating tube colours red, blue, red, blue etc. Or the pillar could have half of the tubes in one colour and the other half in another colour. By creating a pattern system, the pillars have two forms of feedback upon which human swarm rules can be written: 1) individual colour of each tube, 2) overall pattern shown on the pillar.

Each colour is also mapped to a note in the scale of G, and the notes corre-sponding to each tube are played in sequence to add an auditory dimension to the experience. The notes are played using the sonic-pi music generation tool [1] which allows for the panning of the sound based on the physical location of the pillars. In our demonstration, we have two pillars which are set to pan left, and pan right to simulate the spatial effect of the sound. A graphical user interface is provided to allow for the modification of parameters on the fly, as well as enabling simulated testing on the system.

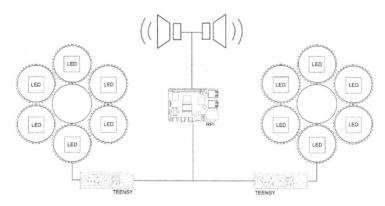

Fig. 5. Top down schematic of Variation 2, six outer tubes per pillar lit by LEDs. The dotted line around each tube represents the touch sensors, located at the top of each tube.

Interaction Rules. In the design of swarm systems, explicit rules of interaction are key for achieving specific outcomes. These systems depend on a transparent cause-and-effect relationship between the participant's actions and the system's responses, which can be orchestrated through a carefully crafted set of rules. Such rule-based interactions differ from non-rule-based systems, where emergent outcomes are shaped by local feedback without predefined directives.

For "The Forest", the interaction is straightforward: when a participant touches a sensor on a tube, it changes colour and emits a corresponding audio note, following the typical sense-act model seen in swarm robotics. This model has been adapted to our installation, requiring participants to react to the pillars' states according to one of three rule sets provided in Table 2. These rules are designed not just for individual action but to guide participants towards collective patterns in light and sound, although the system's inherent variability means convergence is not guaranteed.

The introduction of choice and the variability in human behavior ensure that "The Forest" remains a dynamic audio-visual experience, occasionally finding synchrony before veering off into a new direction. This reflects the essence of a human-swarm interaction, where, despite a limited number of simple rule sets, the diverse actions of participants lead to a continuously evolving installation, a journey through sound and light propelled by collective human engagement.

Table 2. User rule sets used for interacting with the pillars

Rule set A	1. If tube is RED, choose an adjacent tube and change to GREEN
	2. If tube is RED, change to YELLOW and change two additional tubes to YELLOW.
Rule set B	1. If tube is YELLOW, choose an adjacent tube and change to BLUE
	2. If tube is YELLOW, change to RED and change two additional tubes to RED
Rule set C	1. If tube is BLUE, choose an adjacent tube and change to YELLOW
	2. If tube is GREEN, choose an adjacent tube and change to RED

4 Results

This section outlines the testing and assessment of the system. Section 4.1 presents the results of tuning the system with simulated agents. Section 4.2 presents anecdotal description of real-life implementations we ran at various outreach events.

4.1 Simulated Artificial Agents

As an illustration of the "The Forest" concept, we conducted a series of simulations. These simulations were designed to validate that, while the system's behavior is influenced by input rules, it does not limit the creative possibilities-outcomes could range from periodic patterns to chaotic sequences. The goal was to ensure that the system's inherent design allows for a broad spectrum of artistic expressions without undue constraints.

In our experimental setup, we conducted agent-based simulations interfacing with the physical pillars, where software agents acted as surrogates for human interaction. We tested different agent groups, mimicking diverse human person-alities, in configurations of three or six, each following specific rule sets or making random decisions. The rulesets can be found in Table 2. Over the ten experimen-tal scenarios, we varied the number of agents, the rules they followed, and their personal biases to observe the range of art that could emerge from the system. These scenarios are shown in Table 3.

Table 3. The 10 scenarios tested in simulation. Agents follow the rulesets in Table 2. Scenarios (I)–(V) have 3 agents and a single agent is assigned to each ruleset. Scenarios (VI)–(X) have 6 agents and two agents are assigned to each ruleset.

Scenarios	No. Agents	Rule Set	Assignment Per Rule
(I), (VI)	3, 6	Rule 1 and 2	1, 2
(II), (VII)	3, 6	Only Rule 1	1, 2
(III), (VIII)	3, 6	Rule 1 and 2, Bias to Rule 1	1, 2
(IV), (IX)	3, 6	Rule 1 and 2, Bais to Rule 2	1, 2
(V), (X)	3, 6	Random	1

Each simulation ran for 500 turns, allowing agents to interact with the pillars, with random delays introduced to simulate variability in human behavior. We assumed agents would alternate in changing the pillar colours to track the artistic diversity generated by "The Forest." The variations in interaction and emergent art were analyzed using specific metrics to understand the range and nature of the potential outcomes.

1. **Maximum colour connectivity (Max cc)**: colour connectivity between tubes in a pillar is determined by proximity and shared colour. It is quantified by the ratio of colour connections to total tubes in the pillar. A value of 1 indicates uniform colour, 0 signals no adjacent matching colours. Monitoring this connectivity throughout the simulation captures pattern evolution, and the maximum value represents consensus or convergence to a single colour.
2. **Average colour connectivity (Avg cc)**: calculated as the average of the colour connectivity values recorded throughout the simulation.
3. **Average interactions per tube (Avg interactions)**: calculated by divid-ing the total number of interactions involving tubes in a pillar by the number of tubes. This metric offers insight into the average engagement level of tubes within a pillar, thereby indicating the overall interaction dynamics of the forest.

4. **Maximum idle period (Max idle):** the count of turns during which no interaction with a pillar occurred. This metric not only indicates periods of inactivity but also offers insights into the persistence of observed patterns over time.

Table 4 illustrates the outcomes derived from testing 10 scenarios, evaluated across four metrics. Modifying the interaction rules not only affects the results but also influences the degree of engagement. For example, enabling agents to interact with two rules generally fosters consensus, resulting in tubes within a pillar sharing the same colour. The increase in the number of agents correlates with a higher incidence of interactions, although results in a reduction in average colour connectivity. Notably, random agents exhibit the highest interaction and the least idle time, while agents following only one rule display the opposite pattern. This observation shows how the scope of choices provided to an agent impacts their engagement with the pillar. The concept of idle time assumes a distinct significance in the realm of the art experience, suggesting that agents have the opportunity to contemplate a specific pattern for an extended period without any alterations. In this context, Fig. 6a illustrates cycles of convergence to the same colour and divergence to different colours throughout the simulation, Fig. 6d highlights stability in colour connectivity for the majority of the time. This depiction encapsulates two discernible types of artistic experiences.

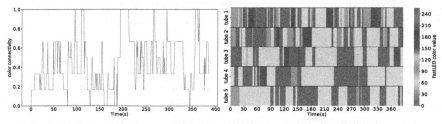

(a) Colour Connectivity in Pillar 2 of Testing Scenario (I)

(b) Light Colours of Pillar 2 of Testing Scenario (I)

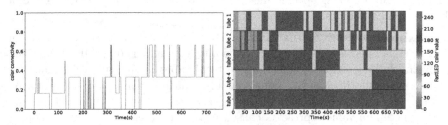

(c) Colour Connectivity in Pillar 1 of Testing Scenario (VII)

(d) Light Colours of Pillar 1 of Testing Scenario (VII)

Fig. 6. Evolution of colour connectivity and light colours throughout the simulations in testing scenarios (I) and (VII) (Color figure online)

Table 4. Evaluation Results from Testing 10 Scenarios Using Artificial Agents

Scenario	Pillar 1				Pillar 2			
	Max cc	Avg cc	Avg i'ns	Max idle	Max cc	Avg cc	Avg i'ns	Max idle
(I)	1	0.45	39.2	18	1	0.43	44.2	17
(II)	0.66	0.25	15.0	23	0.66	0.31	**14.3**	30
(III)	1	0.39	31.0	18	1	0.30	32.7	26
(IV)	1	0.45	49.8	9	1	0.50	54.0	9
(V)	0.66	0.25	67.3	7	0.66	0.23	63.2	7
(VI)	1	0.40	88.7	16	1	0.37	73.7	21
(VII)	0.66	0.26	29.8	**36**	0.66	0.26	27.8	25
(VIII)	1	0.26	62.7	25	1	0.25	68.0	18
(IX)	1	0.46	**103.7**	9	1	**0.47**	89.8	20
(X)	1	0.25	129.0	**6**	0.66	0.25	122.5	9

4.2 Real-Life Demonstration

In this section, we present anecdotal descriptions and observations of events we ran for each of the two variations detailed in Sects. 3.1 and 3.2. We ran two outreach events, at the Bristol Festival of Nature 2022[1] and the FUTURES Festival 2023[2]. A supplementary video to this paper is available[3]. Over the trials, we noticed different types of interactions according to the design of each variation.

Variation 1: User-Led Dynamics in an Interactive Sound Installation. Here, the number of users at any given time ranged between roughly 0–10. Although there was no strictly enforced limit, space constraints served as a natural limit to the number of users. The binary framing with an environmental narrative created a strong impact on users as their own personal experiences influenced how they perceived and interacted with the pillars. The "human swarm" was also highly heterogeneous with a diverse age range and backgrounds for participants. This likely influenced the outcome in how each user chose to explore the installation.

As there were no explicit interaction rules, interactions occurred asynchronously with no direct coordination, often interfering with previous interactions. However, there was indirect coordination in how the resulting soundscape influenced subsequent user interaction. Users would also observe each other to understand how to interact with the installation. The lack of rules also resulted in greater unpredictability and possibilities for user interaction. The open-endedness of interaction was well suited for this particular outreach event as the "human swarm" was highly dynamic in size: users could join and leave at any time. Users were also drawn to interacting with pillars and sound objects depending on aesthetic preferences. Through this simple "voting" dynamic, the

---

[1] https://www.bnhc.org.uk/festival-of-nature/.
[2] https://futuresnight.co.uk/.
[3] https://youtu.be/A9HGEBUOZ7g.

resultant soundscape represented the desires of the collective in the sounds they wished to hear, either natural, man-made or a combination of both.

Variation 2: Convergence and Coordination in Group Interaction.
Here we ran two different scenarios, the first where participants were provided with the rule sets as listed in Table 2. Each set consists of two rules and a user can choose to apply either of their two rules at any given time. Each user is given a rule-set at random and there is at least one of each rule-set in the group. The users are also encouraged to look out for emerging patterns. Figure 7 shows the setup at the outreach event.

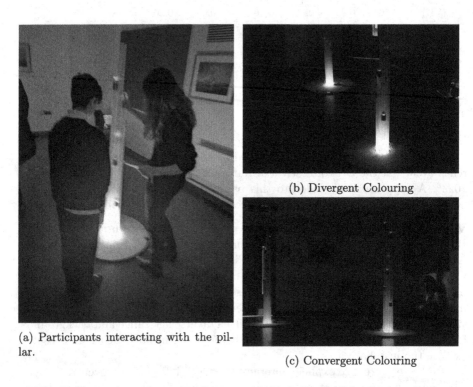

(a) Participants interacting with the pillar.

(b) Divergent Colouring

(c) Convergent Colouring

Fig. 7. Photos from the outreach event with Variation 2 (Color figure online)

We observed that although users were encouraged to look for patterns emerging over time, this prompt was not explicit enough to trigger a collaborative effort in pattern-making. Instead, users would interact with the pillars in an individualistic manner, focused on their own personal experience of the installation. We also observed that users sometimes deviated from the rule sets, following their own intuitions or preferences for interaction. In some cases, this actually led to colour convergence or a harmonic colour output as "deviant" users would bypass the constraints of the rule sets but act with intention to produce their own desired patterns, together with interactions from other users.

Fig. 8. Pre-specified target patterns

In the second scenario, the same rule sets are applied, but an additional goal was given to participants. The participants must aim to create a pre-specified pattern as a group, by following their individual rule sets. The pattern options provided were colour agnostic and are shown in Fig. 8. In this scenario, we found that there was convergence to a pattern in the majority of trials. However, the time to convergence varied depending on the level of collaboration in each trial. In some groups, a leader would emerge to coordinate the interactions of each user leading to faster convergence to a pattern. In other groups, there was much less cohesive teamwork as users would act individually.

The results thus far have been presented in terms of tube colour as this is the most visually practical. As described in Sect. 3.1, each colour has a corresponding audio note mapped to it and as the two pillars evolve in colour, so does the accompanying sound track. The evolution of that sound is in keeping with the colours presented due to the fixed colour-sound mapping model, however, it was observed in the real-life demonstration that it is possible to distinguish between patterns both audibly and visually. Alternating colours for example present themselves as alternating notes in the sequence. This both enriches the experience for the participants and provides another means of sensing through which the sense-act swarm model of rule design can be used. Figure 9 shows a segment of audio that emerged from the real-life demonstration. However, a detailed examination and analysis of the audio produced is beyond the scope of this paper, but it represents an intriguing avenue for future academic research to explore in depth.

Fig. 9. A transcribed snippet of audio taken from the system demonstration

5   Discussion

We have shown how different swarm rules may be designed for two variations of the Forest: in the first variation, implicit interactions allow for a soundscape to emerge over time from individual voting. In the second variation, we tested several scenarios and have shown that the rule sets for each one lead to different emergent outcomes. Each scenario took three factors into consideration: the number of participants, the way in which the three rule sets are distributed, and the heterogeneous nature of human's personal biases in choosing a rule. In simulation, these scenarios demonstrated the ability for the Forest to generate a vast number of emergent art pieces.

From testing rule sets and scenarios in simulation to real-life demonstrations, we observed that our choice of pillar design and scenarios did indeed generate a variety of emergent outputs as expected, particularly for variation 2. However, personal bias of users inevitably resulted in different outputs than seen in simulations. The participants themselves are not perfect agents in the sense that they have the choice to deviate from the rules and constraints given to them. This adds an element of unpredictability to the system - the unpredictability of human nature and subjective bias. Just as human society follows social norms that derive from and are applied by individuals subject to their own psychology and personality, we find that participants follow the swarm rules in much the same way and what emerges is a product of the "human swarm". With just three proposed rules, we have shown that there are a variety of different artistic outcomes and that the system lends itself to an ever-changing, interactive installation.

We can also compare the impact of implicit and explicit constraints from the first and second variations respectively: implicit constraints result in greater open-endedness of user interaction but lower levels of explicit coordination between participants. In the second variation, we observed users communicating directly with each other in order to coordinate their efforts to generate patterns. Furthermore, the addition of colour output made it simpler to identify pattern convergence, which may have contributed to enhanced user collaboration.

Consequently, engaging with the Forest could have a profound and enduring effect on the participants. The Forest may demonstrate the impact of individual actions on an ecosystem and the complex connections between all who inhabit the space. Individuals have agency in affecting the outcome of such complex systems where participants "vote" for the most desirable soundscape or colour patterns through their interactions. The Forest also demonstrates that emergent outcomes are possible through collective art-making and play. This is perhaps a reflection of the evolutionary process of human culture.

6   Conclusion

We have demonstrated how swarm algorithms may guide user interaction with our system, "the Forest", encouraging collaborative art-making and emergent outcomes from individual choices. As both the audience and actors in the Forest, we hope that participants may be able to experience the process of emergence itself by embodying the human swarm.

There is considerable potential for future investigation. This includes additional features such as increasing the number of pillars or dispersing the interaction points across a larger space. Swarm rules can be varied or enforced more strongly by giving users interactive components, such as physical tokens that activate specific components of the system. The narrative could be expanded to enhance how users interact with and perceive the experience, which may involve designing new tree species for the Forest – adding more heterogeneity to pillar design. We may also explore how inter-connectivity of pillars, that is functional dependence between pillars, affects the resulting user experience and degree of collaboration. The resulting experience from changing factors may be evaluated from data collected as part of real-world human trials. Finally, the system could be made available to artists as a source of inspiration or as an instrument for performance.

Acknowledgements. This work has been supported by funding from the Cabot Institute for the Environment, the UKRI Trustworthy Autonomous Systems Node in Functionality (EP/V026518/1) and the Bristol Robotics Laboratory. We would also like to thank Paul O'Dowd, University of Bristol, for his input and feedback during the design process of the Forest.

References

1. Aaron, S.: Sonic pi - performance in education, technology and art. Int. J. Perform. Arts Digital Media **12**(2), 171–178 (2016). https://doi.org/10.1080/14794713.2016.1227593
2. Alhafnawi, M., Hauert, S., O'Dowd, P.: Robotic canvas: interactive painting onto robot swarms. In: Artificial Life Conference Proceedings 32, pp. 163–170. MIT Press One Rogers Street, Cambridge, MA 02142–1209, USA journals-info . . . (2020)
3. Alhafnawi, M., Hunt, E.R., Lemaignan, S., O'Dowd, P., Hauert, S.: Mosaix: a swarm of robot tiles for social human-swarm interaction. In: 2022 International Conference on Robotics and Automation (ICRA), pp. 6882–6888. IEEE (2022)
4. Bisig, D., Neukom, M., Flury, J.: Interactive swarm orchestra an artificial life approach to computer music. In: ICMC (2008)
5. Bisig, D., Schacher, J.C., Neukom, M.: Flowspace-a hybrid ecosystem. In: NIME, pp. 260–263 (2011)
6. Brambilla, M., Ferrante, E., Birattari, M., Dorigo, M.: Swarm robotics: a review from the swarm engineering perspective. Swarm Intell. **7**, 1–41 (2013)
7. Fehr, E., Fischbacher, U.: Social norms and human cooperation. Trends Cogn. Sci. **8**(4), 185–190 (2004)
8. Fischer, E.: The Necessity of Art. Verso Books (2020)
9. Frid, E., Lindetorp, H., Hansen, K.F., Elblaus, L., Bresin, R.: Sound forest: evaluation of an accessible multisensory music installation, pp. 1–12 (2019)
10. Galanter, P.: What is generative art? complexity theory as a context for art theory. In: In GA2003-6th Generative Art Conference (2003)

11. Herremans, D., Chuan, C.H., Chew, E.: A functional taxonomy of music generation systems. ACM Comput. Surv. (CSUR) **50**(5), 1–30 (2017)
12. Improbotics: Theatre lab and AI improv pioneers. https://improbotics.org/
13. Jordà, S., Geiger, G., Alonso, M., Kaltenbrunner, M.: The reacTable: exploring the synergy between live music performance and tabletop tangible interfaces, pp. 139–146 (2007)
14. Martin, K.: Abstract art (1951) in Broadsheet no. 1: Abstract Paintings, Sculptures, Mobiles, London : Lund Humphries
15. Mäs, M., Flache, A., Helbing, D.: Individualization as driving force of clustering phenomena in humans. PLoS Comput. Biol. **6**(10), e1000959 (2010)
16. Moussaïd, M., Kämmer, J.E., Analytis, P.P., Neth, H.: Social influence and the collective dynamics of opinion formation. PLoS ONE **8**(11), e78433 (2013)
17. NTT: Swarm arena (2018). https://ars.electronica.art/futurelab/en/projects-swarm-arena/
18. Oxman, N., Laucks, J., Kayser, M., Uribe, C.D.G., Duro-Royo, J.: Biological computation for digital design and fabrication. In: Computation and Performance-Proceedings of the 31st eCAADe Conference, vol. 1, pp. 585–594 (2013)
19. Rosselló, L.B., Bersini, H.: Music generation with multiple ant colonies interacting on multilayer graphs. In: Johnson, C., Rodríguez-Fernández, N., Rebelo, S.M. (eds.) EvoMUSART 2023. LNCS, vol. 13988, pp. 34–49. Springer, Cham (2023). https://doi.org/10.1007/978-3-031-29956-8_3
20. Schranz, M., Umlauft, M., Sende, M., Elmenreich, W.: Swarm robotic behaviors and current applications. Fronti. Robot. AI, 36 (2020)

Evoboard: Geoboard-Inspired Evolved Typefonts

João Eduardo Batista[1]([✉]) [iD], Fraser Garrow[2] [iD], Carlo Huesca-Spairani[3] [iD], and Tiago Martins[4] [iD]

[1] LASIGE, Faculty of Sciences, University of Lisbon, Lisbon, Portugal
jebatista@fc.ul.pt
[2] School of Informatics, University of Edinburgh, Edinburgh, UK
fg28@hw.ac.uk
[3] University of Alicante, Alicante, Spain
chs17@alu.ua.es
[4] CISUC, DEI and LASI, University of Coimbra, Coimbra, Portugal
tiagofm@dei.uc.pt

Abstract. Type design is a field that deals with the creation of visually appealing designs for the written language. The work of the designer is time-consuming and requires many iterations until the final solution is achieved. Although a human expert is required to validate the final results, this task can be aided by automatic design software. We propose Evoboard, an automatic algorithm that evolves a typefont using a geoboard-inspired representation where each character is a self-intersecting polygon. Evoboard uses a genetic algorithm to optimize the number of vertices of the polygon and their positions in a grid. The evolution of the population is guided by an Optical Character Recognition (OCR) model that aims to maximize the recognition of the polygon as the target character. Thanks to this simple pipeline, both the OCR model and the representation can be easily modified by the user to their needs. We evolve a set of 36 alphanumeric characters that are both highly legible and aesthetically appealing, two important aspects of type design.

Keywords: Typefont · Type Design · Optical Character Recognition · Evolutionary Computation · Evolutionary Design

1 Introduction

Type design is a field that deals with the creation of visually appealing designs for written language. While font design must be able to draw attention by itself, it must always be legible so that the user can convey their intended message [6]. The creation of typefonts is a time-consuming task whose target aesthetic is highly influenced by the target audience and the medium in which it is displayed (e.g., physical posters [14] or smartphones [33]). We propose Evoboard, an automatic typefont creation algorithm that intends to help the designers by making the design task less time-consuming while also providing a diverse range of solutions with a similar aesthetic from which the final designs can be picked.

C. Johnson et al. (Eds.): EvoMUSART 2024, LNCS 14633, pp. 17–32, 2024.
https://doi.org/10.1007/978-3-031-56992-0_2

This iterative algorithm allows the user to select the intended representation (or typefont design) and, from this selection, outputs a set of 36 alphanumeric characters. In the current state of this implementation, the algorithm evolves these 36 characters using a geoboard-based (i.e., polygon-based) design but, the pipeline can be easily modified to accommodate the necessities of the user to other designs and to other objectives besides the creation of typefonts.

2   Related Work

Since many decades ago, the use of generative art, or computer art, has been studied by several authors [4,28]. The earliest reference to the use of software for art, to the authors' knowledge, comes from a statement from the British artist Roy Ascott in 1966 [2]. In this document, Ascott acknowledges that communication sciences would have a strong impact on all application fields, including arts, and artists can either accept the changes and shape the evolution of the field or be carried by them. Nowadays, the use of generative art has reached mainstream media, with the general public using easy-access tools such as DALL-E [23], developed by OpenAI, among many other tools. Although this and many other models based on deep learning are available, the evolutionary computation field has its own share of recent works, with several authors showing interest in the automatic creation of typefonts [1,24,26,27].

Since type design is an art field in itself, the list of possible styles and representations used is endless. Ahn *et al.* [1], Martins et al. [20], Parente *et al.* [24], and Yoshida et al. [34] show a high variety of representations in their work. Other well-known applications within the type design field are the *Alphabet Synthesis Machine* [18] and the works by Ze Wang [32]. Other authors, like Unemi *et al.* [30], focus their work on a single representation and explore their variations in design. Additionally, unlike the previous works, that focus on the Latin alphabet, the work by Unemi *et al.* focuses on the katakana syllabary.

Many works rely on user evaluations to guide the evolutionary process, a well-known issue with interactive evolutionary computation [29]. Besides being a slow process, it leads to inconsistent evaluations and human fatigue.

We use a fully automatic fitness function based on the predictions of Optical Character Recognition (OCR) models. OCR-based fitness functions have previously been shown to be effective in the evolution of type designs. In [21], a Sparse Network of Winnow OCR architecture was used successfully to assign fitness to investigate the effect of migration between islands of characters in the evolutionary process. Despite this, a later work by the same authors [20] investigated the efficacy of OCR-based fitness functions compared to a Root-Mean-Square Error (RMSE) based approach and showed that an RMSE was effective and produced qualitatively better results due to early convergence using the OCR model. Also, we acknowledge other fitness metrics that measure legibility, aesthetics, and semantics, as proposed by Rebelo *et al.* [26] on their work using multi-objective evolutionary computation for layout composition. However, we consider that by providing an OCR model and a representation to the algorithm,

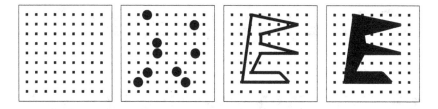

Fig. 1. Representation of the letter E using a 9 × 9 geoboard grid using 9 vertices.

we can obtain legibility from the first and consistent aesthetics from the second, using a single objective fitness function. Other criteria, such as each character being unmistakable from another, something necessary for a typefont to be considered acceptable [22], are implicitly considered by the OCR models minimizing the probability of recognizing a character as any character besides the target. However, since each character is evolved individually, we do not consider other elements such as letterspacing, wordspacing, and linespacing, among others [15].

3 Evoboard: System Overview

We propose *Evoboard*[1] an algorithm that evolved typefonts using a representation based on geoboard, using an approach based on Genetic Algorithms (GA) [16]. We investigate the efficacy of this algorithm as a system for type design when combined with an OCR-based fitness function. Specifically, we consider fitness functions based on OCR of handwritten and typed alphanumeric characters for two types of classifiers, Convolutional Neural Networks (CNN) [17] and Random Forest (RF) [5]. In this section, we begin by discussing the representation itself; we then describe the initialization of the GA and the variation operators and conclude by discussing the OCR-based evaluation.

3.1 Representation

Using a representation inspired by a geoboard (see Fig. 1), Evoboard evolves glyphs using a Cartesian grid, or board, to create characters based on line segments joined over the grid. Using this representation, the genotype of the model consists of an array of genes that encode the glyph through a list of vertices and their respective coordinates on the grid. We investigate both a fixed-length and flexible-length variant of the representation.

For the fixed-length variant, we specify the number of vertices, and only the positions are evolved. In the flexible-length variant, the algorithm evolves both the number of vertices and their positions on the grid. A minimum and maximum number of points is specified upfront. When plotting a glyph, a line is drawn connecting each consecutive position in the list and the first and last positions, and then the shape interior is filled, exemplified in Fig. 2.

---

[1] Implementation available at www.github.com/jespb/Evoboard.

EVOBOARD

Fig. 2. Example of a typefont evolved using Evoboard.

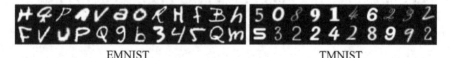

|EMNIST|TMNIST|

Fig. 3. Examples of images from the EMNIST and TMNIST datasets.

3.2 Genetic Algorithm

Initialization: The initial population contains randomly generated individuals. In the flexible-length variant, the length of the individual is a random value between the minimum and maximum length allowed. In the fixed-length variant, the length is determined by an input parameter. After defining the number of genes, each gene is associated with a randomly selected grid position.

Crossover: In the fixed-length variant, the crossover operator used exchange line segments between parents by performing a one-point crossover, where a segment point is randomly selected as a threshold, combining the first section of one parent with the second section of the other, creating a single child. In the flexible-length variant, the crossover operator selects and exchanges a random segment of each parent, considering the array as a loop, and concatenates them, also creating a single child.

Mutation: The mutation operator for Evoboard modifies line segments of the glyph by modifying individual genes in the genotype. Each gene in the genotype has a probability of being modified equal to the mutation rate. If a gene is modified, a new randomly selected grid position is assigned to it.

Evaluation: We suggest that OCR can be used during fitness evaluation to evolve legible characters. If the evolved glyph is recognized by the OCR system, then it should be assigned a good fitness. As such, we use an OCR-based fitness function, where fitness equals the Probability of Recognition (PR) of the evolved glyph as the target character. We investigate different OCR models by combining two different character datasets with two different classifiers to obtain four fitness functions. The datasets and classifiers are discussed below.

3.3 Inducing OCR Models

Character Datasets: The MNIST dataset [8] is one of the best-known and widely used distributions for OCR of handwritten digits. Since we aim to evolve both letters and digits, we use the Extended version EMNIST, which includes upper and lowercase characters [7]. We use a subset of EMNIST by selecting only the digits and uppercase letters, exemplified in Fig. 3.

We also use OCR models trained to detect typeface characters rather than handwritten characters. These characters were selected from a MNIST-style dataset, Typography-MNIST (TMNIST) [19]. In preliminary work, we used OCR models induced using both the full TMNIST dataset, including non-alphanumeric characters, and a subset of TMNIST where we selected only alphanumeric characters, exemplified in Fig. 3. Since using the full dataset results in better glyphs, we use the respective OCR model in the experimental setup.

Classifiers: For OCR, we employ two classifiers, RF and CNN. CNNs learn end-to-end, which automates feature extraction and dimensionality reduction. They are considered to be the state-of-the-art for image classification and recognition tasks. As will be stated in the results section, depending on the application, high performance is sometimes undesirable, and weaker classifiers are preferred. RFs, on the other hand, despite performing well for tabular data and classification tasks, do not perform automatic feature extraction.

4 Experimental Setup

We evolve glyphs for all 26 uppercase letters of the Latin alphabet and 10 numerical digits while exploring fixed and flexible-length Evoboard variations. In this section, we describe the details required to recreate the experiments.

The OCR models induced with the EMNIST dataset used the training and test partitions available in the reference article [7], and the OCR models induced with the TMNIST dataset used 70% of the dataset for training and 30% for testing. The classifiers were trained to predict each character in a multiclass setting (i.e. one classifier is used for all characters). Both datasets are built of 28×28 grayscale images and respective labels. To train the models, we convert each sample to an array of 784 numerical features.

The CNN model comprises two convolutional layers and one dense layer followed by an output layer. The CNNs are induced using the Adam optimizer for 10 epochs. The RF classifier uses a maximum depth of 15 and 50 estimators. All other parameters are set as default based on their sklearn implementations [25]. In preliminary experiments, we do five runs using each OCR model for each Evoboard variation and each alphanumeric character. Then, we focus the experiments on 30 evolutionary runs using the RF OCR models, as will be explained in the results section. In total, we create 80 variants of each character, 20 using CNNs and 60 using RFs. The non-default parameters used are shown in Table 1.

5 Results and Discussion

In this section, we discuss the results obtained when using CNN and RF-based OCR models in the fitness function. We start by commenting on the results obtained using both models and later focus on the RF models, due to a better performance. We discuss the evolution of the PR values over the generations, stating that some characters are easily evolved while others require additional generations. Lastly, we discuss the glyph diversity in each experiment.

Table 1. Parameter settings used in the experiments.

Evoboard		CNN	
Generations	100	Conv. layer 1 filters	32, 3x3
Population size	100	Pooling layer 1 size	2
Fixed length variant:		Conv. layer 2 filters	64, 3x3
No. points	11	Pooling layer 1 size	2
No. positions in x and y	21	Dense Layer units	128
Variable length variant:		Activation func.	Relu
No. points	5 to 25	Output func.	Softmax
No. positions in x and y	25	Loss func.	Sparse CCE
Selection	Tournament	Optimizer	Adam
Tournament size	2	Epochs	10
Elite size	1	Random Forest	
Crossover probability	0.7	Max depth	15
Mutation probability	0.05	No. estimators	50

5.1 Character Recognition Probability as Fitness

We start this section by validating the OCR models, taking into consideration their training and test accuracy in the balanced datasets used to induce them. Then, we discuss their performance as wrapped classifiers in the fitness function.

Convolutional Neural Networks for OCR: When inducing the CNN models for the EMNIST (CNN-E) and the TMNIST (CNN-T) datasets, we obtain 94.31%/98.34% training accuracy and 90.44%/96.51% test accuracy in each dataset, respectively. CNNs are known for being powerful predictive models. However, CNNs are known to suffer from some problems when trying to classify samples that are Out-of-Distribution (OOD), such as classifying a sample with high certainty that a human would clearly understand to be misclassified. A phenomenon sometimes known as *adversarial examples* [13]. Evolved glyphs are drawn from a different data distribution than handwritten or typed characters, making CNNs less suitable to be used in a fitness function since they classify glyphs from early generations as the correct character without giving the GA enough generations to produce something that is recognizable by humans. Recent work also states that models that obtain perfect fitness in early generations may not be suitable to be used in fitness functions [3].

In the experiments using CNN models, the fitness value of the GA individuals reaches its maximum in early generations in nearly all the runs for all 36 characters. However, the evolved designs are, in most cases, not recognizable by humans. In preliminary work, we apply CNN-E and CNN-T on the Evoboard pipeline and perform 5 runs with each of 36 characters. Furthermore, in an attempt to alleviate this issue, we investigated training the CNNs as binary classifiers, rather than a multiclass setting. However, this suffered from the same issue due to the glyphs being OOD from the training datasets. This echoes the findings presented in [20], offering further evidence to suggest that deep learning OCR may present some limitations to this task.

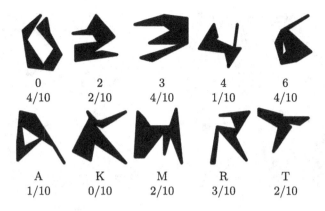

0	2	3	4	6
4/10	2/10	4/10	1/10	4/10

A	K	M	R	T
1/10	0/10	2/10	3/10	2/10

Fig. 4. Best typefonts evolved using a CNN model in the fixed-length variant of Evoboard. Below, we indicate how many runs we consider successful in creating a character that resembles the target out of 10 runs (5 runs using each dataset).

Take Fig. 4 into consideration. This figure contains the best-evolved designs of the characters in which the CNN models obtained the best results. Here, we can see that, even in something as simple as evolving the character 0, only in 4 out of 10 runs we obtain a glyph that we recognize as a 0. Regarding the other glyphs, even if they resemble the target characters, most are something we do not consider to be appealing. As such, the remainder of this section will be focused on the experiments using the RF models.

Random Forest for OCR: When inducing the RF models for the EMNIST (RF-E) and TMNIST (RF-T) datasets, we obtain 95.05%/98.80% training accuracy and 79.24%/92.41% test accuracy in each dataset, respectively. Although these accuracy values seem sub-optimal when compared to the CNN results, by taking into consideration the confusion matrices in Fig. 5, we see that the most common mistakes from the RF-E model come from misclassifying similar characters (e.g., O/0, 1/I/L, 5/S, or 8/B). The RF-T model only shows signs of frequently misclassifying the 0 and O characters.

As previously stated, having a strong predictive model may lead to bad results when generating a character representation. If the PR value of the OCR model easily reaches 100%, the glyph will not evolve to something human-recognizable. In Tables 2 (digits) and 3 (letters), we see the median PR values of each character in each experiment, and the median length of the representations[2].

Taking into consideration the results regarding the evolution of digits and a threshold of 75% to define if we expect good results, we expect the models to have produced good results for all characters using the EMNIST dataset in both experiments with fixed and variable lengths. Although the results for 0/1/5 are below this threshold, we expect them to be mistaken for similar characters.

---

[2] In the fixed-length Evoboard variant, the length is always set to 11.

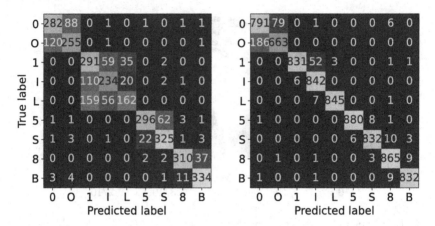

Fig. 5. Subsection of the confusion matrix of the RF models induced using the EMNIST (left) and TMNIST (right) datasets when applied to the test data.

Table 2. Median PR values (%) of the best character evolved in each of 30 runs and median length. For each character, the test cases with the highest values and those with no statistical difference from them are highlighted in green.

	0	1	2	3	4	5	6	7	8	9
Fixed Length										
RF-E	61	58	84	88	88	71	87	96	66	69
RF-T	39	66	81	65	74	48	47	95	32	48
Flexible Length										
RF-E	52	65	85	82	89	58	92	98	66	68
Median Length	11	7	9	10	8	8	9	6	9	7
RF-T	22	79	77	54	78	43	30	100	23	39
Median Length	9	7	9	9	7	8	9	7	10	8

Now taking into consideration the results regarding the evolution of letters, they indicate that RF-E should be successful at evolving all characters except for G and Q. And also B/I/O, which may be mistaken for 8/1/0, respectively. Similarly to the digits, the results using RF-T indicate that these experiments will lead to worse results than those using RF-E.

Regarding the median length of the representations in the flexible-length Evoboard. In many test cases, Evoboard evolves a representation with a short length (6 to 8 points) that outperforms the respective fixed-length experiment, indicating that certain characters like 1, 7, and J, among others, require fewer points to be recognized as the target characters. Characters that require a large number of points tend to have a higher probability of recognition in fixed-length experiments since they only focus on optimizing the positions of the points.

Table 3. Median PR values (%) of the best character evolved in each of 30 runs and median length. For each character, the test cases with the highest values and those with no statistical difference from them are highlighted in green.

	A	B	C	D	E	F	G	H	I	J	K	L	M	N	O	P	Q	R	S	T	U	V	W	X	Y	Z
Fixed Length																										
RF-E	87	62	92	80	93	85	71	78	57	92	90	92	96	86	56	84	70	77	80	96	88	94	93	87	93	93
RF-T	84	38	76	82	66	72	60	70	44	87	81	89	77	96	54	59	59	57	55	85	86	92	48	71	85	91
Flexible Len.																										
RF-E	91	49	94	64	87	85	69	78	67	96	90	98	92	87	53	77	57	77	80	98	86	97	86	86	95	90
Median Len.	8	9	9	9	10	8	10	8	8	6	8	7	8	9	8	11	9	10	8	6	9	8	9	10	7	8
RF-T	93	33	74	48	35	73	43	61	34	93	83	97	65	97	41	45	52	32	48	97	84	98	40	67	92	93
Median Len.	7	10	9	10	9	8	9	9	9	8	9	7	10	9	10	8	10	10	8	8	10	7	8	9	7	8

5.2 Evolution of the Probability of Recognition

By taking into consideration the progression of the PR values (fitness) over the generations, we can have a better idea of the rhythm of the progression of evolving a glyph that resembles a specific character. In this section, we study the evolution of the fitness of the generations for two characters where we obtained good results (T and N), and one character where the results were not satisfactory (B).

Take Fig. 6 (letter T) into consideration. Here, we see that using RF-E, the fitness of the algorithm quickly reaches an optimum with a low standard deviation across the different runs. This indicates that this character can be learned within the first 50 generations. When using RF-T, the fitness continues growing until generation 100 (our experimental limit) and has a large standard deviation value that shows no sign of decreasing over time. This indicates that RF-T may be more demanding on the evolved representation than RF-E, and also that there is a larger diversity of solutions using RF-T since the standard deviation is larger.

Taking Fig. 7 (letter N) into consideration, we see that although Evoboard obtains high PR values, there's a higher standard deviation in these values when compared to the T character. This may be justified by, as previously seen in Table 3, the N character requiring a bigger median length (9 on RF-E and RF-T) than the T character (6 on RF-E and 8 on RF-T). As such, the results show that this character requires more time to optimize the positions of all vertices when compared to the T character, which is quickly evolved.

Lastly, take Fig. 8 into consideration. We found the B character more difficult to obtain in all experiments. Unlike the previous characters, B has a slow evolution, indicating that more generations are needed to evolve its design. In the RF-T experiment with fixed length, we see a more steady growth. However, the median PR value only reaches 38%, implying large difficulties in evolving this character. Although the B and 8 characters are similar, the models successfully produced an 8 character that is recognizable by us more often than the B character. Out of 120 runs (30 for each experiment), we only obtained one B character that we consider satisfactory. This character can be seen in Fig. 11.

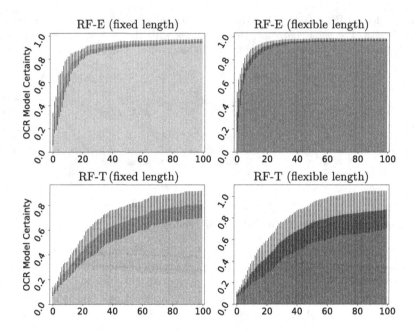

Fig. 6. Mean PR of the T character over the generations and standard deviation.

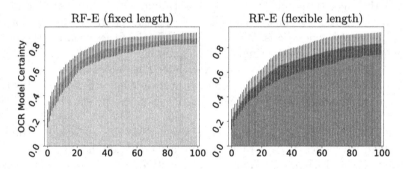

Fig. 7. Mean PR of the N character over the generations and standard deviation.

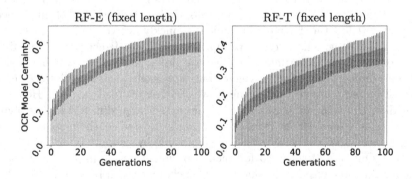

Fig. 8. Mean PR of the B character over the generations and standard deviation.

5.3   Typefont Diversity: The Letter R

Among the characters that consistently evolved with acceptable results, we consider that R had the most diverse solutions across different runs. As such, in this section, we focus our commentaries on the diversity of solutions in this letter. In Fig. 9, we see 10 of the 30 evolved Rs for each experiment. The images were selected based on our personal preference of the evolved Rs.

Taking this image into consideration, we see that using RF-E consistently leads to glyphs that resemble the letter R. However, we also notice some lack of diversity, since most images resemble the letter R with its "hole" filled and little variation to the relative position of each vertex. Using RF-T in the fixed-length Evoboard, we see that all of the selected Rs contain a hole (open or closed) and are more diverse. The PR value of these glyphs as an R ranges from 50 to 75%. Although this value is lower than those obtained using RF-E, we consider the obtained glyph more visually appealing than those evolved using RF-E. Lastly, the Rs evolved using RF-T and the flexible-length Evoboard have lower scores, and most do not resemble Rs. Interestingly, some of the Rs in Fig. 9 are visually similar to Rs, but the model gives them a low PR value. This gives some emphasis on the difference between recognition and similarity and on why CNNs obtain high PR values without producing recognizable designs in many cases. Meanwhile, the glyph in position 7 only contains the right edge of the character and has a PR value of over 50%, indicating that the models can recognize an R by that shape alone. This is interesting since, except for B, no other characters contain a similar right edge.

These commentaries are consistent with what we see in Fig. 10, regarding the evolution of the PR values of the letter R in each experiment. When using RF-E, we see low standard deviation values. This indicates that the glyphs should be consistent across different runs. By opposition, when using RF-T in the fixed-length Evoboard and especially in the flexible-length Evoboard, we notice a larger ratio between the average PR value and the respective standard deviation values, implying that the results are more diverse.

5.4   Evolved Typefont

To conclude the result section, we display the resulting typefont using RF-T with the fixed-length variation of Evoboard. In Fig. 11, we see the best character evolved using RF-T. To select the characters to be displayed, we picked the characters with the best fitness (which, by our standards, has a high visual quality), and if we did not like the decision of the model, we replaced them with our preferred character.

We kept 33/36 of the selected characters and handpicked 3 characters. The RF-E results are not displayed since they were similar in the vast majority of the characters. To generate the same image using RF-E, we would keep 28/36 of the selected characters and handpick 8 characters. This shows that, while using the RF-T as fitness leads to lower PR values, the results are more reliable. We also see that, while this experiment failed to make the outline of the B character, the inner triangles produced an interesting design for this character.

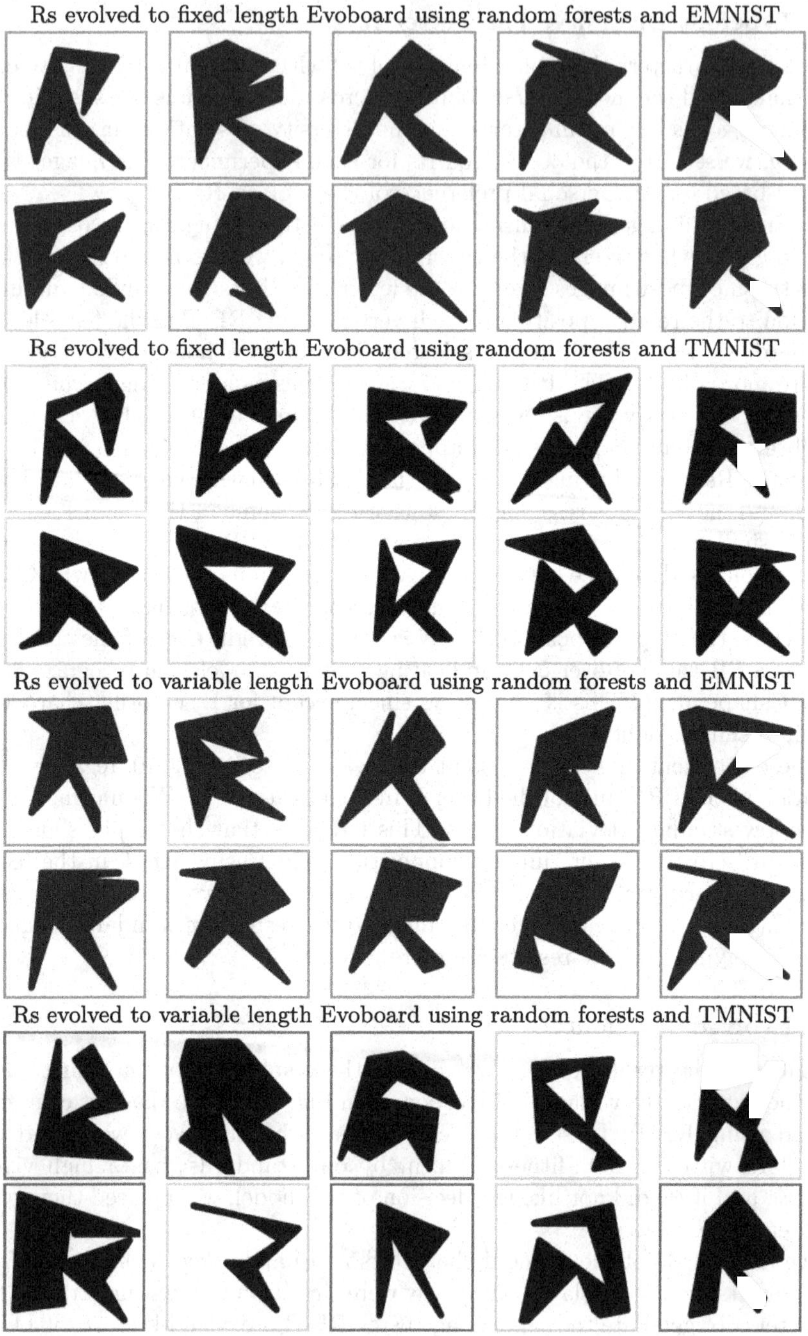

Rs evolved to fixed length Evoboard using random forests and EMNIST

Rs evolved to fixed length Evoboard using random forests and TMNIST

Rs evolved to variable length Evoboard using random forests and EMNIST

Rs evolved to variable length Evoboard using random forests and TMNIST

Fig. 9. Best Rs obtained in each experiment in 30 runs. Green/yellow/red borders represent a probability of recognition of 75–100/50–75/25–50%, respectively. (Color figure online)

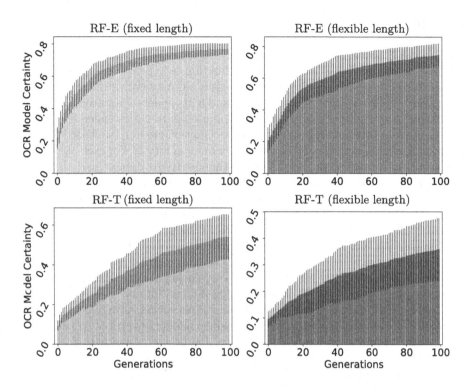

Fig. 10. Mean PR of the R character over the generations and standard deviation.

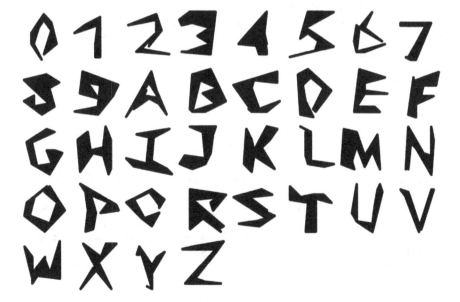

Fig. 11. Set of characters evolved using RF-T and the fixed-length Evoboard.

6 Conclusions

We proposed *Evoboard*, an algorithm that evolves typefonts using a geoboard-inspired representation. Using this representation, a typefont is represented as a self-intersecting polygon. This algorithm uses a GA to evolve the number and position of vertices of this polygon, using an OCR model on its fitness function to measure how similar the current representation is to the intended character.

We use four different optical character recognition models, two based on convolutional neural networks and two based on random forests. For each algorithm, one model is trained using the EMNIST (handwritten) dataset and the other using the TMNIST (typefont) dataset. In preliminary work, we observe that the deep learning approach for OCR tends to fail as a fitness function for this task. This is due to their propensity to incorrectly overpredict with high certainty for out-of-distribution samples, which causes fitness to be maximized in early generations. As such, Evoboard does not create a good representation that is recognizable to us humans. Random forests do not learn the same spurious features as deep learning models, allowing the evolution of typefonts. As such, we focus this work on the results obtained using random forests.

The experimental results are split into four experiments, using models trained in EMNIST/TMNIST and using a fixed/flexible number of vertices. We observed that, when using a fixed number of vertices, the fitness grows much faster and consistently obtains good results in most characters. In characters that require fewer vertices, such as I, J, and L, the flexible-length variant outperforms the fixed-length variant. These results can be explained by stating that it is easier to optimize the position of fewer vertices or the position of many vertices if the algorithm is not simultaneously evolving the optimal number of vertices.

Overall, we consider the results from the experiments quite positive, consistently obtaining good representations for all 36 alphanumeric characters, except for the digit 8 and the letter B. Although these characters were harder to obtain, Evoboard produced interesting results in some evolutionary runs.

Given the overall success in evolving typefonts using Evoboard, there are several variants to be explored in future work, such as: preprocessing the images using feature extraction (e.g., for edge detection); using several polygons, rather than a single polygon containing all vertices, and using different colors for each polygon, allowing the evolved designs to be used for more complex applications.

Acknowledgments. Work funded by FCT through the LASIGE R&D Unit, UIDB/00408/2020 [11] and UIDP/00408/2020 [12] — CISUC R&D Unit, UIDB/00326/2020 [9] and UIDP/00326/2020 [10]; ValgrAI: Valencian Graduate School and Research Network for Artificial Intelligence [31]; and Generalitat Valenciana.

We would also like to thank the Society for the Promotion of Evolutionary Computation in Europe and its Surroundings (SPECIES) for the opportunity to participate in the SPECIES Summer School 2023, which took place in Moraira, Spain, between the 3rd and 9th of September, where this work originated.

References

1. Ahn, Y., Jin, G.: TYPE+CODE II: a code-driven typography. Leonardo **49**(2), 168–168 (2016). https://doi.org/10.1162/LEON_a_01062
2. Ascott, R.: Behaviourist art and the cybernetic vision. Cybernetica **9**, 247–264 (1966)
3. Batista, J.E., Silva, S.: Comparative study of classifier performance using automatic feature construction by M3GP. In: 2022 IEEE Congress on Evolutionary Computation (CEC), pp. 1–8 (2022). https://doi.org/10.1109/CEC55065.2022.9870343
4. Boden, M., Edmonds, E.: What is generative art? Digital Creativity **20**, 21–46 (2009). https://doi.org/10.1080/14626260902867915
5. Breiman, L.: Random forests. Mach. Learn. **45**(1), 5–32 (2001). https://doi.org/10.1023/A:1010933404324
6. Bringhurst, R.: The Elements of Typographic Style. Hartley & Marks, Publishers, Elements of Typographic Style (2004)
7. Cohen, G., Afshar, S., Tapson, J., van Schaik, A.: EMNIST: an extension of MNIST to handwritten letters. CoRR abs/1702.05373 (2017)
8. Deng, L.: The MNIST database of handwritten digit images for machine learning research. IEEE Signal Process. Mag. **29**(6), 141–142 (2012)
9. FCT - Foundation for Science and Technology: CENTRO DE INFORMÁTICA E SISTEMAS DA UNIVERSIDADE DE COIMBRA. https://doi.org/10.54499/UIDB/00326/2020. Accessed 17 Jan 2024
10. FCT - Foundation for Science and Technology: CENTRO DE INFORMÁTICA E SISTEMAS DA UNIVERSIDADE DE COIMBRA. https://doi.org/10.54499/UIDP/00326/2020. Accessed 17 Jan 2024
11. FCT - Foundation for Science and Technology: LASIGE - Extreme Computing. https://doi.org/10.54499/UIDB/00408/2020. Accessed 17 Jan 2024
12. FCT - Foundation for Science and Technology: LASIGE - Extreme Computing. https://doi.org/10.54499/UIDP/00408/2020. Accessed 17 Jan 2024
13. Goodfellow, I., Shlens, J., Szegedy, C.: Explaining and harnessing adversarial examples. In: International Conference on Learning Representations (2015). http://arxiv.org/abs/1412.6572
14. Guo, S., et al.: Vinci: An intelligent graphic design system for generating advertising posters. In: Proceedings of the 2021 CHI Conference on Human Factors in Computing Systems. CHI 2021, Association for Computing Machinery, New York, NY, USA (2021). https://doi.org/10.1145/3411764.3445117
15. Hochuli, J.: Detail in typography (English reprint). B42, Paris, France (2015)
16. Holland, J.H.: Adaptation in Natural and Artificial Systems: An Introductory Analysis with Applications to Biology. Control and Artificial Intelligence. MIT Press, Cambridge (1992)
17. LeCun, Y., et al.: Backpropagation applied to handwritten zip code recognition. Neural Comput. **1**(4), 541–551 (1989). https://doi.org/10.1162/neco.1989.1.4.541
18. Levin, G., Feinberg, J., Curtis, C.: The alphabet synthesis machine. https://www.alphabetsynthesis.com/. Accessed 15 Nov 2023
19. Magre, N., Brown, N.: Typography-MNIST (TMNIST): an MNIST-style image dataset to categorize glyphs and font-styles (2022)
20. Martins, T., Correia, J.A., Costa, E., Machado, P.: Evotype: from shapes to glyphs. In: Proceedings of the Genetic and Evolutionary Computation Conference 2016, pp. 261–268. GECCO 2016, Association for Computing Machinery, New York, NY, USA (2016). https://doi.org/10.1145/2908812.2908907

21. Martins, T., Correia, J., Costa, E., Machado, P.: Evotype: evolutionary type design. In: Johnson, C., Carballal, A., Correia, J. (eds.) EvoMUSART 2015. LNCS, pp. 136–147. Springer, Cham (2015). https://doi.org/10.1007/978-3-319-16498-4_13

22. Muller-Brockmann, J.: Rastersysteme für die visuelle Gestaltung - Grid systems in Graphic Design. Niggli Verlag, Sulgen, Switzerland (1999)

23. OpenAI: DALL-E 3. https://openai.com/dall-e-3. Accessed 14 Nov 2023

24. Parente, J., Martins, T., ao Bicker, J., Hardman, P., Machado, P.: Working with type: approaches on generative and evolutionary typographic creation. In: Proceedings of the Eleventh International Conference on Computational Creativity (2020)

25. Pedregosa, F., Varoquaux, G., Gramfort, A., Michel, V., Thirion, B., Grisel, O., Blondel, M., et al.: Scikit-learn: machine learning in python. J. Mach. Learn. Res. **12**, 2825–2830 (2011)

26. Rebelo, S., Martins, T., ao Bicker, J., Machado, P.: Exploring automatic fitness evaluation for evolutionary typesetting. In: Creativity and Cognition. C&C 2021, Association for Computing Machinery, New York, NY, USA (2021)

27. Rebelo, S.M., Martins, T., Ferreira, D., Rebelo, A.: Towards the automation of book typesetting. Vis. Inform. **7**(2), 1–12 (2023). https://doi.org/10.1016/j.visinf.2023.01.003

28. Richardson, A.: Data-Driven Graphic Design: Creative Coding for Visual Communication. Bloomsbury Publishing, London (2017)

29. Takagi, H.: Interactive evolutionary computation: fusion of the capabilities of EC optimization and human evaluation. Proc. IEEE **89**(9), 1275–1296 (2001). https://doi.org/10.1109/5.949485

30. Unemi, T., Soda, M.: An IEC-based support system for font design. vol. 1, pp. 968–973 (2003). https://doi.org/10.1109/ICSMC.2003.1243940

31. ValgrAI: ValgrAI - Valencian Graduate School and Research Network for Artificial Intelligence. https://valgrai.eu/. Accessed 17 Jan 2024

32. Wang, Z.: Ze Wang's website. https://zewang.info/Generative-Typography. Accessed 15 Nov 2023

33. Yin, W., Mei, T., Chen, C.W.: Automatic generation of social media snippets for mobile browsing. In: Proceedings of the 21st ACM International Conference on Multimedia, pp. 927–936. MM 2013, Association for Computing Machinery, New York, NY, USA (2013). https://doi.org/10.1145/2502081.2502116

34. Yoshida, K., Nakagawa, Y., Köppen, M.: Interactive genetic algorithm for font generation system. In: 2010 World Automation Congress, pp. 1–6 (2010)

Motifs, Phrases, and Beyond: The Modelling of Structure in Symbolic Music Generation

Keshav Bhandari🆔 and Simon Colton$^{(✉)}$🆔

School of Electronic Engineering and Computer Science, Queen Mary University
of London, London, UK
{k.bhandari,s.colton}@qmul.ac.uk

Abstract. Modelling musical structure is vital yet challenging for artificial intelligence systems that generate symbolic music compositions. This literature review dissects the evolution of techniques for incorporating coherent structure, from symbolic approaches to foundational and transformative deep learning methods that harness the power of computation and data across a wide variety of training paradigms. In the later stages, we review an emerging technique which we refer to as "sub-task decomposition" that involves decomposing music generation into separate high-level structural planning and content creation stages. Such systems incorporate some form of musical knowledge or neuro-symbolic methods by extracting melodic skeletons or structural templates to guide the generation. Progress is evident in capturing motifs and repetitions across all three eras reviewed, yet modelling the nuanced development of themes across extended compositions in the style of human composers remains difficult. We outline several key future directions to realize the synergistic benefits of combining approaches from all eras examined.

Keywords: Generative Music · Deep Learning · Computerized Music

1 Introduction

Musical structure embodies the thoughtful organization of fundamental elements such as pitch, harmony, rhythm, and timbre in a composition [50]. Structure gives a sense of coherence, unity, and direction to the piece. On a basic level, notes are organized into motives and phrases. These phrases are then combined into higher-level sections such as verses, choruses and melodic segments. The repetition, variation, and development of musical ideas and themes manifest in the music and create relationships between these sections to form an overarching structure. Understanding and perceiving this musical structure is key to appreciating and enjoying music for listeners [24].

Musical structure is undoubtedly a crucial aspect of people's cognitive process that influences their perception of music [39,59,60]. Surprisingly, research involving infants as young as four and a half years old suggests that musical experience might not be a prerequisite for perceiving musical phrase structures [37]. Studies focusing on rhythm have delved into perceptual elements like timing and tempo [28] and syncopated patterns [22]. For human listeners, perceiving musical structure can be thought of as a pattern recognition task [65]. As we listen, we segment the music into motives, phrases, and sections. We identify similarities, repetitions, variations, and points of departure over time. Through this process, we form a mental representation of the music's structure. However, human memory's limitations become evident when individuals struggle to recall details; instead, they remember specific impactful moments [50] and distinct fragments rather than entire compositions, highlighting the selective nature of musical memory [48,65].

Given this understanding, songwriters strategically incorporate catchy repetitive phrases to enhance memorability and emotional impact for listeners [43,79]. However, Burns [6] argues that repetition gains meaning through its relationship with change. Both endless repetition and constant variation could lead to monotony. Thus, repetition and change are opposite possibilities from moment to moment in music. The tension between them can be a source of meaning and emotion. For example, in pop music, the repetition of verse and chorus sections helps emphasize music ideas, while the contrast between verse and chorus can create more emotional intensity [44]. Similarly, within Western tonal music, artists leverage repetition of harmonic progressions (sequences of chords) to guide listeners through a journey that creates dramatic narratives, conveying a sense of conflict that demands a resolution [39,62].

While structure and predictability are key for many applications involving composition, cognitive science and musical therapy [69], some contexts like video game and social media music require adaptive, event-driven approaches [2]. The music in video games often evolves in response to the player's progress, creating a unique journey for each player. In this scenario, long-term structural plans around musical form might be impractical, as the musical narrative needs to be flexible and responsive to the unfolding events in the game. Similarly, social media music quickly sets moods with catchy hooks rather than intricate structures [23]. Additionally, avant-garde music is another area that deliberately challenges conventional musical structures, requiring artists to explore abstract and unpredictable forms that push listener expectations.

From a computational perspective, automatically analyzing and generating musical structure poses challenges for music research. Unlike other art forms like stories which have explicit plot structures, the "language" of music structure is more abstract and relies heavily on repetition, variation, and development of themes. This makes explicitly modelling musical structure difficult. While well-structured music generation with AI spans both raw audio and symbolic (sheet) music domains, this literature review specifically focuses on research pertaining to musical structures in the symbolic domain. While there has been a plethora of surveys on music generation systems, recent surveys such as [5,7,12,35,42,61] have covered works that use deep learning techniques. Among these, [7] touches upon early deep learning methods built upon recurrent neural networks (RNNs), and generative adversarial networks (GANs), while placing greater emphasis on symbolic methods. Similarly, [5] also surveys early deep learning models until Music Transformer [33]. Furthermore, [42] and [61] have a very narrow focus on multi-track music using GANs and three specific deep learning models respectively. While [35] is comprehensive in their selection of papers by area of application, our literature review differs in two ways. First, we offer a deeper analysis on structure by exclusively focusing on studies that aim to improve musical structure in their methodology. Our selection of papers predominantly spans between 2007 to 2023. Early papers spanning this time frame can be seen in Sect. 2 which provides a foundational overview by covering symbolic methods before delving into later advancements using deep learning in Sect. 3. Second, we examine an emerging "subtask decomposition" approach in Sect. 4 that breaks down the music generative process into smaller steps incorporating neuro-symbolic methods or musical knowledge. Reflecting on this trend, we address future directions for modelling long term musical structure in Sect. 5.

2   Symbolic Methods

In early work on computerized music generation with long-term structure, [16] proposes a reinforcement learning approach on top of a Variable-Length Markov model for musical style imitation and improvisation with long term dependencies. Multiple agents model different musical attributes (e.g. pitch, rhythm, harmony) to capture the anticipatory foundations of musical expectation in a collaborative yet competitive framework. The chosen "behavior" agent for each episode takes actions and updates the others' policies accordingly. Agents leverage *Factor Oracles*, which compactly gather repetitive sequence factors, enabling efficient access to long musical contexts. Results modeling Bach inventions demonstrates the model's sensitivity to phrase boundaries, which clearly segment the score into formal sections.

The authors of [53] propose a constraint programming approach for controlled generation of Markov sequences, with applications to music. They formulate the Markovian property as a cost function represented by stacks of Elementary Markov Constraints and generate sequences by exploring the space of all sequences satisfying both the constraints and optimizing the Markovian cost. A key advantage of this approach is the ability to specify arbitrary structural constraints beyond just Markovian properties and find globally optimal solutions through search rather than greedy methods. These constraints can encode musical structures like repetition, motifs, endings, etc. Different Markovian cost functions can also bias generation towards more structured sequences. In addition, the model performs chunk-wise generation which allows it to maintain structure over long sequences (full songs) rather than just short clips. The chunks can be stitched to maintain longer-term dependencies.

Another notable example of a Markov approach for generating music with long-term repetitive and phrasal structure is the Racchmaninof-Jun2015 algorithm [13]. In this study, the authors develop a Markov model for inheriting long-term repetitive structure from template pieces using the SIACT pattern discovery algorithm from [15]. The beat-relative-MIDI state space representation used in their paper is an improvement over a previous model called Racchmaninof-Oct2010 [14] that uses beat-spacing state space representation in which the spacing difference with respect to the previous note could consist of tonally obscure representations. Using the extracted structural framework of a template piece, the Markov model can generate complete musical textures in a forward as well as backward generative way to form phrases that are perceived as having a beginning, middle and an end.

Herremans et al. [27] present an application of Markov models combined with Variable Neighbourhood Search (VNS) to generate structured music in the style of Bagana, a traditional lyre from Ethiopia. The structure of Bagana pieces are encoded as cyclic patterns capturing repetitive cycles and global form. These patterns are realized by the VNS optimization technique that generates candidate melodies fitting the predefined structure of a template piece. A first order Markov model learned from the Bagana corpus is used in the next step to evaluate how well the melody matches the corpus statistics and musical structure. The VNS then selects the melodies optimizing the objectives to improve over iterations. While the Markov model does not directly generate notes, it provides the statistical basis for the objective functions guiding the VNS to produce melodies adhering to the structure with stylistic consistency. The integration of optimization and statistical learning helps in structuring music in Bagana's style.

Morpheus [26] extends the VNS approach to polyphonic music, with automatic detection of more complex long term patterns in the template piece with respect to a user-specified tension profile. Morpheus comprises 3 components that are core to its framework. First, the authors use the COSIATEC and SIATECComPress [47] pattern detection algorithms to extract repeated note patterns from a template piece. Users can choose between COSIATEC, which captures each note in precisely one pattern, or SIATECCompress, which captures more relationships between different notes, resulting in overlapping patterns. This allows imposing long-term structure in the generated music. Second, these patterns are then constrained in a VNS optimization framework that assigns pitches to generate new music matching a target tension profile. And finally, tension used in the VNS optimization algorithm is quantified using the spiral array model of tonality developed in [25]. While Morpheus has received favorable feedback overall, certain limitations are evident from a structural perspective. These include limited flexibility in pattern detection and the absence of captured variations or transformations, attributed to the pattern recognition algorithms used. Additionally, Morpheus is restricted in developing motifs, primarily repeating patterns without sophisticated development, and lacks a formal quantitative evaluation of structure retention, relying solely on informal listening tests for assessment.

Genetic algorithms, a class of evolutionary computing approaches that are inspired by the principles of natural selection, have also been explored in the concept of evolving musical motifs and structures over generations. For example, in [63], the authors propose an automatic music composition system called Phrase Imitation-based Evolutionary Composition (PIEC) that generates new melodies with structure and form by imitating phrases from a sample melody composed by people. The framework uses a genetic algorithm along with intraphrase and interphrase rearrangement to mimic the ascending or descending melodic progression and note distribution of phrases in the sample melody. Similarly, in [1], the authors use a genetic algorithm to generate variations on a musical piece. To ensure that the generated variations have the underlying melodic skeleton of the original piece, the paper derives a melodic similarity fitness function from the spiral array model proposed in [9].

3   Deep Learning Methods

Symbolic methods, though useful in their versatility, are limited in their development across diverse musical datasets because they rely heavily on specific templates and encode strict musical rules, potentially restricting the generated music. In contrast, deep learning breaks free from these constraints. By learning from extensive datasets, deep learning models understand complex patterns and functions, and exploit locality through the use inductive biases. This enables them to contribute to improved quality in applications such as prompt based conditional generation, unconditional generation with no inputs or assistance, musical infilling as well as accompaniment generation [45].

3.1   Foundational Deep Learning Techniques

Lookback RNN and Attention RNN [66] by Google Magenta were two notable approaches early in the deep learning era that modeled long term structure in the generated melodies. Lookback RNN encoded repeating patterns from 1 or 2 bars ago in the input to allow the RNN model to identify repeating events better in the generated melody. Attention RNN used an earlier version of the self-attention mechanism from [3] to generate melodies with longer dependencies without storing information in the LSTM cell's state. However, while these improvements are noticeable over brief periods, they may not necessarily impact the higher-level structural aspects of the piece.

Building on the Lookback RNN model, the authors of StructureNet [46] propose an enhancement at the generation stage of an RNN based melody generation model that induces repetition and structure in the generated melodies. StructureNet is trained on sequences of structural elements like repeats extracted from a dataset of melodies. When generating new melodies, it runs alongside the melody RNN model, biasing it towards notes that would form repeats similar to those seen in the training data. To evaluate if StructureNet improves structure, the authors compared statistics related to repeats as well as pitch and rhythm distributions between melodies generated with and without StructureNet. The results showed that StructureNet increased the occurrence and lengths of repeats, matching the melodies in the training data more closely, while preserving pitch and rhythm characteristics.

In another study, [8], the authors proposed using WaveNet [51], a generative model based on dilated convolutional neural networks, for generating new melodies that fit a given chord progression. They found WaveNet was able to better learn musical structure compared to its LSTM counterpart. This is because the WaveNet model encodes melodic structure more explicitly through its dilated convolutions across multiple time scales. Each layer doubles the receptive field of the previous layer, allowing it to learn dependencies over larger spans of melody. In contrast, an LSTM relies solely on its internal memory to capture structure, which is less explicit. As a result, WaveNet was able to generate coherent rhythms and melodies over many time steps that exhibited more repetitive patterns and long-range dependencies over the LSTM baseline that was reflective of compositional structure.

3.2 Transformative Deep Learning Approaches

Along the lines of model based enhancements for inducing structure in longer compositions is Music Transformer [33], which uses the transformer model equipped with the relative attention mechanism. This allows the model to learn patterns based on relative distances between events, making it easier to capture motifs and repetition on multiple timescales. However, this approach doesn't explicitly capture abstract structural elements such as how musical phrases and sections transition and develop over time. Additionally, when generating longer music pieces (over a minute), the model is not able to adhere to the initial prompt sequence and produces output with poor musicality and global structure.

In [4], the authors introduced Long-Short Term Universal Transformer (LSTUT), a fusion of transformers and RNNs, aiming to capture both local patterns and extensive musical relationships. This approach stemmed from a fundamental hypothesis: the intertwining of recurrence for short-term structure and attention for long-term dependencies enhances inductive biases, enabling the modelling of music spanning several minutes. However, while their methodology is promising, the evaluation is somewhat limited. Their assessment primarily relies on cross-entropy loss and manual inspection of attention heads to decipher the learned musical features. The lack of a subjective test and absence of generated samples restricts a comprehensive understanding of the model's potential.

Along the lines of feature representation based enhancements is the Pop music transformer [34] in which the authors propose REMI (revamped MIDI derived events) to represent MIDI data following the way humans read them. REMI uses position and bar events to embed an explicit beat/bar grid, tempo events to allow flexible tempo changes, and chord events to represent harmony. This equips models with a sense of rhythm and harmonic structure to better learn musical dependencies. Pop Music Transformer uses REMI with a Transformer-XL backbone. Objective evaluation shows the model generates piano compositions with more consistent rhythm and clearer downbeats compared to baselines that use the MIDI-like representation [52]. Subjective listening tests also indicate REMI results in more pleasing and coherent continuations of given musical prompts. While REMI offers a simple yet effective way to inject musical inductive biases into sequence models through data representation, the input sequence length can be a point of concern for music of longer duration.

The Theme Transformer model [58] effectively tackles the constraints of Music Transformer by incorporating thematic materials into the generated outputs. Initially, the authors employ an unsupervised approach, combining contrastive learning with clustering techniques to extract recurring thematic patterns from 2-bar melodic segments. These themes are then integrated into a transformer encoder, and music generation occurs through a transformer decoder, which cross-attends to the encoder. The architecture of the Theme Transformer incorporates gated parallel attention and theme-aligned positional encoding. This design ensures the sustained influence of the thematic condition over extended generation periods. This is achieved by modulating between the self-attention mechanism based on the decoder's auto-regressive inputs and the cross-attention from the encoder's thematic outputs. Evaluations demonstrate that the Theme Transformer excels in generating music with repeated thematic elements compared to prompt-based Transformers. However, it faces challenges in creating variations and evolving the recurring themes, showcasing areas for potential improvement.

In [32], a novel R-Transformer model is introduced, aiming to tackle the under-explored problem of modelling varied motif repetitions. To facilitate this, a new music repetition dataset comprising over 500,000 labeled motif repetitions across 5 types is constructed from an existing piano music dataset. R-Transformer combines a Transformer encoder for note representation learning and a repetition-aware learner that exploits repetition characteristics based on music theory, enabling controlled generation of designated repetitions. The model is trained using both reconstruction and classification losses. Evaluations show it outperforms baseline models in generating varied high-quality repetitions of motifs and creates subjectively more enjoyable music than prior models, as rated by both musicians and non-musicians. This is attributed to explicitly modelling repetition types and their combinations, which prior generative models overlook.

3.3 Hierarchical Neural Networks

There exists a class of models that draw inspiration from the hierarchical nature of musical elements like notes, chords, bars, phrases, and movements. These models leverage this inherent hierarchy by employing specialized network structures to dissect and generate intricate melodic sequences.

MeloNet [29] is an early example of a small feed-forward based neural network architecture that uses a hierarchical approach to generate melodic variations in Baroque style given a melody input. Specifically, it employs two neural networks operating at different timescales - a "supernet" that predicts motif classes at a higher level based on the melody and previous motifs, and a "subnet" that generates the actual notes based on the motif, harmony, and previous notes.

Similarly, [70] introduces a hierarchical RNN model for monophonic melody generation, employing three Long-Short-Term-Memory (LSTM) sub-networks that operate in a sequential manner: generating bar profiles, beat profiles, and notes. The outputs of higher-level sub-networks guide the generation of finer melody components in lower-level sub-networks. The subjective experiments conducted using this hierarchical approach demonstrate improved melodic quality and long term structure over MidiNet [74] and MusicVAE [55].

Expanding on this concept, Guo et al. [79] employed a three-tiered hierarchical RNN to simultaneously model rhythmic and pitch structures. This approach allowed the model to capture hierarchical musical relationships across varying time scales, evident in the presence of more repeated patterns indicated by a higher compression ratio metric. MusicFrameworks [17] adopted a multi-step hierarchical process, breaking down melody generation into manageable subtasks guided by music frameworks. This method facilitated the use of simpler models trained with less data, enabling the creation of long-term structures, including repetitions, within full-length melodies.

Another notable example of a hierarchical transformer model is HAT (**H**armony-**A**ware Hierarchical Music **T**ransformer) proposed in [77]. HAT leverages musical harmony to jointly model form and texture structure for pop music generation. HAT represents music as tokens with attributes like notes, chords and phrases, and uses Transformer blocks in a three tiered hierarchy to capture structure at different levels. Subjective as well as objective experiments on musical structural attributes show HAT generates pop music with improved coherence in form and stability in texture compared to prior methods. However, there is still room for performance improvement for longer chord progressions and ending sections.

3.4　Long Sequence/Efficiency Models

A class of models hypothesize generating longer sequences is key to improving musical quality and structure, a notion grounded in the observation that training data often surpasses the capabilities of standard full-attention vanilla transformer models. The Compound Word Transformer work [31] groups consecutive music tokens into "compound words" and processes them together for compactness. This allows their model to handle longer sequences to capture long term dependencies and induce structure. In a user study, the authors demonstrated improved structural outcomes during unconditional generation. Similarly, MuseNet [11] adopts the Sparse Transformer [10] architecture, expanding the context window to 4096 tokens. The aim is to generate music with extended structural coherence from a given prompt. Another notable advancement is Museformer [75], a Transformer model for long sequence symbolic music generation that captures musical structure through a fine and coarse-grained attention mechanism. It allows each token to directly attend to selected "structure-related" bars based on statistical analysis of common repetitive patterns in music, providing fine-grained attention for modelling local structures. All other non-structure-related bars are summarized via coarse-grained attention to retain necessary context while greatly reducing computation compared to full attention. This combined approach efficiently generates long, structured musical sequences by capturing repetitive patterns and variations through fine-grained attention, while still incorporating global context. However, one limitation is that the bar selection is based on statistics and may not generalize perfectly to all music. The SymphonyNet [41] approach introduces a permutation invariant byte pair encoding (BPE) representation for symbolic symphony music generation. The BPE representation is proposed to preserve note structure through a compressed encoding scheme by preventing the overflow of long sequence multi-instrument tokens.

3.5   Pre-training Methods

Pre-training involves training a model on a large unlabeled dataset before fine-tuning it for a specific task, aiding the model in learning general representations applicable to downstream tasks. Pre-training can prove advantageous in symbolic music generation tasks due to the intricate, multi-dimensional structure of music, encompassing local and global patterns involving elements like pitch, rhythm and dynamics. Music's high complexity makes it important to learn musical syntax and patterns from vast datasets, enhancing coherence during fine-tuning. However, compared to other domains like text, there are fewer large-scale structured symbolic music datasets available [45,57,73] for task specific problems. Utilizing unlabeled data through pre-training facilitates transfer learning [30], enabling models to leverage discovered patterns and structures when fine-tuning on limited task-specific datasets, maximizing the utility of available labeled data. A few studies [21,40,57,67,73] have explored pre-training in the symbolic music domain aimed at generation with better musical structure and overall coherence.

In the LakhNES project [21], the authors pre-trained a Transformer-XL model using language modelling on a large MIDI dataset. This was then fine-tuned on multi-instrumental music generation. The goal was to leverage the patterns learned during pre-training to generate more coherent long-form music. SongMASS [57] also used a masked language modelling approach to pre-train an encoder-decoder model. They specifically lengthened the masked spans during pre-training to capture longer-term repetitive structures like verses and choruses. This helped generate songs with improved global structure. MuseBERT [67] and MRBERT [40] adapted BERT [20], a language representation model for various downstream music generation tasks by pre-training on piano music segments. They proposed techniques to handle music's polyphonic and multi-dimensional nature compared to text. Recently proposed MelodyGLM [73] introduces a multi-task pre-training framework to model both local patterns and long-term structure in melodies. It uses melodic n-grams and long span masking objectives tailored to music during pre-training. MelodyGLM also constructs a large-scale melody dataset to enable robust pre-training. Both objective and subjective evaluations demonstrate MelodyGLM's ability to generate melodies with improved structure and coherence.

4 Subtask Decomposition

In the context of deep learning, a new method has emerged which involves decomposing the training and generation process into smaller, more manageable steps (typically 2 stages). First, a high-level structure or plan for the music piece is created, outlining its organization and patterns. In the second stage, the actual musical content, such as notes and chords, is generated while adhering to this plan. This separation helps in addressing the challenges of modelling long-term structures and generating details locally. The plan ensures overall coherence, while the detailed music is filled in from the bottom-up, providing a holistic musical composition. There are two notable differences between pre-training and subtask decomposition methods. The former involves the use of the first pre-trained model in the fine-tuning stage whereas the latter involves an entirely separate model in the final stage. Second, the subtask decomposition framework explicitly models musical structure through repeated patterns or an outline of a melodic skeleton in its initial stage. Additionally, while subtasks could be thought of as a hierarchy, what separates them from the hierarchical models in Sect. 3.3 is that the latter trains a single model in a hierarchical fashion. In contrast, subtask decomposition approaches may involve multiple training objectives.

An early design of this framework can be seen in [65] in which the authors use modular planning theory to generate tree like structures at multiple levels which are referred to as sentences, double phrases and phrases. Each level affects the next through object dependencies and constraints. These planning structures are then filled with musical material using a bottom-up Markov based generation technique. Their results show that automated planning can efficiently produce repetition and novelty in melodies, resembling human compositions.

In another study, [68] the authors focus on the problem of drum pattern generation, specifically drum sequences that are rhythmically and structurally compatible with a given melodic track. The proposed model uses a two stage generation framework. In the first stage, a variational autoencoder (VAE) generative adversarial network from [38] is trained to generate a self-similarity matrix (SSM) for the drum track given the SSM of the melodic track. This drum SSM captures the structural information of the music. In the second stage, another VAE model then generates the actual MIDI drum patterns conditioned on the drum SSM and using a bar selection mechanism to encourage self-repetition. Objective and subjective tests conducted against baselines yielded promising outcomes. Objective and subjective tests against baselines showed promising results, highlighting the model's enhanced performance due to its incorporation of global song structure through the drum SSM.

Similarly, PopMNet [71] proposes a two stage process for the generation of pop melodies with well defined long term structure. It represents melody structure as a directed acyclic graph capturing repetitive and sequential relationships between bars. In the first stage, a convolutional GAN model is trained on this graph based structural representation to generate plausible melodic structures. In the second stage, an RNN model is then trained to generate the actual melody conditioned on the structure and chords from the first stage. Human evaluations reveal that melodies generated by PopMNet receive higher ratings in terms of enjoyability, humanness, smoothness, and structure compared to previous models such as AttentionRNN [66], LookbackRNN [66], MidiNet [74], and Music Transformer [33]. The analysis shows that PopMNet melodies exhibit repetition patterns similar to real pop songs, a feature previous models struggled to capture. However, one limitation of the model is its inability to capture complex structural patterns beyond repetitions and sequences.

Melons [80] overcomes PopMNet's limitations by utilizing a more complex multi-edge graph representation of musical structure, incorporating 8 types of bar-level relations compared to PopMNet's focus on repetition and rhythm sequences. Additionally, Melons adopts a Transformer architecture for both structure and melody generation, an upgrade from PopMNet's CNN and RNN models. In cases where no edges exist between bars, an unconditional transformer model generates melodies. Conversely, when relationships exist, a conditional transformer model is employed, considering the previous 8 bars and pairwise bar relationships as context. While Melons outperforms PopMNet in structure and overall musicality, there's still a gap between generated melodies and human-created music. About 80% of the bars in the dataset used in the study cover the proposed relations, suggesting potential room for improvement by exploring additional structural relationships between bars.

In [18], the authors present a modular approach for generating stylistic pop music by imitating a single seed song. This approach uses purely statistical and rule-based methods to capture distinctive melodic, harmonic, rhythmic, and structural attributes from the seed song. A proposed structure alignment procedure helps adhere to the stylistic section lengths and repetition patterns of the seed song. Separate generative models produce melody lines, chord progressions, bass-lines, and overall song structure. The generated songs are evaluated using both objective similarity metrics and subjective listening tests. While results show the model can produce enjoyable, novel music that listeners recognize as similar to the seed song, the authors acknowledge the potential for enhancing musical quality by combining their knowledge-based method with advanced deep learning techniques. They note that much of the model's success stems from making good sequential choices guided by music theory and statistics. However, more sophisticated deep learning sequence models may be able to make even better choices for the next note in a generated sequence. The challenge would be retaining the benefits of the knowledge-based structure and harmony modelling while leveraging the pattern learning strengths of deep networks.

WuYun [76] proposes a hierarchical two-stage skeleton-guided melody generation architecture that incorporates musical knowledge. It first extracts a melodic skeleton of structurally important notes using music theory concepts such as downbeats, accents and tonal tension. In the first stage, a Transformer-XL network is trained on the extracted melodic skeleton to generate new skeleton sequences. The second stage comprises a Transformer encoder-decoder model that generates the full melody conditioned on encoding the skeleton to guide the process. Experiments on the Wikifonia dataset and subsequent subjective evaluations show WuYun produces melodies with improved structure and musicality compared to prior models including Pop Music Transformer [34] and Melons [80] but still falls short with respect to the musical qualities of human compositions.

AccoMontage [78] is a two step neuro-symbolic system for generating piano accompaniments for complete songs given a lead sheet that comprises the melody and chord information of the complete song. AccoMontage combines rule-based optimization for high-level structure and deep learning for local coherence. The system first retrieves candidate accompaniment phrases from a database using dynamic programming to optimize for good transitions and phrase-level fit. It then re-harmonizes the retrieved phrases using neural style transfer to match the target chord progression of the template song. Subjective and objective evaluations show AccoMontage generates more coherent and well-structured accompaniments compared to pure learning-based and template-matching baselines. However, the dependence on manual phrase level annotations in the dataset places constraints in the applicability of the model to other datasets.

The TeleMelody model in [36] applies a two-stage neuro-symbolic approach for lyric-to-melody generation. The first stage trains a lyric-to-template model to generate the symbolic template from lyrics. The template design incorporates musical knowledge and rules revolving around musical elements such as tonality, chord progression, rhythm pattern, and cadence, providing a symbolic representation that bridges lyrics and melodies. The second stage then trains a template-to-melody model in a self-supervised manner by extracting these templates from existing melodies and training a transformer model to reconstruct the melodies from the templates. Evaluations show TeleMelody generates melodies with better structure and musical attributes than previous end-to-end models with less requirement on paired training data. This is possible due to the intermediate template which helps reduce overall difficulty and improve data efficiency.

MeloForm [44] also uses a neuro-symbolic approach to generate melodies with musical form control by combining expert systems and neural networks. MeloForm has two main components. First, an expert system generates synthetic melodies with a given musical form using handcrafted rules. It develops motifs into phrases and arranges phrases into sections with repetitions/variations based on the musical form. Second, a 4 layered transformer encoder-decoder network refines the synthetic melodies from the expert system to improve musical richness without changing the form. The Transformer is trained on the Lakh MIDI Dataset [54] using an iterative masked sequence-to-sequence approach. During refinement, it conditions on the adjacent phrase's rhythm and harmony to predict the pitch of masked phrases while preserving overall structure. The expert system provides flexibility over musical form, while the neural network improves musical coherence. This allows MeloForm to support various musical forms such as verse-chorus, rondo, variational, sonata, etc. Subjective evaluations performed on MeloForm's generated outputs show an improvement over Melons [80] in musical structure, thematic material, melodic richness and overall quality.

Compose & Embellish [72] introduces another two-stage linear transformer based framework to generate piano performances with lead sheets as the intermediate output. The authors extract the monophonic melodic line using the skyline algorithm from [64]. For structural information, the authors employ a structural analysis algorithm from [19] that utilizes edit similarity and the A* search to identify repetitive phrases within the composition. Notably, the first stage involves a pre-training strategy using a separate larger dataset (Lakh MIDI Dataset [54]) for learning new lead sheets, setting it apart from the other two stage models. The second stage fine-tunes the lead sheet model and trains the performance model using the Pop1K7 dataset [31]. Conditioning the performance model on interleaved one-bar segments from the pre-trained lead sheet model enables the generation of complete performance bars with contextual understanding from the latest lead sheet bar.

5 Future Directions

Exploring Structural Encoding: While many music generation systems have used bar level representations such as REMI [34,72] and REMI+ [56] for modelling music with structure, incorporating phrase level representations or meta data may be an interesting direction to explore. Recent work by [49] extends the bar-level REMI encoding to incorporate phrase-level and bar countdown features. Their model demonstrates the value of encoding pop musical structure at the phrase level, aligning more closely with human music perception. However, phrase-based modelling and repetitive motif detection remains an open challenge in algorithmic music generation and composition. One barrier is the lack of standard practices or implementations for extracting musical phrases and structural annotations across genres. Moving forward, the development of robust tools to systematically extract and evaluate musical phrases across corpora would provide the necessary components to assemble hierarchical phrase structures, thereby enabling systems to construct forms at the phrase level. Additional annotations demarcating phrase boundaries within datasets would also facilitate progress in this area.

Mastering Advanced Compositional Techniques: Developing sophisticated long-term structure requires moving beyond basic repetitions and successions. Current models such as [33] and [58] struggle with nuanced variations, hindering the conveyance of complex musical narratives. Advancements demand models to master musical development through compositional techniques such as fragmentation, inversion, augmentation, diminution, stretto, sequence, and modulation across various time frames. Learning how to manifest a motif across the piece in a more structural way by incorporating such melodic and rhythmic transformations may warrant for a similarity conscious generative framework in which the generator understands how similar the last generated phrase is to the initial motif, previous phrase and overall form and has the ability to go back to refine it. Building a pause-think-refine framework along similar lines to [26] would help research advance beyond sequential note-to-note generations that do not explicitly model higher-level motivic structures and transformations.

Integrated Neuro-Symbolic Approaches in Music Generation: Music generation with long term structure is a complex endeavor that requires the coordination of multiple components at the preprocessing stage. For example, preprocessing data to extract melodic/rhythmic/phrase level features into meaningful musical representations provides the groundwork for modelling music, as seen in [44,58,72,76]. However, recent advances in music generation have not yet effectively integrated insights from symbolic methodologies and cognitive musicology to achieve long-term structural coherence. While musical information retrieval has actively researched structural analysis algorithms for decades - spanning tempo estimation, key and modulation detection, cadence identification, and phrase similarity quantification, etc. - modern neural approaches surprisingly overlook these methods, opting for simplistic data preprocessing. Furthermore, cognitive studies elucidating how humans internalize and anticipate musical structure are imperative for designing evaluation protocols. Yet current assessments of generative systems' musicality, both objective and subjective, lack foundations in perceptual research. Effectively melding the methodological tools of symbolic analysis and lessons from music cognition remains an open challenge in paving the way for a truly unified neuro-symbolic approach.

Top Down Structure: Sub-task decomposition studies in Sect. 4 such as [72,76,80] clearly demonstrate the viability of top-down hierarchical plan-then-generate frameworks for algorithmic composition. By separating high-level musical form construction from lower-level sequence generation, these two-stage systems aim to improve global coherence. While we expect this trend to continue, we also hope to see studies that explore alternatives to the existing task breakdown involving pre-training, multi-task learning, unsupervised and contrastive learning methods. What data, features, representations, learning strategies and model architectures the sub-tasks consist of and how many stages of hierarchy make the framework most effective are questions worth looking into.

6   Conclusion

This literature review has charted elements of the evolution of modelling musical structure in AI-powered symbolic composition, from early symbolic systems to contemporary deep learning techniques and subtask decomposition frameworks. While progress is evident in capturing motifs, progressions, and global forms, modelling nuanced development and variation of themes much like human compositions across extended periods remains challenging. Key future directions involve moving beyond rigid structural units, integrating neuro-symbolic approaches at various stages in the system, and further exploring top down structures to shape musical narratives. Continued progress across complementary AI techniques provides hope for achieving the creative development of themes reflective of human composers and musical culture, which may someday be attainable.

Acknowledgements. This work was supported by the UKRI and EPSRC under grant EP/S022694/1. We gratefully acknowledge the use of generative AI in drafting and revising certain sections of this literature review. Finally, we owe much appreciation to our reviewers for their insightful critiques which greatly strengthened the work.

References

1. Alvarado, F.H.C., Lee, W.H., Huang, Y.H., Chen, Y.S.: Melody similarity and tempo diversity as evolutionary factors for music variations by genetic algorithms. In: 11th International Conference on Computational Creativity (ICCC), pp. 251–254 (2020)
2. Amaral, G., Baffa, A., Briot, J.P., Feijó, B., Furtado, A.: An adaptive music generation architecture for games based on the deep learning transformer model. In: 2022 21st Brazilian Symposium on Computer Games and Digital Entertainment (SBGames), pp. 1–6. IEEE (2022)
3. Bahdanau, D., Cho, K., Bengio, Y.: Neural machine translation by jointly learning to align and translate. arXiv preprint arXiv:1409.0473 (2014)
4. de Berardinis, J., Barrett, S., Cangelosi, A., Coutinho, E.: Modelling long-and short-term structure in symbolic music with attention and recurrence. In: Proceedings of The 2020 Joint Conference on AI Music Creativity, pp. 1–11 (2020)
5. Briot, J.P., Hadjeres, G., Pachet, F.D.: Deep learning techniques for music generation–a survey. arXiv preprint arXiv:1709.01620 (2017)
6. Burns, G.: A typology of 'hooks' in popular records. Popular Music **6**(1), 1–20 (1987)
7. Carnovalini, F., Rodà, A.: Computational creativity and music generation systems: an introduction to the state of the art. Front. Artif. Intell. **3**, 14 (2020)
8. Chen, K., Zhang, W., Dubnov, S., Xia, G., Li, W.: The effect of explicit structure encoding of deep neural networks for symbolic music generation. In: 2019 International Workshop on Multilayer Music Representation and Processing (MMRP), pp. 77–84. IEEE (2019)
9. Chew, E., Chen, Y.C.: Real-time pitch spelling using the spiral array. Comput. Music. J. **29**(2), 61–76 (2005)
10. Child, R., Gray, S., Radford, A., Sutskever, I.: Generating long sequences with sparse transformers. arXiv preprint arXiv:1904.10509 (2019)
11. Christine, P.: MuseNet. https://openai.com/research/musenet
12. Civit, M., Civit-Masot, J., Cuadrado, F., Escalona, M.J.: A systematic review of artificial intelligence-based music generation: scope, applications, and future trends. Expert Syst. Appl. 118190 (2022)
13. Collins, T., Laney, R.: Computer-generated stylistic compositions with long-term repetitive and phrasal structure. J. Creative Music Syst. **1**(2) (2017)
14. Collins, T., Laney, R., Willis, A., Garthwaite, P.H.: Developing and evaluating computational models of musical style. AI EDAM **30**(1), 16–43 (2016)
15. Collins, T., Thurlow, J., Laney, R.C., Willis, A., Garthwaite, P.H.: A comparative evaluation of algorithms for discovering translational patterns in baroque keyboard works. In: Downie, J.S., Veltkamp, R.C. (eds.) Proceedings of the 11th International Society for Music Information Retrieval Conference, ISMIR, pp. 3–8 (2010). http://ismir2010.ismir.net/proceedings/ismir2010-2.pdf

16. Cont, A., Dubnov, S., Assayag, G.: Anticipatory model of musical style imitation using collaborative and competitive reinforcement learning. In: Butz, M.V., Sigaud, O., Pezzulo, G., Baldassarre, G. (eds.) ABiALS 2006. LNCS, vol. 4520, pp. 285–306. Springer, Heidelberg (2006). https://doi.org/10.1007/978-3-540-74262-3_16

17. Dai, S., Jin, Z., Gomes, C., Dannenberg, R.B.: Controllable deep melody generation via hierarchical music structure representation. arXiv preprint arXiv:2109.00663 (2021)

18. Dai, S., Ma, X., Wang, Y., Dannenberg, R.B.: Personalised popular music generation using imitation and structure. J. New Music Res. 51(1), 69–85 (2022)

19. Dai, S., Zhang, H., Dannenberg, R.B.: Automatic analysis and influence of hierarchical structure on melody, rhythm and harmony in popular music. arXiv preprint arXiv:2010.07518 (2020)

20. Devlin, J., Chang, M.W., Lee, K., Toutanova, K.: Bert: pre-training of deep bidirectional transformers for language understanding. arXiv preprint arXiv:1810.04805 (2018)

21. Donahue, C., Mao, H.H., Li, Y.E., Cottrell, G.W., McAuley, J.: Lakhnes: improving multi-instrumental music generation with cross-domain pre-training. arXiv preprint arXiv:1907.04868 (2019)

22. Fitch, W.T., Rosenfeld, A.J.: Perception and production of syncopated rhythms. Music Percept. 25(1), 43–58 (2007)

23. Frid, E., Gomes, C., Jin, Z.: Music creation by example. In: Proceedings of the 2020 CHI Conference on Human Factors in Computing Systems, pp. 1–13 (2020)

24. Gabrielsson, A.: The relationship between musical structure and perceived expression (2014)

25. Herremans, D., Chew, E.: Tension ribbons: quantifying and visualising tonal tension. In: Hoadley, R., Nash, C., Fober, D. (eds.) Proceedings of the International Conference on Technologies for Music Notation and Representation - TENOR 2016, pp. 8–18 (2016)

26. Herremans, D., Chew, E.: Morpheus: generating structured music with constrained patterns and tension. IEEE Trans. Affect. Comput. 10(4), 510–523 (2017)

27. Herremans, D., Weisser, S., Sörensen, K., Conklin, D.: Generating structured music for bagana using quality metrics based on Markov models. Expert Syst. Appl. 42(21), 7424–7435 (2015)

28. Honing, H., et al.: Structure and interpretation of rhythm and timing. Tijdschrift voor Muziektheorie 7(3), 227–232 (2002)

29. Hörnel, D.: Melonet i: Neural nets for inventing baroque-style chorale variations. In: Advances in Neural Information Processing Systems, vol. 10 (1997)

30. Hosna, A., Merry, E., Gyalmo, J., Alom, Z., Aung, Z., Azim, M.A.: Transfer learning: a friendly introduction. J. Big Data 9(1), 102 (2022)

31. Hsiao, W.Y., Liu, J.Y., Yeh, Y.C., Yang, Y.H.: Compound word transformer: Learning to compose full-song music over dynamic directed hypergraphs. In: Proceedings of the AAAI Conference on Artificial Intelligence, vol. 35, pp. 178–186 (2021)

32. Hu, Z., Ma, X., Liu, Y., Chen, G., Liu, Y.: The beauty of repetition in machine composition scenarios. In: Proceedings of the 30th ACM International Conference on Multimedia, pp. 1223–1231 (2022)

33. Huang, C.Z.A., et al.: Music transformer. arXiv preprint arXiv:1809.04281 (2018)

34. Huang, Y.S., Yang, Y.H.: Pop music transformer: beat-based modeling and generation of expressive pop piano compositions. In: Proceedings of the 28th ACM International Conference on Multimedia, pp. 1180–1188 (2020)

35. Ji, S., Yang, X., Luo, J.: A survey on deep learning for symbolic music generation: Representations, algorithms, evaluations, and challenges. ACM Comput. Surv. (2023)

36. Ju, Z., et al.: Telemelody: lyric-to-melody generation with a template-based two-stage method. arXiv preprint arXiv:2109.09617 (2021)

37. Krumhansl, C.L., Jusczyk, P.W.: Infants' perception of phrase structure in music. Psychol. Sci. 1(1), 70–73 (1990)

38. Larsen, A.B.L., Sønderby, S.K., Larochelle, H., Winther, O.: Autoencoding beyond pixels using a learned similarity metric. In: International Conference on Machine Learning, pp. 1558–1566. PMLR (2016)

39. Lazzari, N., Poltronieri, A., Presutti, V.: Pitchclass2vec: symbolic music structure segmentation with chord embeddings. arXiv preprint arXiv:2303.15306 (2023)

40. Li, S., Sung, Y.: MRBERT: pre-training of melody and rhythm for automatic music generation. Mathematics 11(4), 798 (2023)

41. Liu, J., et al.: Symphony generation with permutation invariant language model. arXiv preprint arXiv:2205.05448 (2022)

42. Liu, W.: Literature survey of multi-track music generation model based on generative confrontation network in intelligent composition. J. Supercomput. 79(6), 6560–6582 (2023)

43. Livingstone, S.R., Palmer, C., Schubert, E.: Emotional response to musical repetition. Emotion 12(3), 552–567 (2012). https://doi.org/10.1037/a0023747

44. Lu, P., Tan, X., Yu, B., Qin, T., Zhao, S., Liu, T.Y.: Meloform: generating melody with musical form based on expert systems and neural networks. arXiv preprint arXiv:2208.14345 (2022)

45. Makris, D., Zixun, G., Kaliakatsos-Papakostas, M., Herremans, D.: Conditional drums generation using compound word representations. In: Martins, T., Rodríguez-Fernández, N., Rebelo, S.M. (eds.) EvoMUSART 2022. LNCS, pp. 179–194. Springer, Cham (2022). https://doi.org/10.1007/978-3-031-03789-4_12

46. Medeot, G., et al.: Structurenet: inducing structure in generated melodies. In: International Society for Music Information Retrieval Conference (ISMIR), pp. 725–731 (2018)

47. Meredith, D.: Cosiatec and siateccompress: pattern discovery by geometric compression. In: International Society for Music Information Retrieval Conference (ISMIR). ISMIR (2013)

48. Miller, G.A.: The magical number seven, plus or minus two: some limits on our capacity for processing information. Psychol. Rev. 63(2), 81–97 (1956). https://doi.org/10.1037/h0043158

49. Naruse, D., Takahata, T., Mukuta, Y., Harada, T.: Pop music generation with controllable phrase lengths. In: Proceedings of the 23rd International Society for Music Information Retrieval Conference. Bengaluru, India (2022)

50. Ong, B.S., et al.: Structural Analysis and Segmentation of Music Signals. Citeseer (2006)

51. Oord, A.V.D., et al.: Wavenet: a generative model for raw audio. arXiv preprint arXiv:1609.03499 (2016)

52. Oore, S., Simon, I., Dieleman, S., Eck, D., Simonyan, K.: This time with feeling: learning expressive musical performance. Neural Comput. Appl. 32, 955–967 (2020)

53. Pachet, F., Roy, P.: Markov constraints: steerable generation of Markov sequences. Constraints 16, 148–172 (2011)

54. Raffel, C.: Learning-based methods for comparing sequences, with applications to audio-to-MIDI alignment and matching. https://doi.org/10.7916/D8N58MHV

55. Roberts, A., Engel, J., Raffel, C., Hawthorne, C., Eck, D.: A hierarchical latent vector model for learning long-term structure in music. In: International Conference on Machine Learning, pp. 4364–4373. PMLR (2018)

56. von Rütte, D., Biggio, L., Kilcher, Y., Hofmann, T.: Figaro: generating symbolic music with fine-grained artistic control. arXiv preprint arXiv:2201.10936 (2022)

57. Sheng, Z., et al.: Songmass: automatic song writing with pre-training and alignment constraint. In: Proceedings of the AAAI Conference on Artificial Intelligence, vol. 35, pp. 13798–13805 (2021)

58. Shih, Y.J., Wu, S.L., Zalkow, F., Muller, M., Yang, Y.H.: Theme transformer: symbolic music generation with theme-conditioned transformer. IEEE Trans. Multimedia (2022)

59. Stevens, C.J.: Music perception and cognition: a review of recent cross-cultural research. Top. Cogn. Sci. **4**(4), 653–667 (2012)

60. Tan, N., Aiello, R., Bever, T.G.: Harmonic structure as a determinant of melodic organization. Memory Cogn. **9**(5), 533–539 (1981)

61. Tang, H., Gu, Y., Yang, X.: Music generation with AI technology: Is it possible? In: 2022 IEEE 5th International Conference on Electronics Technology (ICET), pp. 1265–1272. IEEE (2022)

62. Temperley, D.: The Cognition of Basic Musical Structures. MIT press, Cambridge (2004)

63. Ting, C.K., Wu, C.L., Liu, C.H.: A novel automatic composition system using evolutionary algorithm and phrase imitation. IEEE Syst. J. **11**(3), 1284–1295 (2015)

64. Uitdenbogerd, A.L., Zobel, J.: Manipulation of music for melody matching. In: Proceedings of the Sixth ACM International Conference on Multimedia, pp. 235–240 (1998)

65. Velardo, V., Vallati, M.: A planning-based approach for music composition. Springer (2015)

66. Waite, E., Eck, D, Roberts, A, Abolafia, D: Generating long-term structure in songs and stories. https://magenta.tensorflow.org/2016/07/15/lookback-rnn-attention-rnn

67. Wang, Z., Xia, G.: MuseBERT: pre-training music representation for music understanding and controllable generation. In: International Society for Music Information Retrieval Conference (ISMIR), pp. 722–729 (2021)

68. Wei, I.C., Wu, C.W., Su, L.: Generating structured drum pattern using variational autoencoder and self-similarity matrix. In: International Society for Music Information Retrieval Conference (ISMIR), pp. 847–854

69. Wigram, T., Gold, C.: Music therapy in the assessment and treatment of autistic spectrum disorder: clinical application and research evidence. Child: Care, Health Dev. **32**(5), 535–542 (2006)

70. Wu, J., Hu, C., Wang, Y., Hu, X., Zhu, J.: A hierarchical recurrent neural network for symbolic melody generation. IEEE Trans. Cybern. **50**(6), 2749–2757 (2019)

71. Wu, J., Liu, X., Hu, X., Zhu, J.: PopMNet: generating structured pop music melodies using neural networks. Artif. Intell. **286**, 103303 (2020)

72. Wu, S.L., Yang, Y.H.: Compose & embellish: well-structured piano performance generation via a two-stage approach. http://arxiv.org/abs/2209.08212

73. Wu, X., et al.: MelodyGLM: multi-task pre-training for symbolic melody generation. arXiv preprint arXiv:2309.10738 (2023)

74. Yang, L.C., Chou, S.Y., Yang, Y.H.: MidiNet: a convolutional generative adversarial network for symbolic-domain music generation. arXiv preprint arXiv:1703.10847 (2017)

75. Yu, B., et al.: Museformer: transformer with fine-and coarse-grained attention for music generation. Adv. Neural. Inf. Process. Syst. **35**, 1376–1388 (2022)
76. Zhang, K., et al.: WuYun: exploring hierarchical skeleton-guided melody generation using knowledge-enhanced deep learning. arXiv preprint arXiv:2301.04488 (2023)
77. Zhang, X., Zhang, J., Qiu, Y., Wang, L., Zhou, J.: Structure-enhanced pop music generation via harmony-aware learning. In: Proceedings of the 30th ACM International Conference on Multimedia, pp. 1204–1213 (2022)
78. Zhao, J., Xia, G.: Accomontage: accompaniment arrangement via phrase selection and style transfer. arXiv preprint arXiv:2108.11213 (2021)
79. Zixun, G., Makris, D., Herremans, D.: Hierarchical recurrent neural networks for conditional melody generation with long-term structure. In: 2021 International Joint Conference on Neural Networks (IJCNN), pp. 1–8. IEEE (2021)
80. Zou, Y., Zou, P., Zhao, Y., Zhang, K., Zhang, R., Wang, X.: Melons: generating melody with long-term structure using transformers and structure graph. In: ICASSP 2022-2022 IEEE International Conference on Acoustics, Speech and Signal Processing (ICASSP), pp. 191–195. IEEE (2022)

The Chordinator: Modeling Music Harmony by Implementing Transformer Networks and Token Strategies

David Dalmazzo[1]([✉]) [ID], Ken Déguernel[2] [ID], and Bob L. T. Sturm[1]([✉]) [ID]

[1] KTH-Royal Institute of Technology, Stockholm, Sweden
{dalmazzo,bobs}@kth.se
[2] Univ. Lille, CNRS, Centrale Lille, UMR 9189 CRIStAL, 59000 Lille, France
ken.deguernel@cnrs.fr

Abstract. This paper compares two tokenization strategies for modeling chord progressions using the encoder transformer architecture trained with a large dataset of chord progressions in a variety of styles. The first strategy includes a tokenization method treating all different chords as unique elements, which results in a vocabulary of 5202 independent tokens. The second strategy expresses the chords as a dynamic tuple describing **root**, **nature** (e.g., major, minor, diminished, etc.), and **extensions** (e.g., additions or alterations), producing a specific vocabulary of 59 tokens related to chords and 75 tokens for style, bars, form, and format. In the second approach, MIDI embeddings are added into the positional embedding layer of the transformer architecture, with an array of eight values related to the notes forming the chords. We propose a trigram analysis addition to the dataset to compare the generated chord progressions with the training dataset, which reveals common progressions and the extent to which a sequence is duplicated. We analyze progressions generated by the models comparing HITS@k metrics and human evaluation of 10 participants, rating the plausibility of the progressions as potential music compositions from a musical perspective. The second model reported lower validation loss, better metrics, and more musical consistency in the suggested progressions.

Keywords: Chord progressions · Transformer Neural Networks · Music Generation

1 Introduction

The domain of music computing has seen a variety of applications of state-of-the-art Natural Language Processing (NLP) models or neural networks, such as

This paper is an outcome of: MUSAiC, a project that has received funding from the European Research Council under the European Union's Horizon 2020 research and innovation program (Grant agreement No. 864189); and a Margarita Salas Grant, UPF, Barcelona, Spain.

© The Author(s), under exclusive license to Springer Nature Switzerland AG 2024
C. Johnson et al. (Eds.): EvoMUSART 2024, LNCS 14633, pp. 52–66, 2024.
https://doi.org/10.1007/978-3-031-56992-0_4

music generated from symbolic music notation [3,8], sound synthesis [4,12], or text-to-sound models such as MusicLM [1]. Nevertheless, compared to melodic sequences, the modeling of harmonic progressions has seen comparatively little research. Hence, we are interested in modeling chords from symbolic music notation by presenting an approach to encode chord progressions (CP). We designed a system to suggest a possible next chord after a CP, a complete CP for a specific bar length, or a parallel plausible CP, where the human user is the leading composer.

When encoding harmony information and chord vocabulary, there is no unified method in the literature. In Western modern popular music, the basic cell of the chord is the triad as a stack of thirds, containing the scale's 1st, 3rd, and 5th notes. The triad has four possible natures: major, minor, augmented, or diminished. A chord can also be expressed in tetrads, adding the 7th scale degree (e.g., Cmaj7: C, E, G, B), having a richer range of possible natures (e.g., major 7th, minor 7th, half-diminished, diminished, dominant, minor-major, among others). Non-triad-based chords are also in the vocabulary, for instance, a stack of fourths or fifths or *suspended* chords where the 3rd is substituted by the 2nd or 4th. Chords can also be extended further by stacking even more thirds, adding scale degrees 9th, 11th, or 13th (e.g. Dmaj7♯11: D, F♯, A, C♯, G). Moreover, chords can have alterations that modify one or several notes and subtractions that eliminate one or several notes. If we consider chord inversions, the range of vocabulary is considerably extended. Hence, modeling symbolic music notation by implementing current NLP architectures to encode form, structure, and harmonic information without losing vocabulary richness is one of the current challenges in the field.

The following three points give the contribution of the proposed system:

A dataset with 70,812 CP from several popular music styles.
A tokenization method to encode chord symbols without reducing its vocabulary.
A positional encoding strategy with domain-specific embeddings of the transformer network to better suit spatial music information.

The following section reviews work applying machine learning to modeling and generating chord progressions. Section 3 presents our application of the transformer architecture to this problem. Section 4 presents the results of our models and analyzes them from a few different perspectives. We conclude with a look at the future development of our system and its integration into a pipeline for music creation. The supplementary material and related code to the publication can be found at the GitHub address: https://github.com/Dazzid/theChordinator.

2   Literature Review

The year 2016 is perhaps the starting point of neural network approaches to model music information, in particular, chord progressions using symbolic music notation. *ChordRipple* [7] presented a Chord2Vec encoder by adapting Word2Vec using datasets derived from the Bach chorales and The Rolling Stone Top 200 songs corpus. The chord vocabulary includes triads, tetrads, and inversions. The authors evaluated the application with music students in a creative exercise and found benefits, but more precise stylistic development was still needed.

A text-based machine-learning model for chord progressions is proposed by Choi et al. [6]. The authors train two Long Short-Term Memory (LSTM) networks using textual representations of chord progression, e.g., "C:7 E:min". One LSTM has a vocabulary of 39 symbols corresponding to the characters which combined create chords, e.g., "E:min" consists of five characters in this vocabulary. The other LSTM has a vocabulary of 1,259 chord symbols. In this case, "E:min" is one vocabulary element. The authors train both systems by using 2,486 scores from The Realbooks and The Fakebooks. They find both models generate progressions that are plausible within jazz styles.

Transformers [13] have shown an improvement over the previous neural networks, enhancing the long-term dependencies by implementing the attentional mechanism as a parallelizable architecture. Most research in the context is focused on adapting or expanding it to specific tasks.

In 2020, Wu and Yang [14] introduced *Jazz Transformer*, a generative composition system including chords, melodies, and form. The model is trained using the Weimar Jazz Database [11], which contains 456 jazz standards, including styles such as Swing, Bebop, Cool, Hard Bob, and Fusion. The data format is MIDI. The authors suggested a set of objective measurements included in the *MusDr*[1] framework that reveals the deficiencies of machine-generated music, such as unpredictable pitch class usage, inconsistent grooving patterns, and chord progressions, or the lack of recurring structures. These metrics demonstrate the limitations of the transformer architecture as a music generator and provide practical quantitative standards for evaluating the efficacy of future automated music composition efforts. Currently, model assessment largely depends on human assessment, thus concluding that besides the training loss and evaluation rating numbers, the music generated still needs to be revised by experts.

The transformer, by default, is not fully suitable as a neural network to encode chord progressions by only providing symbolic music notation as it misses the configuration of voicings and chord construction; thus, in 2021, different publications suggest updating the positional embedding layer of the transformer architecture in order to provide the network with more precise information. Chen et al. [5] propose an encoder-based transformer model to classify chords and provide information on harmonic progressions in the classical music domain. The dataset used is the BPS-FH dataset and the Bach Preludes. They improve the capacity of transformers to encode harmonic features by providing extra information by adding a relative positional encoding, adding a layer of contextual information with the relative function of the chord in the chord sequence.

MusicBERT, by Zeng et al. [15] is a symbolic music understanding model based on the BERT: Pre-training of Deep Bidirectional Transformers for Lan-

---

[1] https://github.com/slSeanWU/MusDr.

guage Understanding architecture, to provide melody completion, accompaniment suggestion, and style classification and recall. They use the Million MIDI Dataset (MMD), which contains 436.631 samples[2], from different styles such as Rock, Rap, Latin, Classic, or Jazz. The data format as a symbolic music notation is shaped as bar-level masking using MIDI. The authors propose adding *musical note embeddings* to the positional embeddings, including time signature, tempo, bar, position, instrument, pitch, duration, and velocity.

MrBert, by Li and Sung [9] is an automatic music generator that creates melodies and rhythm and then creates chord progressions based on the melody. The masked language model is trained with the OpenEWLD dataset (MusicXML), which is translated to music events using the Python library *music21*. The system is based on parallel transformer encoders: one encoding melodies, and the other its rhythmic pattern. Authors argue that chords are generated after the melodies, using a Seq2Seq generation that changes chords when a new symbol outside the chord pattern appears. HITS@k metrics are used to evaluate the compositions. The chord vocabulary is restricted to triads.

Li and Sung [10] propose an encoder-decoder (Seq2Seq) architecture to generate chord progressions given a melody. The data format is music XML. The authors compare three architectures: a) the bidirectional long short-term memory (BLSTM), b) the bidirectional encoder representation from transformers (BERT), and c) the generative pre-trained transformer (GPT2). They evaluated the models using HITS@k [2]. The proposed methodology uses a pre-trained encoder that takes song melodies and couples their relations with the chord progression. The decoder receives the same melodic material and produces suggested chord matchings. The authors argue that the model does not need music theory inference as it relies on the patterns reflected in the symbolic music notation; however, it is not proven the model learns harmonic principles. The transformer is trained using the OpenEWLD dataset and Enhanced Wikifonia Leadsheet Dataset (EWLD), which contain 502 samples in musicXML format, including melodies and chords. 382 chord types were extracted as a vocabulary. Examples of generated music or chord progressions are not available.

3 The Chordinator

We compare two strategies to model chord progressions, maintaining the full range of vocabulary available in the music corpus. All approaches are based on the transformer encoder, changing the token strategy. We also propose complementary embeddings to the transformer to encode chord information by adding a new level of relative embeddings using an array of defined MIDI values added to the positional and tokens embeddings layer. In the second tokenization strategy, we include a style token reference, providing the transformer with contextual information about the style and form.

---

[2] https://github.com/jeffreyjohnens/MetaMIDIDataset.

3.1 Database

The database was formed by exporting 4,300 CP included in the *iReal Pro* application (as songs). The data provides metadata (composer, song name, key, etc.) and information regarding the form, tempo, and chords. The database includes styles such as Jazz, Samba, Ballad, etc. (see the full list in Fig. 1). We filter the CP, keeping only those in 4/4 time signature (around 90% of the dataset), to keep the sequences consistent. We apply data augmentation by transposing all CP into the 18 enharmonic keys, creating 70,812 in total. Furthermore, we prepare the data stream to train a GPT-2[3] transformer architecture with a fixed-length sequence of 512-time steps in the first model and 1024 in the second model. The $< pad >$ token is used to fulfill the defined length.

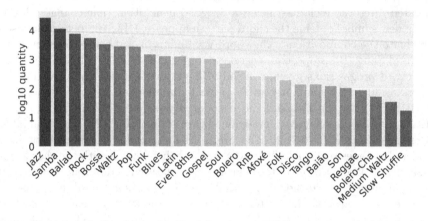

Fig. 1. List of musical styles in the *iReal Pro* dataset. Values are presented in the log10 ratio.

3.2 Model Strategies

Our proposal is based on two main strategies to train a GPT-2-based transformer encoder architecture with chord progressions.

Model 1: It has the entire token vocabulary. All chords are unique elements, producing a vocabulary of size 5,202 unique chords/tokens. The chords are quantized in time, creating a chord per quarter note. When the suggestions are generated, four consecutive chords are equivalent to an entire 4/4 bar.

Model 2: We split the chord units into subsections, simplifying the tokenization range without reducing the vocabulary in the corpus. Starting the chord element with an identifier token '.' (dot), followed by the **root**, **nature**, and **extensions** as shown in Table 1. For instance, '**Dm7 add 9**', is split into '.', '**D**', '**m7**', '**add 9**'. This creates a 1D array of dynamic lengths depending

---

[3] https://github.com/karpathy/minGPT.

on how much information the chord has. Slash chords are also split, following the same structure. For example, 'Cmaj7 / Bb' results in a sequence '.', 'C', 'maj7', '/', 'Bb'. The second root in slash chords becomes the actual root reference. This strategy reduces the vocabulary to a more specific format with 59 Tokens: Song formatting (<start>, <and>, <pad>, '.'), roots ('A' ''A♯' 'Ab' 'B' 'Bb' 'C' 'C♯' 'Cb' 'D' 'D♯' 'Db' 'E' 'Eb' 'F' 'F♯' 'G' 'G♯' 'Gb'), natures ('dim', 'm', 'm6', 'maj7', 'o7', 'sus', 'power'), extensions (additions, subtractions, and alterations), and sharp ('/'). In the cases where the chord is notated with only one note, that chord is considered a major triad.

Extended Model 2: We extend Model 2 in two ways. First, we add a context token at the beginning of the samples with a style specification, meaning we start with a token (<style>) followed by the actual style (such as jazz, pop, blues, samba, etc.). Second, we update the transformer model, adding a MIDI Embedding relative position array into the positional embedding layer. Based on the second strategy, when we have chords divided into root, nature, and extension, the MIDI array is composed of an array of eight numbers where the first is the MIDI root reference, the next three values are used to describe the nature, and extensions could use the next value. The longest sequence of the MIDI array for a chord is seven MIDI values; hence, the eighth MIDI value in the array is an identifier (number 127) to express that there is a slash token.

Table 1. Chord Tokens Format

Chord	Start	Root	Nature	Ext.	Slash	Root_2
C	.	C				
Dm7 add 9	.	D	m7	add 9		
Bbmaj7 ♯11	.	Bb	maj7	♯11		
C♯o7	.	C♯	o7			
Ebmaj7/D	.	Eb	maj7		/	D

3.3   MIDI Embeddings

MIDI Embeddings (ME) are extracted from the symbolic chord representation using the Python libraries *music21* (symbol to notes translation) and *librosa* (notes to MIDI translation). We decided to maintain the chord pitches in only a two-octave range. During training and generation, ME is paired with token inputs, providing only information about the current time-step in the sequence; therefore, when only one note is presented in a specific chord, by default, the ME reads it as a Major chord (triad). When a second token appears next to the original note, the first token is read as a root note (only one MIDI note is then placed), and the nature is updated, adding its related notes to the MIDI array. The same logic is applied to extension and sharp chord, where each step in the sequence updates the ME information, having the complete chord MIDI array only in the last time step of the chord tuple, as shown in Fig. 2. We avoid information leaking following this format.

The formula to update the ME into the positional embedding in the transformer network is defined as follows:

$$midi\_emb = Linear(Emb(midi\_vocab, dm)(m).view(m.size(0), m.size(1), -1), n_{embd})$$

Positional embeddings are added to the input embeddings before the transformer encoder is fed. The positional embedding layer enables the model to distribute information spatially in order to learn specific patterns. The ME takes information from MIDI and adjusts it into an embedding tensor to align with the required architecture format. This, then, is summed with the standard positional encoding layer. It is composed as follows:

1. **Embedding MIDI Values**:
   '$Emb(midi\_vocab, dm)(m)$': This defines an embedding layer where '$midi\_vocab$' is the total number of unique MIDI events and 'dm' is the dimension of the embeddings. '(m)' is the time-step in the chord tuple sequence.
   '$.view(m.size(0), m.size(1), -1)$': The tensor matrix is reshaped to adjust its format to the standard positional embedding. Here, '$m.size(0)$' represents the batch size, while '$m.size(1)$' represents the sequence length. '-1' means that the size in the last dimension is computed so that the total size remains constant.
2. **Linear Transformation**: The reshaped data is passed through a Linear layer. The size of the output data is adapted to match a tensor of shape ($batch\_size, sequence\_length, n_{embd}$). It is then summed with the positional encoding layer.

Input	.	D	○	m7	add 9	/
Token embedding	$W.$	W_D		W_{m7}	W_{add9}	$W_/$
Pos. embedding	$P.$	P_D		P_{m7}	P_{add9}	$P_/$
MIDI embedding	$M.$	M_D		M_{m7}	M_{add9}	$M_/$
Array	0 0 0 0 0 0 0 0	62 0 0 0 0 0 0 0		62 65 69 72 0 0 0 0	62 65 69 72 76 0 0 0	0 0 0 0 0 0 0 127

Fig. 2. MIDI Embeddings: Formatted as an array of eight values for each time-step of the chord tuple. In Model 2, the chord information is split into tuples, and the input layer processes it as a dynamic sequence. The input sequence is translated into tokens embeddings, positional embeddings, and midi embeddings; all those layers share the same tensor matrix shape, and they are summed inside the positional encoding layer, which is passed to the multi-head attention.

3.4   Training the Models

The models are trained on 4 GPUS NVIDIA GeForce RTX 3090 (24576MiB). While we tested many configurations, this manuscript only reports the best versions of both strategies, defined as Model 1 and Model 2.

Model 1 is trained with 120 Epochs, 6 attention heads, 6 layers, 192 embeddings sizes, 64 batch sizes, 192 embeddings, number of workers 6, and a learning rate of 3e-5. Model 1 is prone to overfit after 120 epochs; hence, these configurations are chosen to maintain a small network that adapts to the size of the vocabulary and dataset. Extra attention heads tend to increase the overfitting tendency and do not necessarily lead to better performance.

Model 2 is trained with 250 epochs without showing overfiting tendencies (it could be trained with more epochs). Thus, it has an increment in 8 attention heads and 8 layers. The other features are maintained as they are: 192 embeddings sizes, 64 batch sizes, 192 embeddings, number of workers 8, a learning rate of 3e−5, and a MIDI vocabulary size of 128, which refers to the size of the possible MIDI values. (see table 2).

Table 2. Transformer Format

Configuration	M1: Full Tokens	M2: MIDI Emb
Epochs	120	250
Heads	6	8
Layers	6	8
Embedding	192	192
Batch size	64	64
Learning rate	3e−5	3e−5
Workers	8	8
MIDI_vocab	-	128

3.5 Metrics

HITS@k (k = 1, 3, and 5) [15] is used to evaluate the predictions in both models. HITS@k calculates the proportion of the correct answer given by the candidates, denoted by the letter k. It was calculated as follows:

$$HITS@k = \frac{1}{n} \sum_{i=1}^{n} I(rank_i \leq k) \tag{1}$$

where,

$HITS@k$: This represents the HITS at k metric.
n: Total number of instances or items being evaluated.
$\sum_{i=1}^{n}$: This denotes the sum over all instances.
$I(rank_i \leq k)$: An indicator function that evaluates to 1 if the rank of instance i is less than or equal to k, and 0 otherwise.

Table 3 compares the metrics of Model 1 and 2.

Table 3. Metrics

	Train Loss	Val. Loss	HITS@1	HITS@3	HITS@5
Model 1	0.3699	0.4108	0.7383	0.8294	0.8556
Model 2	0.04982	0.03645	0.9138	0.9707	0.984

4 Results

4.1 Generated Chord Progressions

We generated 20 CP per model in one unique iteration loop where no selection of the CP was performed, and they were shown to the evaluators as they were; no voicing, arrangement, or editing was performed. All CP were constrained with a length of 8 to 24 chords, producing a collection of 40 randomized CP. The Model behind the generated CPs are not identifiable by the evaluators. Ten participants, with an average age of 36 (s.d. $+/-8$), reported using a 1 to 10 scale the plausibility of the chord progressions as a starting point of a conceivable music piece: 1 implies no plausibility as a sequence, and 10 is very plausible to start a potential composition. The overall median score received was **Model 1**: 5.7; **Model 2**: 6.7, as shown in Fig. 3. All suggestions begin with a root token, which can be any scale note; nevertheless, Model 2 includes a random number defining the style from the first six most common styles in the dataset (Jazz, Samba, Ballad, Pop, Bossa, Rock). To generate a sound sample, we implemented the Python library *librosa* and *music21* to generate the chords in MIDI format using a standard Piano. Figure 4 shows examples of generated CP with form tokens in Model 2.

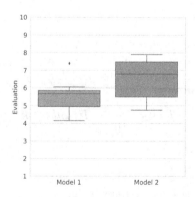

Fig. 3. Boxplot of human evaluation of Model 1 (orange) and Model 2 (blue). (Color figure online)

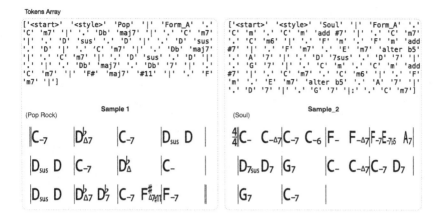

Fig. 4. Two samples from Model 2 using Pop and Soul context token. The tokens array provides information about chords and form, using the token '|' as a bar identification.

4.2   Trigrams

A supplementary dataset comprises three chord cells, including information on the song and composer. Those cells are called trigrams, which provide information on the suggested progressions to check for original versions and sources.

Trigrams are formed by creating groups of subsequent three chords, moving the windowed observation one time step forward. A sequence with Em7, A7, Dmaj7, Ab7 add ♯11, Gm7, are then packed as (Em7, A7, Dmaj7), (A7, Dmaj7, Ab7 add ♯11) and (Dmaj7, Ab7 add ♯11, Gm7). That means four chords are equal if two consecutive trigrams from a suggested progression are found in a particular song. Hence, we can define a threshold of matching trigrams to determine duplication. The trigram analysis can be executed with a list of suggested CP or immediately after a generation. In this case, analyzing 40 CP, the report is too extensive to include in the manuscript. A list of the most 20 trigrams cells found is shown in Fig. 8. The report includes a tag for **Sequence**, which is the CP analyzed; **Location**, which reports the chord position in the sequence; **Trigram**, which is the progressions cell; and a list of **Composer** and **Song** where it was found.

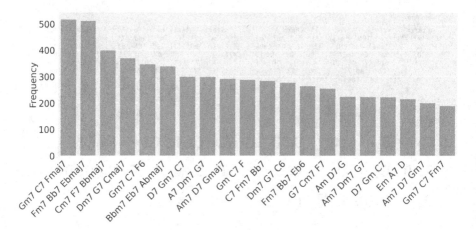

Fig. 5. Most used trigrams found in the dataset. Most progressions are sequences of II-V-I in major and minor functional formation.

Both Models show similar trigram repetitions found in the dataset. However, it does not mean the whole progression is copied; it only provides information about the original trigram cell location. However, it is possible to determine if a whole progression is, in fact, replicated from the training data. No generated CP reported more than two trigram cells from the same song. In Fig. 7, some of the most common trigrams found are shown, and Model 1 is more prone to duplicate cells.

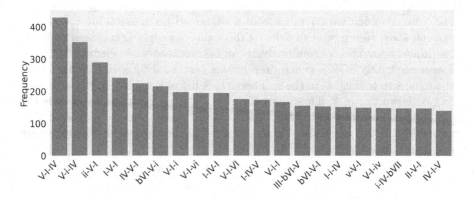

Fig. 6. Most common harmonic functions found in the dataset.

Based on trigrams analysis, we observed that sequences based on ii-V-I progressions and their different versions in major and minor are prominent in the dataset (see Fig. 5). However, the ii-V-I is not the most used progression, as

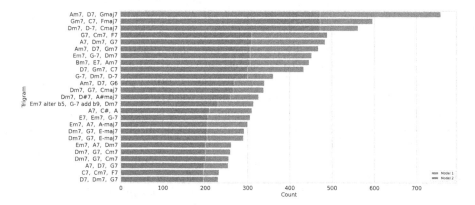

Fig. 7. Model 1 (orange) and Model 2 (blue) report trigrams cells also found in the dataset. However, Model 1 is more prone to replicate common trigram cells found in the dataset (Color figure online)

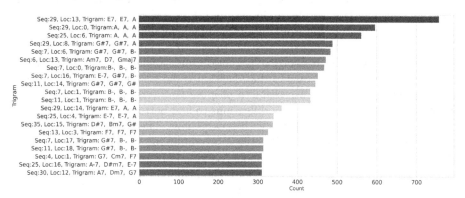

Fig. 8. Trigram analysis provides information about trigram cells found in the dataset. It contains the sequence number; we have 40 CP suggested in this case. The Sequence tab will point to that suggested CP origin. The Location tab provides information on the chord position within the progression; the Trigram tag is the chord cell, and the Count tag expresses how many times it was found in the dataset. The Figure contains 20 of the most trigram cells found

shown in Fig. 6. The most used progressions are combinations of IV-V-I motions and some permutations, commonly audible in the *Blues*, *Rock*, *Pop*, or *Cinematic Music*.

5 Discussion and Conclusion

We have presented two tokenization strategies with an extension of the positional embeddings (MIDI Embeddings) for Model 2. We treat the chord vocabulary with a specific methodology differentiating the root, nature, and extensions, thereby reducing tokens without diminishing the chord vocabulary and its richness. We trained each model on a large database of chord progressions and used it to generate new suggested progressions or chord completion. Our database of 70,812 songs has diverse music styles in popular music. Our evaluation of sequences generated by these models shows that reducing the vocabulary and adding contextual information improves the encoding process of the chord vocabulary. Thus, adding MIDI notes for chord constructions to the Positional Embeddings layer has helped the transformer network stabilize the learning process, allowing more training epochs without overfitting; it also helped to report better HITS@K and validation metrics (see Fig. 3).

Our system contributes to a growing literature on modeling harmonic progressions as a system that can be used as a composer assistant. Future work will explore the usefulness of this system with users performing musical tasks to address further development and usability. We will also explore ways of generating melodies from chord progressions and vice versa. This will be implemented as an accessible web application for others to study.

As expressed in [14], and also based on the outputs generated by Model 1, the vanilla encoder transformer architecture (GPT-2) is not an ideal network for encoding music information from only providing symbolic music notation. Even though it can generate plausible textual tokens, the musicality is still unclear. From the perspective that chord construction and voicing formats are hidden from symbolic music notation and that decision is the responsibility of the musician's knowledge, it is still necessary to propose new versions of related neural networks and test architectural modifications to find better strategies to encodes all musical information related to chord progressions. Therefore, our next model is based on multi-hot encoders (2 octaves) exemplifying the notes in the piano from an expert performer playing chord progressions and an encoder-decoder architecture to suggest, on the one hand, the symbolic music notation but, on the other, the voicings for more musical outputs.

The human evaluation reported a median score of 5.7 for Model 1 and 6.7 for Model 2. Even though there is a difference in appreciation of Model 2 quality, the low score could be directly affected by the test format shown to the participants. Raw MIDI piano sounds were chosen to evaluate the chord progressions, and no voicing was added to avoid human completion from expert performers. The CP were played at 120BPM; in that term, the suggested CPs are expressed in a raw MIDI version that can be highly improved by musicians' knowledge in terms of performance and interpretation and also in sound quality. However, HIT@K metrics reported a difference where Model 2 is much more accurate, particularly in the validation loss. The interpretation of its musicality based on NLP metrics is still open for discussion.

As a further development, to enhance Model 2, when translating symbolic music notation into MIDI chords, we will add an array of plausible voicings repertoire per chord and a distance calculation of note proximities to suggest more natural voicing. By default, the Python library *music21* is unsuitable for this purpose as it reported errors, particularly with slash chord formatting and notes related to suspended chords.

Model 2 reported a tendency over Model 1 to generate more consistent progression in musical terms. The explanation is given by the modeling of the contextual token, which provides a general view of the music style; therefore, Model 1 mixes styles and formats, while Model 2 is more prone to maintain some degree of consistency. For instance, when *Blues* is given, the form and chords are common to the style. Also, Model 1 chooses chord tonalities wrongly formatted in terms of notation; in other words, it mixes enharmonic, e.g., when Ebmaj7 makes sense in the context, it suggests D♯maj7. Further work is also needed to adjust this duality in names as it is relevant to be fixed in a musical context.

Model 2 also suggests form and structures; further development will incorporate that suggestion into the web application.

For future work, we will add a melody generator that is conditioned on a CP and vice versa, where a melody conditions the generation of a CP. A web application as an interface to play with the model and generate suggestions is under development; it will open its usability to a broader audience. We will perform further human evaluations by expert musicians to test the impact and usability of the system.

References

1. Agostinelli, A., et al.: Musiclm: generating music from text. arXiv preprint arXiv:2301.11325 (2023)
2. Ali, M., et al.: Bringing light into the dark: a large-scale evaluation of knowledge graph embedding models under a unified framework. IEEE Trans. Pattern Anal. Mach. Intell. **44**(12), 8825–8845 (2021)
3. Briot, J.P., Hadjeres, G., Pachet, F.: Deep Learning Techniques for Music Generation. Springer, Cham (2019). https://doi.org/10.1007/978-3-319-70163-9
4. Caillon, A., Esling, P.: Rave: a variational autoencoder for fast and high-quality neural audio synthesis. ArXiv e-prints (2021)
5. Chen, T.P., Su, L.: Attend to chords: improving harmonic analysis of symbolic music using transformer-based models. Trans. Int. Soc. Music Inf. Retrieval **4**(1) (2021)
6. Choi, K., Fazekas, G., Sandler, M.: Text-based LSTM networks for automatic music composition. In: Proceedings of 1st Conference on Computer Simulation of Musical Creativity. Huddersfield, UK (2016)
7. Huang, C.Z.A., Duvenaud, D., Gajos, K.Z.: Chordripple: recommending chords to help novice composers go beyond the ordinary. In: Proceedings of the 21st International Conference on Intelligent User Interfaces, pp. 241–250 (2016)
8. Huang, C.Z.A., et al.: Music transformer: generating music with long-term structure (2018). arXiv preprint arXiv:1809.04281 (2018)
9. Li, S., Sung, Y.: MRBERT: pre-training of melody and rhythm for automatic music generation. Mathematics **11**(4), 798 (2023)
10. Li, S., Sung, Y.: Transformer-based seq2seq model for chord progression generation. Mathematics **11**(5), 1111 (2023)
11. Pfleiderer, M., Frieler, K., Abeßer, J., Zaddach, W.G., Burkhart, B. (eds.): Inside the Jazzomat - New Perspectives for Jazz Research. Schott Campus (2017)
12. van den Oord, A., et al.: WaveNet: a generative model for raw audio. ArXiv e-prints (1609.03499) (2016)

13. Vaswani, A., et al.: Attention is all you need. In: Advances in Neural Information Processing Systems, vol. 30 (2017)
14. Wu, S.L., Yang, Y.H.: The jazz transformer on the front line: exploring the shortcomings of ai-composed music through quantitative measures. arXiv preprint arXiv:2008.01307 (2020)
15. Zeng, M., Tan, X., Wang, R., Ju, Z., Qin, T., Liu, T.Y.: Musicbert: symbolic music understanding with large-scale pre-training. arXiv preprint arXiv:2106.05630 (2021)

Deep Learning Approaches for Sung Vowel Classification

Parker Carlson[1,2](✉)[ID] and Patrick J. Donnelly[2][ID]

[1] UC Santa Barbara, Santa Barbara, USA
parker_carlson@ucsb.edu
[2] Oregon State University, Corvallis, USA
patrick.donnelly@oregonstate.edu

Abstract. Phoneme classification is an important part of automatic speech recognition systems. However, attempting to classify phonemes during singing has been significantly less studied. In this work, we investigate sung vowel classification, a subset of the phoneme classification problem. Many prior approaches that attempt to classify spoken or sung vowels rely upon spectral feature extraction, such as formants or Mel-frequency cepstral coefficients. We explore classifying sung vowels with deep neural networks trained directly on raw audio. Using VocalSet, a singing voice dataset performed by professional singers, we compare three neural models and two spectral models for classifying five sung Italian vowels performed in a variety of vocal techniques. We find that our neural models achieved accuracies between 68.4% and 79.6%, whereas our spectral models failed to discern vowels. Of the neural models, we find that a fine-tuned transformer performed the strongest; however, a convolutional or recurrent model may provide satisfactory results in resource-limited scenarios. This result implies that neural approaches trained directly on raw audio, without extracting spectral features, are viable approaches for singing phoneme classification and deserve further exploration.

Keywords: Sung Vowels · Phoneme Classification · Raw Audio · Automatic Speech Recognition · CNN · LSTM · Transformer · VocalSet

1 Introduction

The ability to discriminate between different phonemes is a critical and often studied aspect of automatic speech recognition (ASR) systems. While most modern ASR systems perform strongly for well-resourced natural languages, performance tends to degrade with lower-resource languages. Further, many ASR systems tend to struggle with variations in pronunciation and the rate of speech [16]. These variations in delivery are further exaggerated during singing. Moreover, the numerous fundamental differences between speech and song present many challenges when applying conventional ASR systems to singing.

Parker Carlson–Work done while at Oregon State University.

© The Author(s), under exclusive license to Springer Nature Switzerland AG 2024
C. Johnson et al. (Eds.): EvoMUSART 2024, LNCS 14633, pp. 67–83, 2024.
https://doi.org/10.1007/978-3-031-56992-0_5

A typical ASR pipeline often contains an acoustic front-end which extracts phonemes or other acoustic features from the audio signal, an acoustic model which proposes likely words based on the observed phonemes, and a language model which limits word choice based on the grammatical structure of the language. Before deep learning approaches became widespread, many ASR systems relied on a set of spectral features, such as formants or Mel-Frequency Cepstral Coefficients (MFCCs), which often were modeled with hidden Markov models [8]. Researchers have also trained deep neural networks on MFCC features or raw audio, including bidirectional long short-term memory networks, convolutional neural networks, and more recently, transformer models [2,10,16,28].

While end-to-end neural methods for ASR usually outperform phoneme-based ASR systems for high-resource languages [12], they often struggle to extend to low-resource languages given the absence of sufficient training data [17]. Likewise, researchers investigating singing, such as the task of automatic lyric transcription, also suffer from a lack of annotated training data. When sung, vowels take importance over consonants and are sustained much longer than their spoken counterparts [3]. Moreover, sung vowels inherently have different acoustic properties than spoken vowels [31]. For example, the formant frequencies of sung vowels varies, with the direction and magnitude of the change depending on the vowel and formant [32]. In addition, differences between male and female voices are exaggerated during singing. For instance, the "singer's formant", a clear formant around 3000 Hz that produces a ringing timbre is more common in trained male singers than female singers [33,37]. Furthermore, singers may alter the pronunciation of words or elide certain phonemes, exacerbating the challenges of applying spoken ASR techniques to vocal lines. These differences can be clearly observed in Fig. 1, a comparison of a sung and spoken vowel.

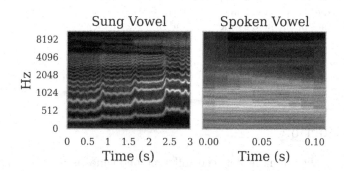

Fig. 1. Sung and sung vowels are significantly different in duration, pitch, and harmonics. Mel-spectrograms visually demonstrate the difference between a spoken and sung vowel **a**. The spoken vowel is taken from the TIMIT corpus [9]; the sung vowel is sung by `female1` with vocal technique `belt`.

In this work, we investigate the use of deep neural networks to classify sung vowels directly from raw audio. We replicate experiments from VocalSet [38] and extend them to investigate vowel classification, evaluate the efficacy of three neural models for sung vowel classification, and examine their performance relative to two spectral models. To our knowledge, this is the first study to to explore neural methods trained on raw audio to the sung vowel classification problem.

2   Related Work

Although most research in the area of automatic phoneme recognition has focused on speech, a few researchers have applied these methods to sung vocal lines. In this section, we review relevant literature in the areas of automatic vowel classification, automatic lyric transcription, and pre-trained musical transformers.

2.1   Automatic Vowel Classification

The task of automatic vowel classification can be viewed as a subset of the phoneme classification problem. Unfortunately, there do not exist many datasets of singing examples that contain annotations at the phonemic level. Therefore, many phoneme classification systems are trained on spoken speech data that may be synthetically altered or augmented to closer resemble aspects of singing [22,23,25]. As more datasets, such as *VocalSet* [38] and *DAMP Sing!* [30], are released, researchers will be less reliant on augmented speech alone.

Vowel classification has been studied in numerous context across many different languages [10,15,20]. When investigating this task, researchers may examine spectral features, such as spectrograms or analysis of vocal formants, either labeled by experts or automatically generated [5,13,21]. Other times, researchers rely upon acoustic features, such as MFCCs, zero-crossing rate, and energy. And these features have been explored through a variety of models, such as k-nearest neighbor, linear and quadratic discriminant analysis, and support vector machines [13,20,21]. However, there remain many unsolved problems to explore. For example, a study by Jha and Rao reported a 10% difference in accuracy between male and female voices when classifying five sung Hindi vowels [15]. Furthermore, vowel classification varies significantly by language, on account of the differing number of vowels and language-specific pronunciation.

2.2   Automatic Lyric Transcription

Automatic lyric transcription (ALT) is an emerging field that intertwines approaches from the fields of speech recognition and music information retrieval. The two primary tasks of ALT are lyric transcription and lyric alignment. In the first task, researchers seek to transcribe sung vocal lines to text. In the second task, researchers attempt to align this transcription to precise time-stamps, in order to determine what is being sung when. Although the field has not received the same attention and resources as ASR, recent work has advanced understanding of both lyric transcription [4,6] and lyric alignment [11,27,34].

There have been many different approaches to ALT, with some systems focusing on classifying sung speech in a monophonic setting [6] and others tackling the more challenging polyphonic setting [11]. In one such study, Mesaros and Virtanen found that their hidden Markov model adapted to gender showed improvements over the baseline model for the lyric transcription task [25].

While much of the previous work in ALT has focused on word-level classification and alignment, there has been some work on phoneme classification and alignment. Spectrograms have been used to aid in phoneme alignment with sung speech [34], as well as classifying words sung in Thai [18]. Additionally, MFCCs have been used to classify sung Hindi vowels with Gaussian mixture models [15].

2.3   Pre-trained Musical Transformers

For many text classification problems, the dominant paradigm is a two-stage process that begins with pre-training a large transformer that captures a general understanding of a language, followed by additional training to fine-tune it for a domain-specific problem [7]. Due to the success of pre-training and fine-tuning transformers for text-based tasks, there has been many recent approaches on developing a similarly capable model for acoustic, speech, and musical understanding [14,36]. One such state-of-the-art model, MERT [24], introduced musical pre-training objectives and applied them to HuBERT [14], a popular model for speech representation. The authors found their model generalized well to fourteen tasks that required a broad understanding of musical features. In this study, we fine-tune MERT to investigate the task of sung vowel classification.

3   Methodology

In this section, we discuss the dataset used in this study and provide a detailed description of our model architectures for our vowel classification experiment.

3.1   Dataset

To explore this task, we utilize VocalSet [38], a dataset of 10.1 h of recordings of twenty professional singers performing a variety of vocal techniques on five different Italian vowels (/a/, /e/, /i/, /o/, /u/). Henceforth, we refer to these vowels as a, e, i, o, and u respectively, following convention established in VocalSet. The dataset contains examples of eleven male and nine female singers, comprising the seven different common voice types (bass through soprano).

We preprocess our dataset following the approach of Wilkins et al. [38], making informed choices for any missing details. For each recording, we automatically remove silence at the beginning, middle, and end of recordings using Sox.[1] Next, we split the recordings into three second examples sampled at 44.1 kHz. Because of the presence of short (less than one second) examples, we opt to discard any

Table 1. Examples are nearly uniformly distributed by vowel in our dataset.

Vowel	Examples
a	1148
e	1147
i	1149
o	1190
u	1187

examples with a duration less than three seconds of audio. Lastly, we standardize each example to have a zero mean and unit variance.

Although VocalSet provides suggested train and test splits [38], we choose to evaluate our models using k-fold cross-validation in order to test our models on every example. We randomly partition the singers into four folds of five singers, while ensuring that each fold has an approximately equal number of male and female singers.[2] We exclude spoken excerpts in order to focus on sung vowels, as well as the trillo technique because each example is shorter than three seconds in duration. These exclusions represent less than 2.5% of all available data. After preprocessing our dataset we have 5821 examples, totaling 4.85 h. The distribution of examples per vowel is shown in Table 1.

VocalSet Replication. In their original paper, Wilkins et al. explored VocalSet in the context of vocal technique classification and singer identification [38], but curiously, they did not explore the sung vowel classification task. To contextualize our results and provide a vowel classification baseline for VocalSet, we replicate their convolutional model and evaluate on the tasks of vocal technique classification and singer identification explored in their initial publication.

3.2   Neural Models

We select three neural models of varying architectures to evaluate in the task of sung vowel classification. We evaluate a convolutional model, a recurrent model, and a pre-trained transformer. Each of these model architectures have been successfully applied to raw audio analysis in various contexts [1,24,26].

Convolutional. For our convolutional model, we select a three layer convolutional neural network (CNN) that is identical to Table 1 of Wilkins et al. [38] in order to provide an identical baseline for our vowel classification results. We direct readers to the original VocalSet paper [38] for a detailed model description.

---

[1] https://sox.sourceforge.net/sox.html.

[2] Each split of singers is either (two male, three female) or (three male, two female).

Recurrent. To consider a recurrent neural model of similar depth to our convolution model, we stack three Bidirectional Long Short-Term Memory (BiLSTM) layers with eight units each. After each of the first two BiLSTM layers, we apply a 20% dropout during training. The last BiLSTM layer is followed by a dense layer with 32 units and a ReLU activation function, then a final dense layer with five units and a softmax activation. Both dense layers have a L2 regularization of 1e-3. We train this model for 200,000 batches of size 64 using RMSProp [35] with a learning rate of 1e−4 and momentum of 0.99 using TensorFlow.[3]

Transformer. As our transformer model, we consider MERT, a state-of-the-art pre-trained acoustic transformer that has been shown to generalize well to a variety of musical tasks [24]. We begin from MERT_v0_public, a 95M parameter version of MERT pre-trained on music4all, a dataset of 910 h of publicly available music [29]. MERT is pre-trained with acoustic and musical masked language model objectives, which predict k-means clusters from log-Mel spectral features and reconstruct masked audio from a constant-Q transform respectively. For each of the transformer layers, we take the average of all embeddings across the time dimension, creating thirteen total embeddings that capture the musical information of the entire example. We empirically consider each embedding as well as combinations of them, following the authors' suggestion,[4] finding that a sum of the first seven embeddings yields the best results. This combined embedding is used as input to a dense layer with 256 units and ReLU activation, followed by a dense layer with five units and a softmax activation function. We fine tune the entire model for five epochs using the Adam optimizer [19], cross-entropy loss, and a learning rate of 3e−5 using PyTorch[5] and Transformers.[6]

3.3 Spectral Models

We first explore two existing spectral models that have been successfully applied to vowel classification to serve as a baseline for our raw audio neural models.

Formant Model. We follow recent work from Korkmaz et al. to classify spoken Turkish vowels using a formant-based classifier [20]. We divide each example into frames of 512 samples (11.6 ms) with a 256-sample overlap (5.8 ms) and apply a Hamming window to each frame. Next, we apply linear predictive coding with eight filters to extract three formants per frame. Frames in which no formants were found are discarded (21.2% of data), and frames missing the second or third formant have these values imputed as zero. We use these features to train a k-nearest neighbor classifier, with $k = 1$, using the Manhattan distance. We do not aggregate classification results from the frames back to three-second examples.

---

[3] tensorflow.org.

[4] github.com/yizhilll/MERT/blob/main/scripts/MERT_demo_inference.py.

[5] pytorch.org.

[6] github.com/huggingface/transformers.

MFCC Model. We follow Jha and Rao's approach to classify sung vowels using MFCCs and Gaussian mixture models (GMM) [15]. We split the audio into 1024 sample (23.2 ms) frames with a 512 sample overlap, resulting in 87 frames per second. From each frame, we extract 13 MFCCs. For each vowel, we fit a single mixture GMM with a full covariance matrix. Classification is performed by maximizing the likelihood of the test vectors with respect to each model. Each example is assigned to a vowel according to the minimum Mahalanobis distance between the MFCC vector and each GMM. Following Jha and Rao [15], an example is considered to be correctly classified if more than half of the MFCC vectors are assigned to the correct vowel. If fewer than half of the MFCC vectors predict the same vowel, the example is classified as an error.

4   Results

In this section we discuss the findings of the experiments outlined in the previous section. First, we briefly examine the performance of the convolutional model on the two VocalSet replication tasks. Next, we thoroughly compare the performance of the five vowel classification models in a variety of settings.

Table 2. Results for our VocalSet replication by task. Top-1 Accuracy was not reported in [38]; MERT only reported Top-1 Accuracy [24]. M. and F. denote the accuracy of classifying male and female singers, respectively.

MODEL	PREC.	RECALL	ACCURACY				
			Top-1	Top-2	Top-3	M.	F.
Vocal Technique							
VocalSet [38]	0.676	0.619	-	0.801	0.867	-	-
Conv. (Ours)	0.628	0.607	0.619	0.753	0.814	-	-
MERT [24]	-	-	0.756	-	-	-	-
Singer ID							
VocalSet [38]	0.473	0.516	-	0.638	0.700	0.684	0.351
Conv. (Ours)	0.698	0.660	0.673	0.808	0.863	0.701	0.632
MERT [24]	-	-	0.780	-	-	-	-

4.1   VocalSet Replication

Our approach underperforms on the vocal technique classification task and outperforms on the singer identification task, relative to VocalSet [38] (Table 2). While a discrepancy in performance between male and female singers still exists, it is less than observed by Wilkins et al. During dataset preprocessing, we observed that our dataset sizes to be slightly different than reported by VocalSet. We estimate to have approximately 22% more data for our vocal technique classification task, and 38% less data for our singer identification task. We attribute these differences in dataset size to differing parameters in preprocessing. We follow Wilkins et al. in training for a fixed number of batches, rather than epochs. As such, when batch size is fixed, the number of epochs is inversely proportional to the number of examples. Because our vocal technique classification underperforms compared to Wilkins et al., despite having more training data, and conversely our singer identification outperforms with less data, our results suggest that our models are undertrained. Nevertheless, our results are similar to those observed by Wilkins et al., and we continue to our vowel classification task.

4.2   Sung Vowel Classification

Our results for the vowel classification task are presented in Table 3. We found all neural models to achieve accuracies above 68%. However, we found that both the formant model and MFCC model to failed to converge, demonstrating accuracies corresponding to chance. We note that the MFCC model predicted the vowel u for 99.5% of examples, and the formant model's predictions appear to follow a uniform distribution. These results underscore the challenge of the sung vowel classification task. Existing spectral methods for spoken vowels failed to generalize, and existing spectral methods for sung vowel classification failed to replicate. On account of the inability of our spectral models to classify sung vowels, in our subsequent analysis, we focus only on the three neural models.

Table 3. Our results on the vowel classification task. Because our spectral models fail to converge, we do not evaluate male or female accuracy for these methods.

Model	Precision	Recall	Accuracy		
			Total	Male	Female
Neural					
Convolutional	0.723	0.719	0.718	0.749	0.674
Recurrent	0.685	0.685	0.684	0.714	0.640
MERT	0.799	0.797	0.796	0.827	0.752
Spectral					
Formants	0.200	0.200	0.201	-	-
MFCCs	0.041	0.199	0.203	-	-

Of the three neural models, we find that MERT performed best, outperforming both our convolutional and recurrent models with an accuracy of 79.6%. While our convolutional model generally outperformed our recurrent model, we find that our recurrent model performs better in classifying the vowel a as shown in Fig. 2. Furthermore, for certain singers and vowels, our recurrent model also outperforms our convolutional model, as shown in Figs. 6 and 7.

Across all vowels, MERT demonstrated the best performance. However, it was not universally better. In some cases, MERT performed notably worse than

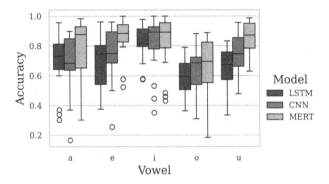

Fig. 2. The neural models performed decently, although all models had a high variance in accuracy across singers and vowels. Overall, MERT performed best. Each data point represents the model's performance on a single singer.

both other models, such as `male4`'s i, `female2`'s o, and `female3`'s o, shown in Figs. 6 and 7. These drops in performance suggest that an ensemble of neural methods may perform well. However, in some cases, all models struggled, such as `male3`'s a and to a lesser degree `female5`'s o, also shown in Figs. 6 and 7.

One important consideration is the disparity of computational resources required by the different models. Our convolutional and recurrent models are of similar sizes, with approximately 112,000 and 70,000 parameters respectively, whereas MERT has 95 million parameters. While pre-trained transformers have attractive properties, for instance, the ability to generalize to new tasks with minimal training, the vast difference in model size compared to our traditional neural models potentially makes them a poor choice in computationally-limited scenarios, such as real-time phoneme classification systems.

To varying extents, we find that all three models made similar errors, shown in Fig. 5. Like Jha and Rao [15], we find our models struggle to distinguish between the back vowels o and u. We also notice that our models struggle to distinguish between the a and o vowels. To a lesser degree, our models struggle with the central vowels e and i as well. In addition to these shared mistakes, the recurrent model also occasionally misclassifies u as the vowel i.

Some of these particular confusions are perhaps to be expected. For the spoken pronunciation of e and i, the placement of the tongue is more closed and towards front compared to the other vowels. On the other hand, for the pronunciation of a and o, the tongue is more open and towards the back. It is likely that these similarities between vowels are further exaggerated when sung.

Our models demonstrate similar relative performance in classifying vowels across most vocal techniques as shown in Fig. 3. Notably, they struggle more with classifying `inhaled` vowels and vowels performed with the `lip_trill` technique. It is perhaps unsurprising that these models struggled to identify vowels performed with these two extended techniques. The `inhaled` vowels are quite breathy, which serves as additional noise that muffles the vowel. Vowels performed with a `lip_trill` have a unique timbre that may dampen formants and

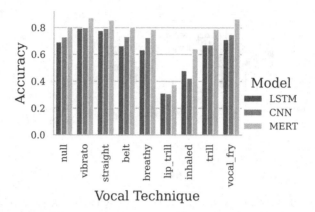

Fig. 3. Certain vocal techniques are more challenging for all models. Vowel classification accuracy was measured for each model and vocal technique. Portions of the dataset that have not been assigned a vocal technique are represented by "null", for instance the techniques `fast_forte` and `slow_forte`.

other acoustic features that the models rely upon to recognize the vowel. This timbre is shown visually in Fig. 4. Anecdotal evidence from the authors suggests that it is difficult for humans to audibly discern vowels sung with the `lip_trill` technique, but this merits further empirical investigation.

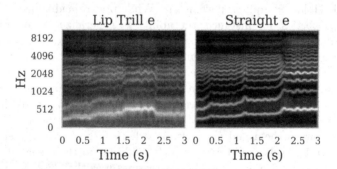

Fig. 4. An example of the vowel e performed with the `lip_trill` and `straight` vocal techniques by `female1`. The `lip_trill` vowel is both noisier and has less defined harmonics compared to the e performed with the `straight` technique.

We notice that our models perform approximately 10% better in terms of accuracy on classifying vowels sung by male singers (58.7% of the dataset) compared to female singers (41.3%). This replicates findings from Jha and Rao, who also observed a 10% difference in accuracy between male and female singers [15], though their dataset was gender-balanced. It remains unclear if our results are caused by the imbalance of the training data in VocalSet, or because of the differences in frequencies between the male and female singing voice.

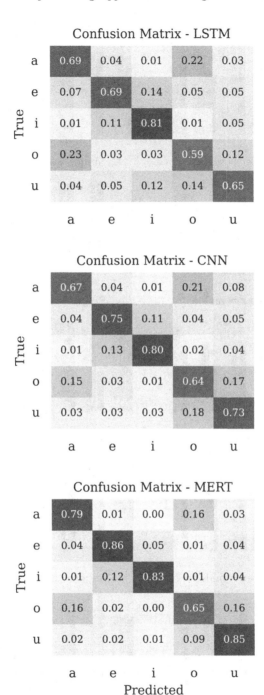

Fig. 5. Each model made similar types of errors as shown in the confusion matrices above. The relative frequency of each error varies by vowel and model.

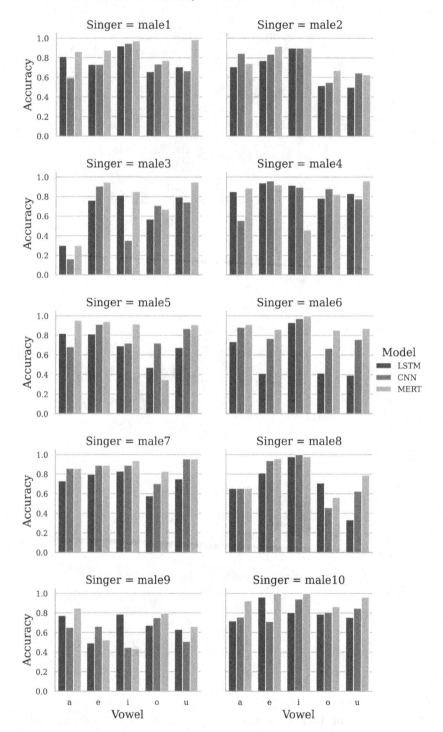

Fig. 6. Our model accuracy for each singer and vowel. Certain singers and vowels were significantly more challenging for some or all models.

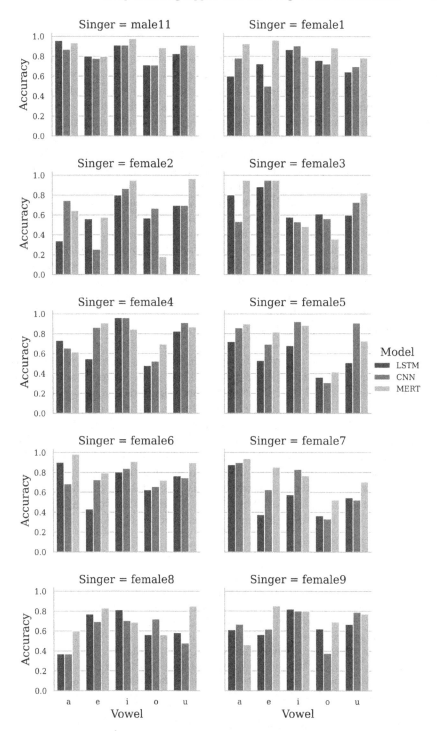

Fig. 7. Our model accuracy for each singer and vowel. Certain singers and vowels were significantly more challenging for some or all models (continued).

5   Conclusion

In this work, we demonstrate neural networks' potential to classify sung vowels in a monophonic vocal settings. By training on raw audio alone, we avoid time-consuming spectral analysis and feature extraction operations which complicate real-time analysis and prohibit use in live systems. Although these results are encouraging, there remain many avenues for improvement and further research.

As future work, we plan to explore additional models, to improve both accuracy and efficiency. We will examine neural models trained on spectral features (MFCCs), as well as an ensemble of neural models. We will also test larger convolutional and recurrent models in the 1M to 10M parameter range to understand the impact of model size, and to investigate if we can achieve comparable performance as MERT with a moderately sized model. We also intend to compare against other state-of-the-art audio and musical transformers.

One major limitation of our approaches for sung vowel classification is the lack of annotated singing data. We only evaluate our models on five Italian vowels because of the composition of VocalSet, which is insufficient for use in automatic lyric transcription systems. We encourage researchers to create more resources that enable classifying a broader spectrum of vowels. Additionally, we plan to evaluate our models on an existing speech dataset, such as TIMIT [9], to test if our models are able to generalize to spoken vowels.

Another limitation of our work is the variation of our models' performance. While our models performed decently overall, for some singer-vowel combinations, performance would degrade. In addition, our models consistently underperformed by about 10% for female singers. One important area for future work is developing models that perform equally for both male and female singers.

Sung vowel classification is a challenging and understudied subset of the phoneme classification problem. Existing spectral methods that we examined for both spoken and sung vowels failed when applied to the robust VocalSet. In this work, we evaluate three neural models and establish baselines for sung vowel classification using the VocalSet dataset. We hope this work inspires future exploration on the challenging sung vowel classification problem.

References

1. Avramidis, K., Kratimenos, A., Garoufis, C., Zlatintsi, A., Maragos, P.: Deep convolutional and recurrent networks for polyphonic instrument classification from monophonic raw audio waveforms. In: ICASSP 2021–2021 IEEE International Conference on Acoustics, Speech and Signal Processing (ICASSP), pp. 3010–3014 (2021). https://doi.org/10.1109/ICASSP39728.2021.9413479
2. Baevski, A., Zhou, Y., Mohamed, A., Auli, M.: wav2vec 2.0: a framework for self-supervised learning of speech representations. In: Advances in Neural Information Processing Systems, vol. 33, pp. 12449–12460 (2020). https://doi.org/10.48550/arXiv.2006.11477
3. Burrows, D.: Singing and saying. J. Musicology 7(3), 390–402 (1989). https://doi.org/10.2307/763607, http://www.jstor.org/stable/763607

4. Dabike, G.R., Barker, J.: Automatic lyric transcription from karaoke vocal tracks: resources and a baseline system. In: Proceedings of Interspeech, pp. 579–583. International Speech Communication Association (ISCA) (2019). https://doi.org/10.21437/Interspeech.2019-2378
5. De Wet, F., Weber, K., Boves, L., Cranen, B., Bengio, S., Bourlard, H.: Evaluation of formant-like features on an automatic vowel classification task. J. Acoust. Soc. Am. **116**(3), 1781–1792 (2004). https://doi.org/10.1121/1.1781620
6. Demirel, E., Ahlbäck, S., Dixon, S.: Automatic lyrics transcription using dilated convolutional neural networks with self-attention. In: Proceedings of the International Joint Conference on Neural Networks (IJCNN), pp. 1–8 (2020). https://doi.org/10.1109/IJCNN48605.2020.9207052
7. Devlin, J., Chang, M.W., Lee, K., Toutanova, K.: BERT: pre-training of deep bidirectional transformers for language understanding. In: Burstein, J., Doran, C., Solorio, T. (eds.) Proceedings of the 2019 Conference of the North American Chapter of the Association for Computational Linguistics: Human Language Technologies, Volume 1 (Long and Short Papers). pp. 4171–4186. Association for Computational Linguistics, Minneapolis, Minnesota (2019). https://doi.org/10.18653/v1/N19-1423, https://aclanthology.org/N19-1423
8. Gales, M., Young, S.: Application of hidden Markov models in speech recognition. Found. Trends Sig. Process. **1**(3), 195–304 (2008). https://doi.org/10.1561/2000000004
9. Garofolo, J.S., et al.: Timit acoustic-phonetic continous speech corpus (1993). https://doi.org/10.35111/17gk-bn40
10. Graves, A., Fernández, S., Schmidhuber, J.: Bidirectional LSTM networks for improved phoneme classification and recognition. In: Duch, W., Kacprzyk, J., Oja, E., Zadrożny, S. (eds.) ICANN 2005. LNCS, vol. 3697, pp. 799–804. Springer, Heidelberg (2005). https://doi.org/10.1007/11550907_126
11. Gupta, C., Yılmaz, E., Li, H.: Automatic lyrics alignment and transcription in polyphonic music: does background music help? In: Proceedings of the International Conference on Acoustics, Speech and Signal Processing (ICASSP), pp. 496–500. IEEE (2020). https://doi.org/10.1109/TASLP.2022.3190742
12. He, Y., et al.: Streaming end-to-end speech recognition for mobile devices. In: IEEE International Conference on Acoustics, Speech and Signal Processing (ICASSP), pp. 6381–6385 (2019). https://doi.org/10.1109/ICASSP.2019.8682336
13. Hillenbrand, J., Gayvert, R.T.: Vowel classification based on fundamental frequency and formant frequencies. J. Speech Lang. Hear. Res. **36**(4), 694–700 (1993). https://doi.org/10.1044/jshr.3604.694
14. Hsu, W.N., Bolte, B., Tsai, Y.H.H., Lakhotia, K., Salakhutdinov, R., Mohamed, A.: Hubert: self-supervised speech representation learning by masked prediction of hidden units. IEEE/ACM Trans. Audio, Speech, Lang. Process. **29**, 3451–3460 (2021). https://doi.org/10.1109/TASLP.2021.3122291
15. Jha, M.V., Rao, P.: Assessing vowel quality for singing evaluation. In: National Conference on Communications (NCC 2012), pp. 1–5. IEEE (2012). https://doi.org/10.1109/NCC.2012.6176860
16. Karpagavalli, S., Chandra, E.: A review on automatic speech recognition architecture and approaches. J. Sig. Process. Image Process. Pattern Recogn. **9**(4), 393–404 (2016). https://doi.org/10.14257/ijsip.2016.9.4.34
17. Kermanshahi, M.A., Akbari, A., Nasersharif, B.: Transfer learning for end-to-end ASR to deal with low-resource problem in Persian language. In: 26th International Computer Conference, Computer Society of Iran (CSICC), pp. 1–5 (2021). https://doi.org/10.1109/CSICC52343.2021.9420540

18. Khunarsal, P., Lursinsap, C., Raicharoen, T.: Singing voice recognition based on matching of spectrogram pattern. In: Proceedings of the International Joint Conference on Neural Networks (IJCNN), pp. 1595–1599 (2009). https://doi.org/10.1109/IJCNN.2009.5179014

19. Kingma, D.P., Ba, J.: Adam: a method for stochastic optimization. In: 3rd International Conference on Learning Representations (ICLR) (2015). https://doi.org/10.48550/arXiv.1412.6980

20. Korkmaz, Y., Boyacı, A., Tuncer, T.: Turkish vowel classification based on acoustical and decompositional features optimized by genetic algorithm. Appl. Acoust. **154**, 28–35 (2019). https://doi.org/10.1016/j.apacoust.2019.04.027

21. Korkmaz, Y., Boyaci, A.: Classification of Turkish vowels based on formant frequencies. In: Proceedings of the International Conference on Artificial Intelligence and Data Processing (IDAP), pp. 1–4 (2018). https://doi.org/10.1109/IDAP.2018.8620877

22. Kruspe, A.M.: Training phoneme models for singing with "songified" speech data. In: Proceedings of the 16th International Conference on Music Information Retrieval (ISMIR), pp. 336–342 (2015). https://archives.ismir.net/ismir2015/paper/000034.pdf

23. Kruspe, A.M.: Application of automatic speech recognition technologies to singing. Ph.D. thesis, TU Technische Universität, Ilmenau, Germany (2018). https://www.db-thueringen.de/receive/dbt_mods_00035065

24. Li, Y., et al.: MERT: acoustic music understanding model with large-scale self-supervised training (2023). https://doi.org/10.48550/arXiv.2306.00107

25. Mesaros, A., Virtanen, T.: Automatic recognition of lyrics in singing. EURASIP J. Audio, Speech, Music Process. **2010**, 1–11 (2010). https://doi.org/10.1155/2010/546047

26. van den Oord, A., et al.: Wavenet: a generative model for raw audio (2016). https://doi.org/10.48550/arXiv.1609.03499

27. Ou, L., Gu, X., Wang, Y.: Transfer learning of wav2vec 2.0 for automatic lyric transcription. In: Proceedings of the 23rd International Conference on Music Information Retrieval (ISMIR), pp. 891–899 (2022). https://archives.ismir.net/ismir2022/paper/000107.pdf

28. Palaz, D., Collobert, R., Doss, M.M.: Estimating phoneme class conditional probabilities from raw speech signal using convolutional neural networks. In: Proceedings of Interspeech (2013). https://doi.org/10.21437/Interspeech.2013-438

29. Santana, I.A.P., et al.: Music4all: a new music database and its applications. In: International Conference on Systems, Signals and Image Processing (IWSSIP), pp. 399–404. IEEE (2020). https://doi.org/10.1109/IWSSIP48289.2020.9145170

30. Smule, I.: DAMP-MVP: Digital Archive of Mobile Performances - Smule Multilingual Vocal Performance 300x30x2 (2018). https://doi.org/10.5281/zenodo.2747436

31. Story, B.H.: Vowel acoustics for speaking and singing. Acta Acust. Acust. **90**(4), 629–640 (2004)

32. Sundberg, J.: Articulatory differences between spoken and sung vowels in singers. STL-QPSR, KTH **1**(1969), 33–46 (1969)

33. Sundberg, J.: Articulatory interpretation of the "singing formant". J. Acoust. Soc. Am. **55**(4), 838–844 (1974). https://doi.org/10.1121/1.1914609

34. Teytaut, Y., Roebel, A.: Phoneme-to-audio alignment with recurrent neural networks for speaking and singing voice. In: Proceedings of Interspeech. pp. 61–65. International Speech Communication Association (ISCA) (2021). https://doi.org/10.21437/interspeech.2021-1676

35. Tieleman, T., Hinton, G.: Lecture 6e rmsprop: divide the gradient by a running average of its recent magnitude. In: COURSERA: Neural Networks for Machine Learning, vol. 4, pp. 26–31 (2012). https://www.cs.toronto.edu/~tijmen/csc321/slides/lecture_slides_lec6.pdf
36. Verma, P., Berger, J.: Audio transformers: transformer architectures for large scale audio understanding. adieu convolutions (2021). https://doi.org/10.48550/arXiv.2105.00335
37. Weiss, R., Brown, W., Jr., Moris, J.: Singer's formant in sopranos: fact or fiction? J. Voice **15**(4), 457–468 (2001). https://doi.org/10.1016/s0892-1997(01)00046-7
38. Wilkins, J., Seetharaman, P., Wahl, A., Pardo, B.: Vocalset: a singing voice dataset. In: Proceedings of the 19th International Conference on Music Information Retrieval (ISMIR), pp. 468–474 (2018). https://doi.org/10.5281/zenodo.1193957

Investigating the Viability of Masked Language Modeling for Symbolic Music Generation in abc-notation

Luca Casini$^{(\boxtimes)}$ ⓘ, Nicolas Jonason ⓘ, and Bob L.T. Sturm ⓘ

KTH Royal Institute of Technology, Stockholm, Sweden
{casini,njona,bobs}@kth.se

Abstract. The dominating approach for modeling sequences (e.g. text, music) with deep learning is the causal approach, which consists in learning to predict tokens sequentially given those preceding it. Another paradigm is masked language modeling, which consists of learning to predict the masked tokens of a sequence in no specific order, given all non-masked tokens. Both approaches can be used for generation, but the latter is more flexible for editing, e.g. changing the middle of a sequence. This paper investigates the viability of masked language modeling applied to Irish traditional music represented in the text-based format abc-notation. Our model, called abcMLM, enables a user to edit tunes in arbitrary ways while retaining similar generation capabilities to causal models. We find that generation using masked language modeling is more challenging, but leveraging additional information from a dataset, e.g., imputing musical structure, can generate sequences that are on par with previous models.

Keywords: Symbolic Music Generation · Masked Language Models · Irish Traditional Music

1 Introduction

This paper investigates the viability of masked language models (MLMs) in the context of symbolic music generation and compares them with models based on an autoregressive approach. We can define MLMs as sequence models where the learning objective consists in the parallel prediction of masked tokens in a sequence based on all of the unmasked ones. This is opposed to causal modeling, where the task is to predict the following token given the context preceding it. The MLM paradigm was popularized by Google's BERT (Bdirectional Encoder Representations from Transformers) [5], designed mostly for feature extraction and natural language understanding tasks. BERT-like models have

This paper is an outcome of MUSAiC, a project that has received funding from the European Research Council under the European Union's Horizon 2020 research and innovation program (Grant agreement No. 864189).

proved very effective at this but are generally found lacking when it comes to language generation in comparison to their causal counterparts, of which OpenAI's GPT and its successors are notable examples [2]. Still, the attractive aspect of MLMs is their flexibility when it comes to the generation process, as the lack of constraints in the generation order allows editing existing sequences. Some researchers have shown that masked language models can be adapted for generation [16]. Arguably, MLMs can be thought of as a generalization of causal language models where the masking is not strictly left-to-right and should in theory be capable of generalizing the same task. In this paper we describe our exploration of these ideas for symbolic music editing and generation.

We focus our experiments on modeling Irish traditional dance music expressed using abc-notation [15]. There are a number of reasons for choosing this task. First, being text-based makes it easier to compare and adapt natural language processing (NLP) techniques. There is also a large dataset of music available in this format.[1] The structure of music in this tradition is also very regular and codified [1], making it easy to verify when a model is learning sensible patterns and musical concepts. Finally, past research efforts on the same tasks offer a point of comparison for the results we obtain with our model, most notably the two autoregressive models *folk-rnn* [13] and *Tradformer* [3].

The remainder of the paper is structured as follows. The next section discusses related work. Section 2 presents the particulars of our model, called abcMLM, including its architecture, training and masking procedure. Section 3 presents a number of application scenarios for abcMLM and evaluates some of its generated output. Section 4 concludes the paper and points to future developments of this work. The code for this work can be found on GitHub, along with a link to a demo page https://github.com/mister-magpie/abcMLM.

1.1 Related Work

The model of abcMLM is based on the transformer architecture, which represents the current state of the art for sequence modeling [14]. We now discuss a number of relevant works that share an architecture similar to ours.

Tradformer [3] is a decoder transformer in the style of GPT, trained on Irish traditional music. The model surpassed previous state-of-the-art methods in the task and also showed flexibility by being able to learn Swedish folk music from a very small dataset with transfer learning. Given the shared objective, this model serves as a point of comparison for the results with abcMLM in terms of generation quality, while also considering the limitations it has because of its autoregressive approach.

Music Transformer [8] is a large-scale transformer decoder model trained on a MIDI-like representation of performances of classical piano music. The model is capable of dealing with very long sequences, using a purposefully designed attention layer, and can generate piano music from scratch, or accompaniments for melody prompts.

---

[1] For example, the tens of thousands of tunes at http://thesession.org.

LakhNES [6] is a similar model that instead focuses on video game music characterized by 3 voices and a noise channel, and uses a decoder based on transformer-XL.

An example of a masked music model is MuseBERT [17], where the authors use a transformer encoder to learn how to reverse a corruption process that includes both masking and swapping tokens in a sequence. The inputs use a matrix-like representation where each row corresponds to a note and columns are attributes like onset time, duration and pitch. The resulting model can be fine-tuned for tasks like texture generation, chord analysis and accompaniment refinement.

1.2 Data Representation

In this paper, we apply masked language modeling to a specific type of music and format, namely Irish traditional music expressed with abc-notation [15]. Abc-notation uses a very compact text-based representation of music, and appears in large online communities of folk musicians.

In the abc standard,[2] letters from A to G correspond to the respective pitches, with octave shifts indicated by appended commas or ticks. Accidentals are denoted by specific prefix characters. Duration is indicated by an integer number that symbolizes a multiple of a base unit, while preceding it with a slash corresponds to a shorter note (e.g., /2 means half). Measure lines are denoted by vertical beams (pipes), which can be coupled with a colon for repetitions or a number to indicate first and second endings. Each abc-notated tune begins with series of fields that can specify various attributes like title, author, origin, key, and meter.

As an example, the abc-notation of a version of the tune *"The Connachtman's Rambles"* (transposed to C) is shown in Fig. 1. Training a MLM on this transcription means making it predict any of the masked elements. Figure 2 shows masks as "□" as an example. A well-trained model should generate a plausible beginning for the tune.

```
M:6/8
K:Cmaj
|:EGG cGG|AGG cGF|EGG ced|cAA AGF|
EGG cGG|AGG cde|fed ced|1 cAA AGF:|2 cAA A3||
|:eaa ege|edc dcd|eaa ege|edc d3|
eaa ege|edc cde|fed ced|1 cAA A3:|2 cAA AGF||
```

Fig. 1. A transcription of the tune *The Connachtman's Rambles* transposed to C major.

---

[2] See https://abcnotation.com/wiki/abc:standard:v2.1.

```
M:6/8
K:Cmaj
|:□□□□□□|□□□□□□|EGG ced|cAA AGF|
EGG cGG|AGG cde|fed ced|1 cAA AGF:|2 cAA A3||
|:eaa ege|edc dcd|eaa ege|edc d3|
eaa ege|edc cde|fed ced|1 cAA A3:|2 cAA AGF||
```

Fig. 2. An example of masking applied to the same tune. Given this input, a masked language model would predict the twelve masked tokens in parallel.

2   abcMLM Implementation

The model we implement and train, abcMLM, is a standard transformer encoder, as shown in Fig. 3. Each element of an input sequence is embedded in an n-dimensional space and summed with a learned positional encoding. These transformed inputs pass through a dropout layer and then through a stack of transformer encoder layers. These layers are identical to the ones in Tradformer [3], with which we share most hyperparameters to allow for direct comparison. Similarly to Tradformer, we also noticed that performance scales with the size and depth of the model. In our case we increased the model size to 256. Finally, the output is processed by a final linear layer that yields a vector of the vocabulary size for each position in the sequence. This is used to compute the loss function for the masked positions, following the architecture of BERT and the other masked language models.

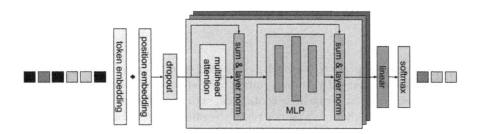

Fig. 3. Graphical summary of the model architecture.

2.1   Training

The dataset we use is the same as for Tradformer [3], which consists of more than 23,000 abc-notated transcriptions of Irish traditional music. Our inputs begin by specifying meter and mode, as they are relevant for the generation process and help condition the model. The total number of tokens in the vocabulary is 129 after including special tokens for masking, padding, and sequence start/end. We limit the maximum sequence length to 256 tokens, as opposed to 512 for Tradformer. This is due to memory limitations but does not affect our conclusions. In fact, it only causes the removal of around 2,000 transcriptions, since the average training sequence length is 169.

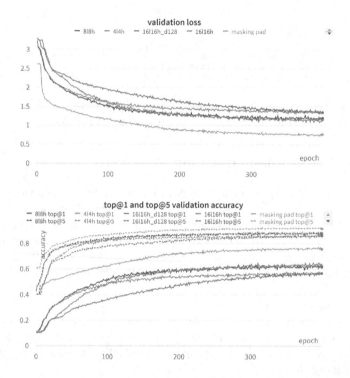

Fig. 4. Validation loss and accuracy metrics for different model with 4, 8 and 16 layers and attention heads. The best model with a model size 128 performs as bad as the 4 layers model.

Convergence appears to be much slower compared to autoregressive models. This is not surprising, as the training target of simultaneously predicting multiple tokens, often done with very little context depending on the mask size, is harder than next-token prediction. The inherent randomness in the masking procedure also makes batches very heterogeneous, and consequently forces us to increase the batch size as much as possible. We found 256 to be a good compromise between speed and learning capabilities – which is 4 times the value used by Tradformer. Figure 4 shows validation metrics for the different model variants.

2.2 Masking

The main difference with respect to BERT and similar models is in the masking procedure that constitutes the training inputs. Whereas these models mask a fixed amount of the input (usually 15%), we want to learn a good representation even for completely masked sequences. We took inspiration from existing work in image generation using masked vision transformers [4], masking the training input sequences as follows. According to some function $\gamma : \mathbb{R} \in [0, N]$ we determine how much of the non-padded length-N sequence to be masked. We define

$\gamma(t) = Ncos(t)$ with t uniformly sampled in the interval $[0, \frac{\pi}{2}]$, rounded to the closest integer. This number of masks is then applied across the sequence in random positions (without replacement). Although we followed [4], γ could probably be a simple linear function with minimal impact. This approach also appears to translate well in the domain of music generation using pianoroll representation [9].

After experimenting with generation, we found that the model has a hard time predicting the end of sequence token in positions that are not the last one. We hypothesize that this behavior is dependent on the fact that the padding tokens are never masked and ignored by the attention layer, resulting in a model that associates mask tokens to musical content only and puts the end-of-sequence token on the last masked position. We tried training a model with masking allowed on any position. This resulted in a deceptive increase in accuracy metrics and a lower loss score. Given that for most sequences almost half of the tokens are padding, predicting their presence on the right side of the input is very easy and their large number easily skews the metrics.

3   Example Applications

We now demonstrate and discuss the main use cases of abcMLM. First, we discuss the main goal of this work, which is editing. After that we describe our experience with generation and how we arrived at using structural information from the dataset in order to generate novel and convincing tunes.

3.1   Editing a Known Tune

The main contribution of abcMLM is demonstrating the viability of this class of models for editing existing material. We tried masking measures in existing tunes and replacing endings of sections with new first and second endings. In our qualitative evaluation we found that in almost all instances the model is able to generate correct and good sounding variations. We experienced a small amount of redundancy, with around one in ten variations being a duplicate.

As an example, we show a few generated endings for "The Connachtman's Rambles", presented in the introduction. We masked the tokens of the final two endings of each part, and in a central portion of the tune (bars 3–4 and 11–12). Each entry in Table 1 is an independent run of the model, the left column is for endings, and the right column is for central measures. Each of the outputs is a good variation, capturing melodic patterns that are coherent with the tune and style. Is especially nice to notice how the model has picked up this cross-relation between the A and B sections where endings appear to mirror each other (the practice of Irish music entails repeating a tune several times alternating between sections sequentially).

Table 1. Editing endings for "The Connachtman's Rambles"

Ending edits (bar 7-8 and 15-16)	Central edits (bar 3-4 and 11-12)
|1ccccGF:||2cGcc3|...|1cGcc3:||2cGccGF|	EGG cGG|AGA AGF|...|ccc def|edd d3|
|1cAAAGF:||2cAdc3|...|1cAdc3:||2cAGAGF|	EGG EGG|AGG FED|...|egg geg|aaa a3|
|1ccccGF:||2cGcc3|...|1cGcc3:||2cGccGF|	EGG cGG|AGA AGF|...|ccc def|edd d3|
|1cAAAGF:||2cAdc3|...|1cAdc3:||2cAGAGF|	EGG AGG|AGG cAG|...|ecg aeg|gec d3|
|1cAGAGF:||2cAGA3|...|1cAGA3:||2cAGAGF|	EGG cGG|FEF DED|...|eaa egg|edc d3|
|1cAAAGF:||2cAGA3|...|1cAGA3:||2cAGAGF|	EGG AGG|GFF FGF|...|edc dcc|dBB d3|
|1cedcAG:||2cedc3|...|1cedc3:||2cedcAG|	EGG cGG|efe dcF|...|eAA cBc|def̂ g3|
|1cAAAGF:||2cGEc3|...|1cGEc3:||2cGEDGF|	EFF FFF|DDC DEF|...|Bfe ede|cAG c3|
|1cdccGF:||2cdcc3|...|1cdcc3:||2cdccGF|	EEE cGE|DDD DCD|...|eed cde|ddd d3|
|1cAAcGF:||2cGGc3|...|1cGGc3:||2cGGAcG|	EGG EGG|AGG FED|...|egg geg|aaa a3|

In Fig. 5 we also superimposed a few edits from the table to the score to give a visual idea of the same results.

Fig. 5. Examples of edits on a known tune. Each color is a different output.

3.2 Infilling Structures of Odd Length

We also experiment what the model produces when the input presents a structure it has not seen – for example, the number of measures is not the typical eight, or each bar contains an odd number of tokens (e.g., five tokens for a 6/8 measure in a jig). The first example in Fig. 6 shows some referencing of melodic material and its variation, which is typical of Irish traditional music. In Fig. 7 instead, we have an example of the model infilling five masked tokens in each of seven bars. The model effectively ignores the 6/8 m and employs the > token (stealing time from the following note for the preceding one), resulting in outputs that play more like polkas than jigs. While not always faithful to the meter, those types of outputs are musically interesting and can contribute in creative directions.

```
M:6/8
K:Cmaj
GCEGEG | A2AGEG | GCEGEG | A2AGEG | GCEGEG | A2AGEG | GEDCDC : |
E2EGEG | A2AAcA | GEGCEG | A2AAcA | GEGCDE | GEDCDE | DCDCDC : |
```

Fig. 6. Infilling an odd-length structure of 7 bars with 6 tokens each in 6/8 time.

```
M:6/8
K:Cmaj
c>dec | AFG>F | EGC>D | EDD>G | c>dec | AFG>F | EGC>z : |
| : cGEG | cGE>G | AGF>G | AGE>G | cGE>G | AFG>F | EGC>z : |
```

Fig. 7. Infilling the same odd-length structure but providing 5 tokens per bar instead. The model ignores the 6/8 times and outputs what looks like a polka.

3.3 Predicting All Masked Tokens in Parallel

abcMLM outputs a vector parameterizing a categorical distribution over the vocabulary for each position in the sequence. Since the model only learns from the masked tokens we only consider their respective predictions. The most naive use of the models is thus to let it predict a completely masked sequence. Lack of context results in nonsense apart from the obvious header tokens.

Following a masking schedule as done in MaskGIT and doing iterative re-sampling both fail to completely solve this problem, as abc-notation is very easily broken by an erroneous token in the sequence. We also experimented with unmasking a few tokens at a time but eventually ended up with generating them one by one as described in the next section. Any in-between strategy only trades speed for accuracy (which is not great to begin with).

3.4 Generating One Token at a Time

To address the lack of context, we generate tokens sequentially. This is done by predicting one masked token and iterating the process until the sequence is complete. We experiment with ordering the prediction randomly, left-to-right and right-to-left. We find that this generation strategy allows the model to output coherent music since the context is slowly filled in, and the last few predictions become more obvious. The few mistakes that are still present seem to be mostly due to the fact that the model is not able to ignore superfluous mask tokens, resulting in bars longer or shorter than usual. This type of mistake could be easily detected and fixed by a musician using the model as a tune-writing assistant. Different ordering strategies do not appear to result in significant differences in the final output.

All examples in Fig. 8 have counting errors, and show an important limitation of the model. Since <pad> tokens are ignored and never masked, the model has learned that the last masked token is the end of sequence and thus the final length of the output will always correspond to the number of masked tokens. Both folk-rnn [13] and Tradformer [3] do not have this limit and seem to be able to learn when to stop generating, creating outputs of appropriate lengths.

```
M:4/4
K:Cmin
|:G2GFGECE|ECB,ED2DC|B,2C2GB,CD|B,CECB,2DB,|
G2GAFECE|B,2ECB,CDE|F3GBBGG|1G2CB,C2GA:||2G2CB,C2(3DFE
|:C2(3DCB,CDE|F(3GAGFDE|B,CDEC(3ECE|FDGFEC|
B,4DCEG|(3GFEFGECB,2cB|AGGFFG(3FAc|FBGFGFEG:||2FBGFGccB
|:ecdedccc|AccBcccd|(3efgegefgffe|dccGc2cd|
edcBcc|AccGGCB,C|EFGFGFFG|1AFGF3BGG|F2G2C4:|
```

```
M:6/8
K:Cmin
ECCDEG|cBcGEC|E2GGFE|DCDFED|ECCDEG|cBcGEC|E2GGFE|1ECB,C2D:||2ECB,C2F
|:GcccBc|dcAABA|GBBFBB|dcdedc|GcccBc|dcAABA|GBBGFE|1ECB,CEF:||2ECB,C2d
|:ecAGEC|ededcB|GBBBB|FBBfdB|eccGEC|ecAGEC|ededcB|1ECB,C2:||2ECB,C2d|
ecAGEC|ededcB|GBBdcB|dcBedB|edcdcB|c/2d/2edcBG|1ECB,C2d:||2|2ECB,C2G|
```

```
M:6/8
K:Cmin
DBBAGA|B2dfdB|cBcFAc|GcBAcA|DBBAFA|B2dfdB|g2efed|1c3CEG:||2c3ccA|
GccFA2|FAcAcA|GBBFDF|BcedcB|GccFAc|e3ecG|e3efd|gfedcB|1c3cc'2:||2c3c2d
|:e2ggef|c'2ggec|b2ffdf|b2dfdG|e2efdG|e2edef|gfedcB|1c3c2=B:||2c3c2B
|:edcdBG|e2gfdf|b2bfdB|e2efdB|efgc'3|efg=ab3|gfed2B|1c3c2F:||2c3c3|
```

Fig. 8. Autoregressive Generation with different orders: Random (top), left-to-right (middle) and right-to-left(bottom)

We tried addressing the model's tendency to fill the maximum length by masking pad tokens. The resulting model is indeed able to predict the end-of-sequence token in arbitrary positions, but does so in a very peculiar way that impacts the quality of generated tunes. Left-to-right generation (top) appears biased towards generating the end of sequence token very early, usually after 8 measures and especially after a repetition sign. On the other hand, with right-to-left generation (middle) sequences seem to be of a more typical length. But this time the model skips the end of sequence token altogether, often leaving a hanging measure at the end. Random generation (bottom) seems to be an acceptable compromise, exhibiting the same behavior of left-to-right generation but to a lesser extent and producing tunes of the appropriate length more often than not.

3.5   Leveraging Length and Structure from Existing Tunes

To address this problem we draw a sample from the test dataset and use its length to decide the number of mask tokens in the prompt. This solution prevents most of the mistakes we previously observed, most likely due to the fact that the model becomes able to better gauge the position of bar tokens. However, the model still occasionally produces counting errors, especially when generating reels. Figure 9 shows a jig and a reel generated with this strategy.

Since the most common problem appears to be linked to measure tokens, we take the initial idea a step further by prompting the model with the structural tokens from an existing tune. By structural tokens, we mean any token that is not pitch-related and provides information about measures, duration, repetitions, and so on. This strategy provides a "scaffold" that the model can

```
M:6/8
K:Cdor
FDCBCC | GDCDFG | FDCGFG | FDCFBG | FDCBAG | FDCDFG | GFGFDG | FDB , CDF : |
| : G2GB3 | cBGFDF | G2c3 | dBdfBc | d2cB2c | cBcdfdc | B2GFD | DB , B , C3 : |
```

```
M:4/4
K:Cdor
| : C2G2cBGF | B , 2DB , FB , DF | Gcc2c2dB | G2Bdc2BG |
C2G2cBGF | B2cBF2 (3def | g2fdedcd | 1BGFDC3B , : | | 2BGFDC3d
| : c3dB3c | dBBcdBFd | fdgdfdce | BcdBc2df |
gededcBc | BGc2dB2ddef | g2fdc3 | 1bgfddccd | 2BGFDC3B , : |
```

Fig. 9. Autoregressive generation in random order with and end-of-sequence token placed according to a sample from the dataset.

fill with melodic content, and which results in generated material that is more appropriate. Figure 10 shows a jig and a reel and the original tune from the the dataset where we sampled the structure. Figure 11 shows staff notation of the same tunes.

```
T:original tune 5828
M:6/8
K:Cdor
| : df | eccBcc | G3GAB | c3cBG | F3FGB | c3BcB | G3GBG | F3FGB | c3c : |
| : df | eccfed | edecde | d^cdB=cd | e2fg2f | eccBcc | G3GAB | 1F3FGB | c3c : | | 2c3cBG | F3F |
```

```
T:based on test 5828
M:6/8
K:Cdor
| : cd | efegg2 | D3Gcd | e3gfe | d3d2d | c3dGB | G3Bcd | c3c2c | c3d : |
| : ga | b2agfe | B2z2za | b2g2B2 | a2fc2a | b2ggfe | d3Bcd | 1e3g2e | d3d : | | 2e3c2c | c3z |
```

```
T:original tune 21352
M:4/4
K:Cmin
C2GCBCGc | BGFDB , CDB , | C2GCBCGc | BGFDGCC2 : |
CccdedcB | GBFDB , CDB , | CccdedcB | GBFDGCC2 | CccdedcB | GBFDB , CDF | fdecdBc=A | BGFDGCC2 |
```

```
T:based on test 21352
M:4/4
K:Cmin
C2EGcGEG | c2ecdBB2 | c2gcegcB | AGFDECC2 : |
cdefg2fg | b2bfdBB2 | cdefg2fg | fefdc2c2 | cdefg2fg | b2bfdBB2 | cdeBAGFG | AGFDECC2 |
```

Fig. 10. Autoregressive generation in random order leveraging structural tokens (barlines and duration) based on an existing tune in the test dataset.

A quick self-evaluation was performed by the authors (who have musical proficiency and are familiar with the style), generating 10 jigs and 10 reels and comparing them to the same type of tunes generated by the Tradformer. While this test's validity is limited, it helps us to gauge the capabilities of the models. To avoid bias tunes were shuffled so the generating system were unknown. Each tune was scored between 1 and 5. Table 2 reports the average for each model.

Fig. 11. Music notations for the tune in Fig. 10.

Table 2. Evaluation for 20 samples (10 reels and 10 jigs) from abcMLM using lengths or structures sampled from the dataset vs. Tradformer.

Model	abcMLM + length	abcMLM + structure	Tradformer
Jigs	2.8	4.0	4.1
Reels	2.2	3.6	3.7
Avg.	2.5	3.8	3.9

4   Conclusions and Future Work

We have introduced abcMLM: a masked language model for symbolic music notation, specifically Irish traditional folk music represented in abc-notation. The rationale behind exploring this modeling approach was enabling user to not only generate symbolic music with neural networks but also to edit it. The resulting model can also be used for generating new tunes altogether, but in order to do it reliably it required more work compared to causal left-to-right models. We found that training abcMLM is quite challenging, and different factors come into play compared to causal models. Namely, the added randomness of the masking procedure needs to be counteracted by an increased batch size for the model to converge effectively. Training is thus more memory intensive and slower.

Nevertheless, the model is very effective in its intended task of editing existing material. The generated material is often appropriate and rarely redundant according to our qualitative evaluation. When it comes to the generation of complete tunes, abcMLM struggles in comparisons to similar causal models as is the case for similar text generation models. Most mistakes happen in situations where the number of masks to be predicted deviates from what one would expect. This often happens as the model fills the context in a random way. Using structural information from existing tunes alleviates this problem almost completely. This can take the form of using the length of an existing tune, or imputing measure structure and note duration. Overall, we can say that abcMLM is slightly less accurate than its direct competitors, but it gains the capability of predicting tokens in a much more flexible way, enabling creative uses that are beyond the reach of autoregressive models.

One important limitation of a masked language model is the fact that predictions are bound to the presence of mask tokens. Since there is no concept of empty tokens, the model will end up filling random tokens in situations where nothing is needed but mask tokens remain. A way out could be to learn the insertion operation. The model would then predict a position and a token until the sequence is full or the model only suggests empty tokens [12]. Furthermore, we could also include a second stage where deletions are performed. If the two stages end up undoing each other, we can stop the generation process [7,11]. The generalization of these ideas is to implement and learn all Levenshtein operations, resulting in something conceptually identical to diffusion for discrete sequences [10].

While outside of the scope of this paper, encoder models like BERT are also useful for feature extraction, clustering and classification. Using abcMLM in machine listening tasks is definitely something to explore. Increasing the size of the dataset with other abc music datasets available online is also a viable extension. Although not small, the dataset we are using is far from the size of datasets used in NLP, and additional information could only benefit the performance of the model. In this context adding information about the type of music could also provide better latent representations.

References

1. Breathnach, B.: Folk Music and Dances of Ireland: A comprehensive study examining the basic elements of Irish Folk Music and Dance Traditions. Ossian (1971)
2. Brown, T., et al.: Language models are few-shot learners. Adv. Neural. Inf. Process. Syst. **33**, 1877–1901 (2020)
3. Casini, L., Sturm, B.L.T.: Tradformer: a transformer model of traditional music transcriptions. In: Raedt, L.D. (ed.) Proceedings of the Thirty-First International Joint Conference on Artificial Intelligence, IJCAI-22, pp. 4915–4920. International Joint Conferences on Artificial Intelligence Organization (7 2022). https://doi.org/10.24963/ijcai.2022/681, aI and Arts

4. Chang, H., Zhang, H., Jiang, L., Liu, C., Freeman, W.T.: Maskgit: masked generative image transformer. In: Proceedings of the IEEE/CVF Conference on Computer Vision and Pattern Recognition, pp. 11315–11325 (2022)
5. Devlin, J., Chang, M.W., Lee, K., Toutanova, K.: Bert: pre-training of deep bidirectional transformers for language understanding. arXiv preprint arXiv:1810.04805 (2018)
6. Donahue, C., Mao, H.H., Li, Y.E., Cottrell, G.W., McAuley, J.: Lakhnes: improving multi-instrumental music generation with cross-domain pre-training. arXiv preprint arXiv:1907.04868 (2019)
7. Gu, J., Wang, C., Zhao, J.: Levenshtein transformer. In: Advances in Neural Information Processing Systems, vol. 32 (2019)
8. Huang, C.Z.A., et al.: Music transformer. In: International Conference on Learning Representations (2019). https://openreview.net/forum?id=rJe4ShAcF7
9. Jonason, N., Sturm, B.L.: erl-j/masked-generative-music-transformer: Original release (2023). https://doi.org/10.5281/zenodo.7703864
10. Reid, M., Hellendoorn, V.J., Neubig, G.: DiffusER: diffusion via edit-based reconstruction. In: The Eleventh International Conference on Learning Representations (2023). https://openreview.net/forum?id=nG9RF9z1yy3
11. Ruis, L., Stern, M., Proskurnia, J., Chan, W.: Insertion-deletion transformer. arXiv preprint arXiv:2001.05540 (2020)
12. Stern, M., Chan, W., Kiros, J., Uszkoreit, J.: Insertion transformer: Flexible sequence generation via insertion operations. In: International Conference on Machine Learning, pp. 5976–5985. PMLR (2019)
13. Sturm, B.L., Santos, J.F., Ben-Tal, O., Korshunova, I.: Music transcription modelling and composition using deep learning. In: Proceedings Conference on Computer Simulation of Musical Creativity. Huddersfield, UK (2016)
14. Vaswani, A., et al.: Attention is all you need. In: Advances in Neural Information Processing Systems, vol. 30 (2017)
15. Walshaw, C.: The abc music standard 2.1 (2011). https://abcnotation.com/wiki/abc:standard:v2.1. Accessed 13 Mar 2023
16. Wang, A., Cho, K.: BERT has a mouth, and it must speak: BERT as a Markov random field language model. In: Proceedings of the Workshop on Methods for Optimizing and Evaluating Neural Language Generation. ACL (2019). https://doi.org/10.18653/v1/W19-2304
17. Wang, Z., Xia, G.: Musebert: pre-training music representation for music understanding and controllable generation. In: ISMIR, pp. 722–729 (2021)

MoodLoopGP: Generating Emotion-Conditioned Loop Tablature Music with Multi-granular Features

Wenqian Cui[✉], Pedro Sarmento, and Mathieu Barthet

School of Electronic Engineering and Computer Science, Queen Mary University of London, 327 Mile End Rd, Bethnal Green, London E1 4NS, UK
cuiwenqian.app@gmail.com, {p.p.sarmento,m.barthet}@qmul.ac.uk

Abstract. Loopable music generation systems enable diverse applications, but they often lack controllability and customization capabilities. We argue that enhancing controllability can enrich these models, with emotional expression being a crucial aspect for both creators and listeners. Hence, building upon LooperGP, a loopable tablature generation model, this paper explores endowing systems with control over conveyed emotions. To enable such conditional generation, we propose integrating musical knowledge by utilizing multi-granular semantic and musical features during model training and inference. Specifically, we incorporate song-level features (Emotion Labels, Tempo, and Mode) and bar-level features (Tonal Tension) together to guide emotional expression. Through algorithmic and human evaluations, we demonstrate the approach's effectiveness in producing music conveying two contrasting target emotions, happiness and sadness. An ablation study is also conducted to clarify the contributing factors behind our approach's results.

Keywords: Controllable Music Generation · Symbolic Music Generation · Deep Learning · Transformers · Guitar Tablatures · Guitar Pro

1 Introduction

The significance of repetitive, loopable aspects in music structures is evident, especially in loop-centric genres like electronic dance music [11]. Prior works have explored loop generation in both symbolic [1,11,12] and audio domains [16,39], with some having specific focuses, such as drum instruments [2,35]. However, increasing the degree of control in loop-based music generation systems is needed to address creative requirements, with agency over the emotions conveyed by the music standing out due to their direct influence on the listener's experience and engagement. Emotion-controllable music offers potential applications in live performances, soundtracks, gaming [17,19], virtual/augmented reality (VR/AR), and even in personalized music generation and the data-driven musification in the context of smart cities [27].

C. Johnson et al. (Eds.): EvoMUSART 2024, LNCS 14633, pp. 97–113, 2024.
https://doi.org/10.1007/978-3-031-56992-0_7

We utilize LooperGP [1], an advanced loopable symbolic music generation system that can effectively produce coherent and original loops with specified lengths, keys and time signatures, as our baseline. It extracts repeatable sections in music using a correlative matrix approach to derive the training data. This symbolic tablature generation system is trained on the DadaGP dataset [28]—a large-scale compilation of Guitar Pro format tablatures combining musical notes with playing techniques, dramatically elevating the expressiveness in the generated music. Such expressiveness can be harnessed for better emotional representation in music.

To guide our model in generating music conveying specific emotions, we add control tokens to the start of the symbolic token sequences, inspired by the GTR-CTRL model [29]. Our study mainly targets happiness and sadness, which are associated to two quadrants in the two-dimensional valence/arousal space based on Russell's model of affect [26]. Happiness and sadness are representative emotions from the high valence and high arousal quadrant (first quadrant) and the low valence and low arousal quadrant (third quadrant), respectively. Hence, our system operates under the assumption that music with high valence and high arousal expresses happiness, while music with low valence and low arousal expresses sadness. However, we acknowledge the bias of these assumptions and recognize that they may not hold in all contexts.

Even though earlier works have used valence and arousal scores as controls [31,32], we posit that it might not fully harness the model's potential for conditional generation. Motivated by the findings in psychological research [6,8,36], which explored the intrinsic musical features contributing to conveying distinct emotions, we integrate specific musical elements, notably tempo and mode, during both training and inference to enhance conditional generation capabilities. This is to investigate if the features highlighted by music psychology studies can also be advantageous to AI generative systems, and we note that the approach is not bounded by happy and sad emotions, for it can be extended to other emotions by leveraging correlated musical features.

While many features are significantly associated to music's emotional expression, they often remain static throughout a piece. Given music's dynamic nature, representing its essence with a single attribute is limiting. To address this, we introduce an approach integrating multi-granular features at both song and bar levels for emotion-conditioned generation. Specifically, we utilize tonal tension—metrics capturing tonal attributes—as bar-level features, based on their known correlation with musical emotions [3,7].

We trained our model on DadaGP [28], a dataset specializing in Guitar Pro format guitar tablatures, with an encoder/decoder framework to convert symbolic tokens into Guitar Pro files. The Transformer-XL [5] model is employed for sequence generation. Our results highlight the significance of both song and bar-level features in emotion-conditioned music, validated through algorithmic evaluations and a listening test. To summarize our contributions: 1) We improved on LooperGP, a generation system that creates loopable music, by incorporating a control for emotion; 2) We incorporated features from music psychology research in the emotion control process alongside emotion labels; 3) We investigated enriching the emotional control process by integrating both song-level and bar-level features.

2 Related Work

2.1 Emotion-Conditioned Symbolic Music Generation

To generate symbolic music with specific emotions, one common method is to insert emotion control tokens at the start of the sequence as conditions [32]. This conditioning method is widely used in various tasks or domains. Sarmento et al. [29] use tokens to condition the instruments and genres of generated music, and Keskar et al. [20] use control tokens to generate sentences with target attributes.

For other conditioning methods, Tan et al. [33] use low-level musical features to infer high-level features to perform music style transfer. Ferreira et al. [9] uses genetic algorithms to condition mLSTM to generate video game soundtracks with certain emotions. Huang et al. [15] use the tile function to condition the CVAE-GAN architecture. Grekow et al. [10] generate music with certain emotions by random sampling the 20-dimension latent space of CVAE. Instead of using discretized values, Sulun et al. [31] use continuous-valued valence/arousal scores to condition a transformer to generate music, which is classified as the dimensional approach in [37].

Emotions can also be inferred from other modalities. Tan et al. [34] uses image-music pairs with the same emotion to train and condition the music generation model, and Madhok et al. [22] uses the emotion vector classified using image to condition the music generation model.

2.2 Emotion-Related Features in Music

The emotions perceived or felt upon listening to music have been extensively studied in literature, with researchers focusing on intrinsic features such as tempo and mode [6,36]. Dalla et al. [6] designed an experiment where the infants were asked to point to happy or sad faces after listening to music, and they found that fast tempo and major mode are related to happy music, while slow tempo and minor mode are related to sad music[1]. Webster et al. [36] further investigated the combined effect of tempo, mode, and texture, showing that fast tempo, major mode, and simpler melodies result in happier music, while slow tempo, minor mode, and thicker texture result in sadder music.

Juslin et al. [18] examined how five acoustic cues regarding tempo, energy, and articulation, relate to the emotions of happiness, sadness, anger, and fear. Blood et al. [3] uses positron emission tomography (PET) to measure the relationship between musical emotions and the level of musical dissonance. Fernández-Soto et al. [8] investigates the tempo and rhythmic unit to four emotional semantic scales. Yang et al. (2023) [38] highlighted that music emotion

---

[1] Only major and minor modes were considered in this study.

perception was a multimodal phenomenon that depended on less frequently studied features such as musical structure, performer expression, and stage setting, and was affected by individual factors such as musical expertise.

2.3 Tonal Tension

Tonal tension, as described in [13], quantifies the emotional and mental fluctuations induced by tonality in music. It is derived from the spiral array theory, representing pitch classes, chords, and keys in a helical three-dimensional space [4]. Tonal tension comprises three elements: cloud diameter, cloud momentum, and tensile strain [13]. Cloud diameter gauges the maximal distance between any two notes within a cloud, while cloud momentum represents the distance between the centres of effect of two clouds of points, and tensile strain is the tonal distance between the centres of effect of a cloud of notes and the key. These metrics effectively quantify the tonality of a piece, and as such, are suggested as useful control tokens for emotion-conditioned music generation. By utilizing the varying values of tonal tension, more nuanced guidance is expected to be provided in the music generation process.

2.4 DadaGP and Guitar Tablature Generation

DadaGP [28] is a symbolic music generation dataset comprising 26181 guitar tablatures. It also contains an encoder/decoder to transform the guitar tablatures into symbolic tokens, which can be directly used to train sequence-to-sequence models. DadaGP covers 739 musical genres with a main focus on rock, metal, and their sub-genres.

DadaGP serves as a dataset for the generation of guitar and other instruments' parts in a tablature format. There are many works focus on guitar tablature generation, with most of them targeting a specific application. Sarmento et al. [28] trained a Transformer-XL model on the DadaGP dataset to generate guitar music in tablature. Sarmento et al. [30] focus on mimicking the style of four iconic guitarists by analyzing features from DadaGP. Loth et al. [21] trained on a subset of DadaGP to generate progressive metal music. McVicar et al. [23] focuses on generating guitar solo tablatures using MusicXML data.

3   Methodology

In this paper, we aim to enhance emotion-conditioned music generation by utilizing both song-level and bar-level features. We adopt Russell's model of affect [26], associating emotions to valence and arousal values in a two-dimensional space, to determine the level of happiness and sadness expressed by a piece. According to the model, high valence and arousal values correspond to happy emotion, while low valence and arousal values correspond to sad emotion. Therefore, we made an assumption that music with higher valence and arousal values is more likely to convey happy emotion, and vice versa.

To make the model generate music with target emotions, control tokens are added at the start of the token sequence. While only using valence and arousal is intuitive, we aim to explore the benefits of integrating other features. In this work, we classify the features into three categories: emotion labels and music psychology features as song-level features, and tonal tension as bar-level features.

In the following subsections, we will illustrate how we obtained the feature labels and incorporated the music loop information. Finally, we will summarize the pipeline of this work, covering data preparation, model training, and model inference.

3.1   Emotion Labels

To get the emotion labels for each song in the DadaGP dataset, we query the Spotify Web API using the artist name and the song title to retrieve the valence and energy values, where energy here serves as a surrogate to arousal, as in [24]. The matching was carried out using the SpotiPy Python library. As a result of the matching process, a total of 16,173 songs were successfully annotated. The values obtained for valence and energy are continuous, but to use them as control tokens for the generative system, they must be discretized. This involves dividing them into two categories - high and low values. To determine the threshold for this division, the median valence and arousal values of all the pieces in the dataset are calculated and used. For instance, valence-high and valence-low are the tokens used for valence. Based on this categorization, music with high valence and high arousal is classified as happy music, while music with low valence and low arousal is considered sad music. In our case, considering ranges between 0 and 1, the thresholds for valence and arousal are 0.433 and 0.846, respectively.

3.2   Music Psychology Features

In this work, we focus on two features studied in the music psychology literature—tempo and mode. The mode of the music can also be found using the Spotify web API, and it is split into two classes: major mode and minor mode. Although tempo can also be found through Spotify web API, it is not extracted in this work because every song in DadaGP already has a token representing its tempo, and it is the necessary information for the decoder.

3.3   Bar-Level Features

We utilize tonal tension (i.e., cloud diameter, cloud momentum, and tensile strain) as the bar-level features, employing the midi-miner package [25] for feature calculation from musical scores. Similarly, we discretized those values to use them as control tokens. We discretized bar-level features into four levels, using the first quartile, median, and third quartile of the data distribution as separating thresholds. The reason we use four levels instead of two levels to represent bar-level features is to support more combinations. Since the bar-level features are added for each bar of the music, they represent the "state" of that bar, and their purpose is to guide the music generation process. Hence, we would want the number of possible combinations of the features to be relatively large, so that they can represent more creative possibilities. In this study, we used three bar-level features. If these features had two levels, there would be a total of 8 possible combinations. However, if the features had four levels, the number of combinations increases to 64.

Since values can be derived for each bar of the music, we chose to append those values at the start of each bar of the piece, right after the `new_measure` token) token. Therefore, the resulting sequence for every bar is: `new_measure`, `cloud_diameter`, `cloud_momentum`, `tensile_strain`, and then the rest of the tokens in this bar.

3.4 Keeping the Loop Information

The aim of this work is to generate emotion-conditioned and loopable music. Therefore, another part that should be integrated is to make the model generate coherent musical loops. Inspired by LooperGP [1], we use the same loop extraction method [14] and the "Barred Repeats" method proposed in the paper, as it is shown the best result in the LooperGP paper.

3.5 The Overall Pipeline

In this section, we delineate our project pipeline, comprising data pre-processing, model training, and inference.

First, we query the Spotify web API to get the song-level features for every piece, including valence, arousal, and mode. We then perform the correlative matrix approach and the "Barred Repeats" method in LooperGP [14] to get the loops used to train the model. The whole process of the loop extraction results in a further shrink of the dataset size to 13,466. Bar-level features, i.e., tonal tension, are then derived using the midi-miner package [25].

After getting all the necessary features, the next step is to prepare the dataset for training. In the training process, every piece of music is represented by a token sequence. After discretizing all the features above, we add the control tokens to the corresponding positions within the sequence. We put the emotion labels and the music psychology features at the very beginning of the sequence, and put the tonal tension values right after every `new_measure`[2] token.

A Transformer-XL model [5] is employed for the symbolic music generation task, predicting the next token in a sequence. Then, during inference, we use the control tokens to serve as a prompt to steer the model to generate music. Specifically, we want the model to be able to generate happy and sad music, so we use different prompts to make the model generate music with different emotions. We use the prompt sequence [`valence:high`, `arousal:high`, `mode:major`, `time_signature:4`] to generate happy music, and use the prompt sequence

---

[2] `new_measure` is the token representing the start of a new bar.

[valence:low, arousal:low, mode:minor, time_signature:4] to generate sad music. The thresholds for tempo are determined heuristically. We set an upper threshold of 150 BPM and a lower threshold of 100 BPM during inference, and sample the generated tempo to be higher or equal to 150 BPM for happy music and sample the tempo to be lower or equal to 100 BPM to generate sad music. Moreover, we allow the model to freely generate tonal tension without specifying values, assuming it learns to utilize them to guide the generation during inference.

Moreover, a time signature token is added during training. Although this was intended to ensure consistent metre, the model occasionally generates music with varying time signatures. Hence, post-processing steps are employed to regularize the output to $4/4$ m.

4   Experiments

As mentioned in the previous sections, we use both the song-level and bar-level features as control tokens to train the model and then use them as prompts in the inference process. We also conducted an ablation study to determine the contributing factors to the system.

The experiment settings include the following: the Transformer-XL model serves as the backbone of the symbolic music generation task, and we perform the next-token prediction task with cross-entropy loss. We trained each model for 100 epochs, with batch size being 8, learning rate being 0.0002, and AdamW as the optimizer.

5   Evaluations

In this section, we discuss the evaluation methods used in this work. This includes algorithmic approaches and a human-involved approach. The evaluation mainly focuses on the two essential aspects of this project, which are emotion and loop. Therefore, there are three main methods in our evaluation system, including training a neural network model to classify the emotions of the generations, a loop extraction algorithm to determine the number of loops in the generated music, and a subjective listening test to get human feedback on the generations in terms of loops and emotions.

All the evaluation processes are based on the model generations from the epoch 20 checkpoint. This is carefully chosen by ourselves to balance the music quality and the model's ability to generate emotion-specific music. It is mainly based on music quality and variability since the most important aspect of music generation is the music itself. We found out that generations from earlier epochs would result in poor music quality, since the model has not learned the general music composition rules, and the generations from later epochs would result in serious overfitting of the training set since the variability of the model is poor and most of the generations are memorized from the training set. Based on the above criteria, we choose the final checkpoint from epoch 20.

Different methods are evaluated for happy/sad music generated from the trained Transformer-XL model. During inference, the model generates 1000 pieces of happy music and 1000 pieces of sad music, and the 2000 pieces of music are evaluated using the algorithmic approaches. A Type I error α of 5% is used in the statistical analyses.

5.1   Emotion Identification

We utilize the evaluation approach proposed in GTR-CTRL [29], which is to train a neural network model for classifying the emotions expressed by the generated music. Based on GTR-CTRL, this BERT-style classifier effectively categorizes the attributes of the generated music from the symbolic tokens. We expand on the concept of using language classifiers for evaluation and extend it to include emotion classification.

We trained separate models for valence and arousal to measure the level of happiness and sadness in each piece, with both models sharing the exact same GPBERT architecture. We also use the median scores to discretize valence and arousal into binary classification labels. We trained the models on two parts of the data, including the original DadaGP data and the processed DadaGP data containing only the loops. We believe the loop subset of DadaGP might be a biased data source because it only contains parts of the music, whereas the valence and arousal scores by Spotify are derived from the entire piece of music, not just the loop part.

The training configurations are also the same as GTR-CTRL, which includes 768 tokens per song and the GPBERT layer, self-attention layer, feed-forward layer as the model architecture. We trained the models for 10 epochs and chose the best-epoch checkpoint for inference. The best result was achieved at epoch 6 for valence with a 70.89% accuracy and epoch 2 for arousal with an 81.21% accuracy. This result shows that the GPBERT model is slightly better at classifying arousal than valence.

We then use the trained models to classify the generated symbolic music. In this scenario, we want the happy music to have higher valence and arousal scores, and the sad music to have lower valence arousal scores. During the GPBERT model inference, the softmax operation would first calculate a score for every data to indicate its probability of having a high valence/arousal label (pre-argmax score)[3], and the argmax operation would give every piece a binary classification result (post-argmax label). There are two metrics calculated in the table. high valence/arousal percentage (HVP or HAP) calculates the percentage of music having high valence/arousal post-argmax label, and mean valence/arousal score (MVS or MAS) calculates the mean valence/arousal score from the pre-argmax score. Note that they are all on a scale from 0.0 to 1.0, and we use 0.5 to separate high valence/arousal from low valence/arousal during inference. In theory, happy music would have higher high valence/arousal percentage and mean valence/arousal score, and

---

[3] There is also a score for low valence/arousal in the final layer.

Table 1. Comparison between our model (MoodLoopGP) and LooperGP in happy-sad emotion score difference. HVP and HAP stand for high valence percentage and high arousal percentage, and MVS and MAS stand for mean valence score and mean arousal score.

Settings	HVP	MVS	HAP	MAS
MoodLoopGP - Happy	0.6573	0.6553	0.5731	0.5107
MoodLoopGP - Sad	0.2025	0.2165	0.0307	0.0797
MoodLoopGP - Difference	**0.4548**	**0.4388**	**0.5424**	**0.4310**
LooperGP - Happy	0.3666	0.3784	0.1414	0.1828
LooperGP - Sad	0.3425	0.3652	0.1308	0.1756
LooperGP - Difference	0.0241	0.0132	0.0106	0.0072

sad music would have lower scores. Therefore, we then calculate the difference of `high valence/arousal percentage` and `mean valence/arousal score` between happy music and sad music groups, and a larger difference means a better model in making music convey happiness and sadness.

The classification score results from epoch 20 are displayed in Table 1, along with a comparison between our work and LooperGP, which is used as a baseline. It is important to note that in LooperGP, no control tokens were used when generating either happy or sad music. In fact, there was no difference between the two settings at all, as this was done to align with MoodLoopGP. However, there indeed exists a slight variation in the classification score between different trials, but it is smaller enough to be discarded.

When comparing the performance between models, all four metrics verify that MoodLoopGP can effectively generate music with target emotion when providing the corresponding prompt, and the metrics differences between happy and sad music generated by MoodLoopGP are up to 54%. It also shows that the training process creates an unbiased improvement over happy and sad music, which is validated by the fact that the absolute difference between `MoodLoopGP - Happy` and `LooperGP - Happy` and the difference between `MoodLoopGP - Sad` and `LooperGP - Sad` is roughly the same. Additionally, although the metrics difference in `MoodLoopGP - Difference` group for valence and arousal are roughly the same, it seems that the valence scores are more balanced compared to arousal as nearly all the arousal scores are below 0.5.

We also conducted an ablation study to investigate the contributing factors of our approach. We took out one group of features in each trial and then compared the performance. The information is divided into three categories: 1) Emotion Labels (EL): Valence and Arousal tokens. 2) Music Psychology Features (MPF): Tempo and Mode tokens. 3) Tonal Tension (TT): Cloud Diameter, Cloud Momentum, and Tensile Strain tokens.

The results in Table 2 demonstrate that all the features are important for achieving the best performance. When any of the features are removed, the performance drops significantly. It should be highlighted that when the Emotion

Table 2. Ablation study results of the emotion evaluation. "All" means the proposed model (MoodLoopGP), other settings mean all the features are added but the specified one, where EL, MPF, TT stand for Emotion Labels, Music Psychology Features, Tonal Tension, respectively.

Settings	HVP	MVS	HAP	MAS
All - Happy	0.6573	0.6553	0.5731	0.5107
All - Sad	0.2025	0.2165	0.0307	0.0797
All - Difference	**0.4548**	**0.4388**	**0.5424**	0.4310
Missing EL - Happy	0.5900	0.5573	0.2000	0.2494
Missing EL - Sad	0.2100	0.2313	0.0200	0.0702
Missing EL - Difference	0.3800	0.3260	0.1800	0.1792
Missing MPF - Happy	0.5232	0.5247	0.4283	0.4385
Missing MPF - Sad	0.1835	0.2120	0.0444	0.1054
MPF - Difference	0.3397	0.3127	0.3839	0.3331
Missing TT - Happy	0.5624	0.5596	0.5726	0.5310
Missing TT - Sad	0.1242	0.1438	0.0401	0.0831
Missing TT - Difference	0.4382	0.4158	0.5325	**0.4479**

labels are missing, the HAP and MAS drop by roughly 30%, and the HVP and MVS drop the most when the Music Psychology Features are missing. This illustrates that the Emotion Labels seem to contribute more to the arousal and Music Psychology Features seem to contribute more to the valence. Additionally, tonal tension seems to contribute more to the valence than arousal, as both the HAP and MAS Happy/Sad scores between the All and Missing TT groups are roughly the same, whereas relatively large differences are obtained for the HVP and MVS Happy/Sad scores. Removing tonal tension yields the highest difference between Happy and Sad for MAS, however it is close to the difference obtained when all features are used.

5.2 Loop Extraction

Following the evaluation approach from LooperGP [1], we use the same loop extraction method to evaluate the average number of loops per generation. The same parameters are used to implement the loop extraction algorithm, including Minimum Repetition Notes = 4, Minimum Repetition Beats = 2, Minimum Loop Bars = 4 and Maximum Loop Bars = 4. A detailed explanation of the parameters can be found in [1]. We compare our model with the baseline model, which is a Transformer-XL trained on the raw DadaGP dataset instead of the loop subset in order to demonstrate the effectiveness of our model's loop generation ability, and both groups are evaluated on 2000 generations of the corresponding model.

Table 3. Comparison of the average number of loops per generation between Mood-LoopGP and the Transformer-XL model trained in DadaGP paper.

Model	Loops Found	Average Number of Loop
MoodLoopGP	757	0.3789
Transformer-XL-DadaGP	522	0.2702

Table 3 shows the loop extraction evaluation result. MoodLoopGP can generate 45% more loops than the baseline, which demonstrates the advantage of the loop extraction algorithm is successfully kept in MoodLoopGP. We also performed a Wilcoxon Signed-Rank Test to examine the difference between MoodLoopGP and the baseline model. The result ($Z = 2990.0$, $p < 1e\text{-}40$) shows that there is a significant effect of the model type on the number of loops generated.

5.3 Subjective Evaluation

To evaluate the performance of the model from the listener's perspective, we conducted a listening test to study the generated music from the following three aspects: music quality, loop coherence, and the conveyed emotions. We recruited 11 participants, 7 male and 4 female, and approximately 2/3 of them had previously received training in music theory or musical instruments.

There were 60 musical excerpts in the listening test, and they were from three groups of 20 generations with each having 10 happy excerpts and 10 sad excerpts:

- Model generations prompted with all extra information: The model with all information added in the initial prompt to guide the generation. This serves as the expected model.
- Model generations prompted with all information but tonal tension: This is to evaluate the contributions of the bar-level features and demonstrate the benefits by leveraging multi-granular features.
- Human-composed music: Human-composed music is added to serve as the baseline to investigate the difference between human and machine-composed music.

All the excerpts were randomly chosen from their group and were taken from the first four bars of the music to form a loop. Each loop is repeated several times to derive the final piece. The number of repeated times was varied between pieces with different tempos to create pieces having lengths of roughly 30 s. The chosen 60 excerpts were also randomized to prevent order bias during the listening test.

Additionally, all the pieces were rendered from guitar pro tablatures, which do not have dynamic information. This makes the resulting music sound rigid and different from human-performed music. To address this problem, we told the listeners to only focus on the composition part of the music rather than the performing part of the music.

After listening to every excerpt, the listeners were asked to answer the following questions:

1. Have you heard the music in this excerpt before? (Prior to this survey) (Y/N)
2. Do you think the music is composed by a human or a machine? (Human/Machine)
3. Do you like the excerpt? (7-point Likert scale)
4. Does the loop in this excerpt sound coherent to you? (7-point Likert scale from dislike to like)
5. What emotion do you think this excerpt conveys? (7-point Likert scale from sad to happy)

The first question investigates if participants have heard the music before to evaluate biases from prior listening experiences. The second and third questions evaluate music quality based on the assumption that human-composed music and music preferred by listeners indicate higher quality. The fourth question evaluates the quality of the generated music as loops, and the fifth question evaluates it from the emotion's perspective.

Table 4. Percentage of heard and not heard music reported by the participants.

Composition Type	Heard	Not Heard
Machine-Composed: All Information	1.82%	98.18%
Machine-Composed: Without Tonal Tension	2.73%	97.27%
Human-Composed Music	5.45%	94.55%

Table 5. Turing Test: Percentage of music identified as human-composed or machine-composed.

Composition Type	Human	Machine
Machine-Composed: All Information	27.73%	72.27%
Machine-Composed: Without Tonal Tension	27.27%	72.73%
Human-Composed Music	50.45%	49.55%

Table 4 displays the results of the first question, showing that the participants had mostly not heard any of the three music groups prior to the experiment. Table 5 presents the results of the Turing test, indicating that 27% of machine-composed music was classified as human-composed, a lower percentage than the human music group. Surprisingly, only half of the human-composed music was correctly identified, possibly because the listeners are still biased by the loss of dynamic information and the use of virtual instruments.

Table 6. Results for all the Likert scale questions, including the listener's preference, loop coherence (LC), Happy Emotion Scores (HES), and Sad Emotion Scores (SES).

Average Score	Preference	LC	HES	SES
Machine-Composed: All Information	−0.3045	0.1591	0.2091	−0.2818
Machine-Composed: Without Tonal Tension	−0.2045	0.1682	−0.2909	−0.3182
Human-Composed Music	0.6000	0.9091	0.7091	−0.2636

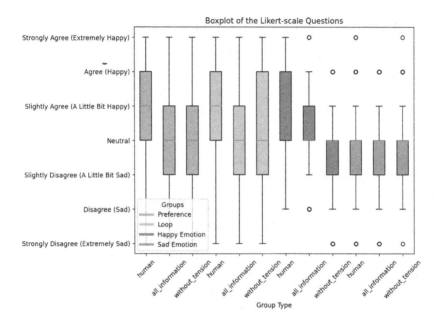

Fig. 1. Boxplot of the Likert-Scale Questions among different groups. The Happy Emotion and Sad Emotion groups correspond to pieces classified as happy and sad, respectively.

Table 6 shows the mean scores for Questions 3 to 5 on a 7-point Likert scale. The left-most answer is assigned −3, the right-most answer is assigned 3, and the stride is 1. This is to place the neutral answer (i.e., 0.) in the middle so that positive mean scores indicate positive ratings from the participants. Figure 1 shows the boxplot of the Likert-scale questions. Human-composed music consistently outperforms machine-composed music, indicating the gaps between human-composed and machine-composed music. Loop coherence scores are positive but close to 0 for all generated music groups, indicating loop coherence to listeners is not very strong, and we observe a slight difference in the median loop coherence scores between human and machine groups. The Happy Emotion and Sad Emotion Scores are obtained from results to the emotion question (Question 5) for pieces classified as happy and sad, respectively. The human group achieves the best result in HES and HES-SES difference. The difference obtained for the

Table 7. Friedman test result for the three groups of music in the listening test.

Question	$\chi^2(2)$	p-value
Preference	51.66	6.06e−12
Loop	36.00	1.53e−8
Emotion	22.41	1.36e−5

Table 8. Wilcoxon Sign Rank Test result between human-composed music and MoodLoopGP with tonal tension.

Question	Z	p-value
Preference	3002.50	1.38e-10
Loop	4179.50	4.59e-7
Emotion	4236.50	8.46e-3

Table 9. Wilcoxon Sign Rank Test result between MoodLoopGP with and without tonal tension.

Question	Z	p-value
Preference	4551.50	0.42
Loop	6372.50	0.70
Emotion	3860.50	1.11e−2

Happy and Sad pieces for the machine-composed groups (all_information only) indicates that the generated music is successful in varying the emotional expression from sad to happy. The boxplot highlights that participants had difficulty differentiating happy and sad music in the `Without Tonal Tension` group, but were able to do so in the `All Information` group. Therefore, Tonal tension likely helped in generating human-perceivable happy and sad music. This is supported by the Wilcoxon Signed Rank Test results, which will be covered later (Tables 8 and 9).

In order to gain a better understanding of the outcomes of the Likert-scale questions, we conducted a Friedman test among the three groups of music. The results, as presented in Table 7, indicate that the source of generation (human, machine with all information, and machine without tension) has a significant impact on the listener's preference, loop, and emotional perspective ($p<1e−4$ for all three questions). We also carried out multiple pairwise comparisons using the Wilcoxon Sign Rank Test with a Bonferroni-corrected α level ($\alpha/3 =.0167$). We found significant differences between the `Human` and `All Information` groups for preference ($Z = 3002.50$, $p<.01$), loop coherence ($Z = 4179.50$, $p<.01$), and emotion ($Z = 4236.50$, $p<.01$), confirming that human-produced music outperforms the machine-generated one. Additionally, we were interested in exploring the effect of bar-level features (i.e., Tonal Tension) in the generation process. We did not find significant differences between the `All Information` (including tonal tension) and `Without Tonal Tension` groups for the preference ($Z = 4551.50$, $p = .42$) and loop coherence ($Z = 6372.50$, $p = .70$) indicating that conditioning based on tonal tension may not contribute to improving preference and loop coherence. However, we found a significant difference between the All Information and Without Tonal Tension groups ($Z = 3860.50$, $p<.0167$) for emotion showing that adding tonal tension in the conditioning improves the generation of emotion-specific music.

6   Conclusion

In this paper, we present MoodLoopGP, a novel approach for emotion-conditioned and loopable music generation utilizing multi-granular musical features. Through the integration of both song-level attributes (emotion labels, tempo, mode) and bar-level attributes (tonal tension), our model demonstrates an enhanced capacity to generate music conveying specified emotions of happiness and sadness while keeping the model's ability of music loop generation. It is supported by the empirical evaluations conducted, including algorithmic emotion classification, loop extraction, and a subjective listening test. Our work demonstrates that incorporating music psychology features can enrich conditional generative models, and our multi-granular conditioning strategy offers a promising direction for more fine-grained control over emotion-specific music generation.

Acknowledgement. This work is supported by the EPSRC UKRI Centre for Doctoral Training in Artificial Intelligence and Music (Grant no. EP/S022694/1).

References

1. Adkins, S., Sarmento, P., Barthet, M.: LooperGP: a loopable sequence model for live coding performance using guitarpro tablature. In: Johnson, C., Rodríguez-Fernández, N., Rebelo, S.M. (eds.) EvoMUSART 2023. LNCS, vol. 13988, pp. 3–19. Springer, Cham (2023). https://doi.org/10.1007/978-3-031-29956-8_1
2. Alain, G., Chevalier-Boisvert, M., Osterrath, F., Piche-Taillefer, R.: Deepdrummer: generating drum loops using deep learning and a human in the loop. In: The 2020 Joint Conference on AI Music Creativity (2020)
3. Blood, A.J., Zatorre, R.J., Bermudez, P., Evans, A.C.: Emotional responses to pleasant and unpleasant music correlate with activity in paralimbic brain regions. Nat. Neurosci. **2**(4), 382–387 (1999)
4. Chew, E., et al.: Mathematical and computational modeling of tonality. AMC **10**(12), 141 (2014)
5. Dai, Z., Yang, Z., Yang, Y., Carbonell, J., Le, Q., Salakhutdinov, R.: Transformer-XL: attentive language models beyond a fixed-length context. In: Proceedings of the 57th Annual Meeting of the Association for Computational Linguistics, pp. 2978–2988. Association for Computational Linguistics, Florence, Italy (2019). https://doi.org/10.18653/v1/P19-1285, https://aclanthology.org/P19-1285
6. Dalla Bella, S., Peretz, I., Rousseau, L., Gosselin, N.: A developmental study of the affective value of tempo and mode in music. Cognition **80**(3), B1–B10 (2001)
7. Daynes, H.: Listeners' perceptual and emotional responses to tonal and atonal music. Psychol. Music **39**(4), 468–502 (2011)
8. Fernández-Sotos, A., Fernández-Caballero, A., Latorre, J.M.: Influence of tempo and rhythmic unit in musical emotion regulation. Front. Comput. Neurosci. **10**, 80 (2016)
9. Ferreira, L.N., Whitehead, J.: Learning to generate music with sentiment. In: Proceedings of the 20th International Society for Music Information Retrieval Conference, pp. 384–390 (2019)

10. Grekow, J., Dimitrova-Grekow, T.: Monophonic music generation with a given emotion using conditional variational autoencoder. IEEE Access **9**, 129088–129101 (2021)
11. Han, S., Ihm, H., Lee, M., Lim, W.: Symbolic music loop generation with neural discrete representations. Proceedings of the 23th International Society for Music Information Retrieval Conference (2022)
12. Han, S., Ihm, H., Lim, W.: Symbolic music loop generation with VQ-VAE. arXiv preprint arXiv:2111.07657 (2021)
13. Herremans, D., Chew, E., et al.: Tension ribbons: quantifying and visualising tonal tension. (2016)
14. Hsu, J.L., Liu, C.C., Chen, A.L.: Discovering nontrivial repeating patterns in music data. IEEE Trans. Multimedia **3**(3), 311–325 (2001)
15. Huang, C.F., Huang, C.Y.: Emotion-based AI music generation system with CVAE-GAN. In: 2020 IEEE Eurasia Conference on IOT, Communication and Engineering (ECICE), pp. 220–222. IEEE (2020)
16. Hung, T.M., Chen, B.Y., Yeh, Y.T., Yang, Y.H.: A benchmarking initiative for audio-domain music generation using the freesound loop dataset. Proceedings of the 22th International Society for Music Information Retrieval Conference (2021)
17. Hutchings, P.E., McCormack, J.: Adaptive music composition for games. IEEE Trans. Games **12**(3), 270–280 (2019)
18. Juslin, P.N.: Cue utilization in communication of emotion in music performance: relating performance to perception. J. Exp. Psychol. Hum. Percept. Perform. **26**(6), 1797 (2000)
19. Kalansooriya, P., Ganepola, G.D., Thalagala, T.: Affective gaming in real-time emotion detection and smart computing music emotion recognition: implementation approach with electroencephalogram. In: 2020 International Research Conference on Smart Computing and Systems Engineering (SCSE), pp. 111–116. IEEE (2020)
20. Keskar, N.S., McCann, B., Varshney, L.R., Xiong, C., Socher, R.: Ctrl: a conditional transformer language model for controllable generation. arXiv preprint arXiv:1909.05858 (2019)
21. Loth, J., Sarmento, P., Carr, C., Zukowski, Z., Barthet, M.: Proggp: from guitarpro tablature neural generation to progressive metal production. The 16th International Symposium on Computer Music Multidisciplinary Research (2023)
22. Madhok, R., Goel, S., Garg, S.: Sentimozart: music generation based on emotions. In: ICAART (2), pp. 501–506 (2018)
23. McVicar, M., Fukayama, S., Goto, M.: Autoleadguitar: automatic generation of guitar solo phrases in the tablature space. In: 2014 12th International Conference on Signal Processing (ICSP), pp. 599–604. IEEE (2014)
24. Panda, R., Redinho, H., Gonçalves, C., Malheiro, R., Paiva, R.P.: How does the spotify api compare to the music emotion recognition state-of-the-art? In: 18th Sound and Music Computing Conference (SMC 2021), pp. 238–245. Axea sas/SMC Network (2021)
25. Ruiguo-Bio: Ruiguo-bio/midi-miner: Python midi track classifier and tonal tension calculation based on spiral array theory (2023). https://github.com/ruiguo-bio/midi-miner
26. Russell, J.A.: A circumplex model of affect. J. Pers. Soc. Psychol. **39**(6), 1161 (1980)
27. Sarmento, P., Holmqvist, O., Barthet, M., et al.: Ubiquitous music in smart city: musification of air pollution and user context (2022)

28. Sarmento, P., Kumar, A., Carr, C., Zukowski, Z., Barthet, M., Yang, Y.H.: DadaGP: a dataset of tokenized guitarpro songs for sequence models. In: Proceedings of the 22th International Society for Music Information Retrieval Conference, pp. 610–618 (2021)
29. Sarmento, P., Kumar, A., Chen, Y.H., Carr, C., Zukowski, Z., Barthet, M.: GTR-CTRL: instrument and genre conditioning for guitar-focused music generation with transformers. In: Johnson, C., Rodríguez-Fernández, N., Rebelo, S.M. (eds.) Evo-MUSART 2023. LNCS, vol. 13988, pp. 260–275. Springer, Cham (2023). https://doi.org/10.1007/978-3-031-29956-8_17
30. Sarmento, P., Kumar, A., Xie, D., Carr, C., Zukowski, Z., Barthet, M.: Shredgp: guitarist style-conditioned tablature generation. In: Proceedings of the 16th International Symposium on Computer Music Multidisciplinary Research (CMMR) 2023. (2023)
31. Sulun, S., Davies, M.E., Viana, P.: Symbolic music generation conditioned on continuous-valued emotions. IEEE Access 10, 44617–44626 (2022)
32. Takahashi, T., Barthet, M.: Emotion-driven harmonisation and tempo arrangement of melodies using transfer learning
33. Tan, H.H., Herremans, D.: Music fadernets: controllable music generation based on high-level features via low-level feature modelling. Proceedings of the 21th International Society for Music Information Retrieval Conference (2020)
34. Tan, X., Antony, M., Kong, H.: Automated music generation for visual art through emotion. In: ICCC, pp. 247–250 (2020)
35. Tripodi, I.J.: Setting the rhythm scene: deep learning-based drum loop generation from arbitrary language cues. arXiv preprint arXiv:2209.10016 (2022)
36. Webster, G.D., Weir, C.G.: Emotional responses to music: interactive effects of mode, texture, and tempo. Motiv. Emot. 29, 19–39 (2005)
37. Williams, D., Kirke, A., Miranda, E.R., Roesch, E., Daly, I., Nasuto, S.: Investigating affect in algorithmic composition systems. Psychol. Music 43(6), 831–854 (2015)
38. Yang, S., Reed, C.N., Chew, E., Barthet, M.: Examining emotion perception agreement in live music performance. IEEE Trans. Affect. Comput. 14(02), 1442–1460 (2023). https://doi.org/10.1109/TAFFC.2021.3093787
39. Yeh, Y.T., Chen, B.Y., Yang, Y.H.: Exploiting pre-trained feature networks for generative adversarial networks in audio-domain loop generation. In: Proceedings of the 23th International Society for Music Information Retrieval Conference (2022)

Weighted Initialisation of Evolutionary Instrument and Pitch Detection in Polyphonic Music

Justin Dettmer[1]([🖂]) [iD], Igor Vatolkin[1] [iD], and Tobias Glasmachers[2] [iD]

[1] Chair for Artificial Intelligence Methodology, RWTH Aachen University, Aachen, Germany
{dettmer,vatolkin}@aim.rwth-aachen.de
[2] Institut für Neuroinformatik, Ruhr-University Bochum, Bochum, Germany
tobias.glasmachers@ini.rub.de

Abstract. Current state-of-the-art methods for instrument and pitch detection in polyphonic music often require large datasets and long training times; resources which are sparse in the field of music information retrieval, presenting a need for unsupervised alternative methods that do not require such prerequisites. We present a modification to an evolutionary algorithm for polyphonic music approximation through synthesis that uses spectral information to initialise populations with probable pitches. This algorithm can perform joint instrument and pitch detection on polyphonic music pieces without any of the aforementioned constraints. Sets of tuples of (instrument, style, pitch) are graded with a COSH distance fitness function and finally determine the algorithm's instrument and pitch labels for a given part of a music piece. Further investigation into this fitness function indicates that it tends to create false positives which may conceal the true potential of our modified approach. Regardless of that, our modification still shows significantly faster convergence speed and slightly improved pitch and instrument detection errors over the baseline algorithm on both single onset and full piece experiments.

Keywords: Evolutionary music approximation · Joint instrument and pitch recognition · Algorithm optimisation

1 Introduction

Automatic music transcription of polyphonic audio recordings is one of the most complex tasks in the field of Music Information Retrieval [2]. A robust algorithm which produces score from audio signals is extremely beneficial for many scenarios and applications: learning to play instruments, analysis of musical genres, rearrangement of music pieces for specific purposes, music recommendation, etc. Among several steps required for successful music transcription, instrument and pitch detection are maybe the two most prominent and not easy to solve. Several instruments playing at the same time, different playing techniques, applied effects, and varying distributions of overtones make these tasks very challenging. This work focuses on joint instrument and pitch detection in polyphonic music. Instead of supervised classification, which requires training of models based on preferably large annotated datasets, or also source separation, which is not always straightforward because of the unknown number of playing sources, we focus here on an analysis-by-synthesis approach like in [11]. This means that we try to re-create a close approximation of the input original sound by the combination of "basic" entities, which are instrument samples in our case. We employ such a method in this work through evolutionary approximation of the given piece, as originally proposed in [33]. The advantage of an evolutionary algorithm over the state-of-the-art Convolutional Neural Networks (CNNs) is that it does not requires training on large datasets, allows for simple addition or removal of instrument classes without a need for re-training, and the general approach remains the same independently of the number of playing sources.

Past work has shown that this rather new approach works well for feature generation in musical genre recognition [33] but is still lacking in its instrument and pitch detection capabilities. [12] has shown improvements after a systematic analysis of different input features and distance measures for the design of fitness evaluation functions. As another idea to improve the original method, we propose a technique to generate initial evolutionary algorithm solutions based on probabilities for each pitch to exclude unlikely pitches from the initial population. We further investigate the algorithm's capabilities in finding individuals with good fitness when approximating pieces that are composed of instrument recordings not contained in our sample library.

We test this modified algorithm on two synthetically created and thus perfectly annotated datasets: one containing single-onset examples and another consisting of complete musical pieces. The results show a significant reduction of instrument and pitch detection errors after initialisation, and smaller errors in the convergence area after a large number of generations. Further, we analyse the relationship between individual fitness and the associated errors.

2 Related Work

While earlier studies on instrument detection have applied classifiers like Gaussian mixture models or support vector machines using manually engineered audio features as input [4,25], more recent state-of-the-art approaches are based on CNNs [15,22] or audio spectrogram transformers [13] with Patchout [20]. The performance of neural networks can be further improved by means of Neural Architecture Search (NAS) [8], as applied in [10], or by knowledge distillation from an ensemble of teacher models to a student network [31].

Such supervised approaches require large annotated datasets for training; however, the existing datasets typically feature either only a small range of instruments, contain a smaller selection of musical pieces, or both: e.g., MedleyDB [3] has only 122 tracks and OpenMIC [17] contains annotations of only 20 instrument classes with many missing labels. Because annotation of large datasets requires a lot of human effort, some works have introduced synthetically generated audio [24,30] which has also some limitations, as is does not contain "real-world" recorded music pieces. On the other hand, heavy optimisation techniques such as NAS involve a lot of computing resources and may sometimes lead to model overfitting.

Unsupervised methods like source-filter models and non-negative matrix factorisation [16] tackle polyphonic music by first dividing the input signal into separate parts that aim to isolate the instruments before applying classification models on the isolated signals, but have also their limitations, in particular for polyphonic music with a high and unknown number of playing instruments.

Pitch detection occurs in different problem contexts as well, and can be hindered by non-harmonic components of instrument sounds which were investigated in [23]. Chromagrams, or pitch profiles predict pitch classes without an exact estimation of the tone height. An improved version which removes timbre information was proposed in [28]. A representation that stores the strengths of all half-tones used in Western music is the semitone spectrum which can be extracted with the Non-Negative Least Squares (NNLS) chroma method [26].

A particular challenge for a robust pitch detection in polyphonic sounds is the task of fundamental frequency (f0) estimation [19]. For this task, deep convolutional neural networks were successfully applied in [32].

3   Background

In this section, we will formally describe the problem of joint instrument-pitch detection and explain the core concepts behind the Evolutionary Algorithm (EA) introduced in [33] upon which we propose our modifications in Sect. 4.

3.1   Joint Instrument-Pitch Detection

In order to talk about instrument and pitch detection accurately, we must first define the underlying combinatorial problem. For our purposes, a given musical piece will be divided into non-overlapping windows that span from one onset to another. An onset is a moment in time at which an instrument $i \in I$ begins playing a sound at pitch $p \in P$, where I is the set of all 51 possible instruments contained in our dataset, and P is the set of the 88 pitches commonly found on a piano claviature (a0 to c8). For any such onset-onset window, we are given a target set X_t of tuples (i, p), containing the instrument and pitch information for that onset. Additionally, some of the instruments are represented with several playing styles or different instrument bodies $s \in S_i$, e.g., piano samples come from five different styles of the MUMS dataset [6] as well as six sound bodies from the Native Instruments Komplete collection [29] (Alicias Keys, Gentleman, Giant hard, Giant soft, Grandeur, and Maverick). Our task is to find a candidate set X_c that is equal to the target set X_t.

The number of elements in X_t is not known a-priori, and may differ between individual onset-onset windows of a musical piece. With a large set of possible instruments I, this combinatorial problem quickly reaches immense complexity, with up to $(|I| \cdot |P|)^n$ possibilities for a single onset, where $n \in \mathbb{N}$ is the assumed maximum possible number of tuples in X_t. This uncertainty introduces further complexity when working with polyphonic music. On top of this labelled information, we denote Y_{sig} as the musical audio signal of such an onset-onset window.

3.2 Evolutionary Approximation of Music

EAs are a type of search algorithm that simulate principles from evolution theory in order to iteratively explore a search space [7]. In our case, the search space is discrete and is comprised of all possible combinations of pitch-instrument tuples within the assumed maximum number of tuples n. We define an individual X_c as part of a population of size μ. In the baseline implementation, the initial population is created randomly. The individual is created by drawing from weighted probabilities for the number of tuples and uniform probabilities for the instruments and pitches per tuple. In addition, a valid instrument style $s \in S_i$ is drawn from the available styles of the instrument i.

In the evolutionary loop, every individual is graded with a fitness value, as described below in Sect. 3.3. Then, we randomly choose $\lambda \in \mathbb{N}$ individuals from the population, copy them, and apply one or multiple of the following mutations: (i) changing the instrument and style of a tuple, (ii) changing the pitch of a tuple, (iii) removing or adding a new tuple. The number of applied mutations n_{mut} is given by the equation

$$n_{mut} = \lfloor G \cdot \alpha + \beta \rfloor, \tag{1}$$

where α and β are hyperparameters (we refer to [33] for details), and G is drawn from a Gaussian distribution with mean 0 and standard deviation 1. The result is clipped in the range $[1, u_{bound}]$, where u_{bound} is also a hyperparameter that sets the maximum possible number of applied mutations.

The new generation of individuals is then also graded with fitness values and finally, the λ individuals with the worst fitness are removed from the population. Additionally, a mechanism for step size adaptation slowly lowers mutation strength in each generation by multiplying u_{bound} by a parameter ζ. Unless otherwise specified, we use the following parameters for our experiments: $\mu = 10$, $\lambda = 1$, $\alpha = 5$, $\beta = 10$, $u_{bound} = 10$, and $\zeta = 0.9954$.

3.3 Evaluating Fitness

Our fitness function was chosen after a set of experiments investigating the performance of different options in [12], where this choice performed the best. It operates on the signal that is generated when combining the instrument recordings contained in an individual's set of tuples. We simply sum the underlying signals saved in our sample library for this combination. The resulting signal X_{sig} is then compared to the target signal Y_{sig} by computing the average of each Short-Time Fourier Transform (STFT) bin in the magnitude spectrum and then calculating the COSH distance [14] as given in Eq. 2. This distance measure can be interpreted as the symmetric variant of the Itakura-Saito divergence [18] and is used to compare signals within the realm of speech processing [1]. Our implementation of the COSH distance is

$$D_{\text{COSH}}(X_{sig}, Y_{sig}) = \frac{1}{2N} \sum_{i=1}^{N} \left(\frac{x_i}{y_i} - \log \frac{x_i}{y_i} + \frac{y_i}{x_i} - \log \frac{y_i}{x_i} - 2 \right), \qquad (2)$$

where x_i and y_i are indices in the averaged bins of the magnitude spectrum for X_{sig} and Y_{sig} respectively, and N is the number of STFT bins which is typically 1024.

As we want to closely approximate the target signal, our aim is to minimise this distance function. A zero distance indicates identical signals that we expect to only encounter when using target signals that were created from our sample library. Because full music pieces contain hundreds or even thousands of onsets, and individual onset approximation will require too much computing resources, the evolutionary individuals optimise the complete track at the same time. First, the distances between an individual (a candidate solution) and all onsets are estimated and sorted in ascending order. Then, the fitness is calculated as the mean distance to $\phi = 5\%$ of onsets with the smallest distance to the candidate solution. This way, it is aimed to assign a similar fitness to different individuals which may well approximate distinct parts of the piece characterised by a set of rather similar onsets (e.g., intro, verse, or chorus).

4   Initialisation with Pitch Probabilities

The baseline implementation of the algorithm initialises its population randomly from uniform probabilities. Every instrument and pitch has the same probability of being drawn. As the target signal is available to us, we can narrow down the distribution of candidate pitches by eliminating pitches that do not prominently appear in the spectrum of the signal. Through the Constant-Q Transform (CQT) [5], an alternative to the STFT that already operates in the logarithmic space, we can calculate magnitude bins for all of our 88 available pitches. For two piano notes (c4, g5), the CQT creates the spectrogram displayed in Fig. 1. One can clearly see that many pitches do not appear in the spectrogram at all, while the most prominent peaks are located around the correct pitches of c4 and g5. There are further maxima located in the octaves above the played notes, as well as the adjacent half tones.

Instead of relying on the CQT to set in stone which pitches we consider, we use it to create a probability distribution across our 88 pitches from which our algorithm will draw upon initialization. To do this, we must turn the two-dimensional result from applying the CQT across the target signal into a one-dimensional probability vector. We investigated two methods of dimensionality reduction: summation and the maximum method. In summation, we calculated the sum of all corresponding CQT frequency bins across the temporal axis. In

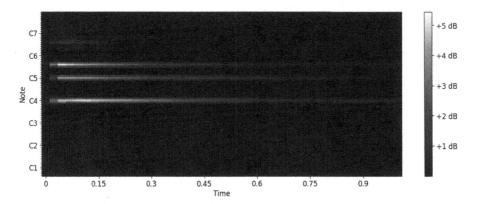

Fig. 1. A CQT magnitude spectrogram of only the harmonic elements from two piano notes (c4, g5) played together.

the maximum method, we applied the maximum function across that same axis, keeping only the largest value per frequency bin. As an additional preliminary step, we applied Harmonic Percussive Source Separation (HPSS) with median filtering as described in [9] in its librosa [27] implementation. Finally, all elements of the resulting vector are divided by the sum of elements to enforce that all probabilities sum up to exactly 1.

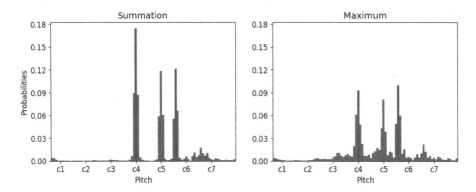

Fig. 2. Extracted pitch probabilities when summing or applying the maximum function on across the temporal axis of the CQT after applying HPSS. The correct pitches of c4 and g5 are highlighted in red. (Color figure online)

Figure 2 shows the resulting probabilities for each pitch in the example signal used for Fig. 1. We can see clear peaks around the correct pitches of c4 and g5 with both methods, however their associated probabilities are below 0.2 in both cases. Although the unlikely probabilities are barely visible in the spectrogram, their accumulation in the probability vector steals probability mass from the desired pitches. To address this, we add an additional step in which we set all but the highest k values to zero before creating the probability vector. An example with $k = 5$ is shown in Fig. 3, where roughly 50% of the probability mass is concentrated on the two correct pitches, with the rest being distributed upon reasonable alternatives. We also observe that the summation method creates a higher peak at the lower of the two correct pitches, while the maximum method benefits the higher pitch. Qualitatively, it is hard to clearly prefer one method over the other; however, we choose to use the summation method in the remainder of this work, as it should prove to be more robust to single peaks in the CQT, and is slightly faster to calculate than the maximum function.

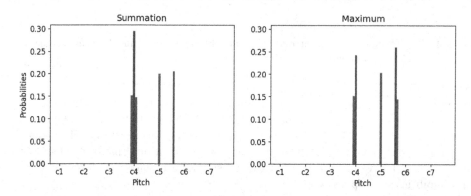

Fig. 3. Extracted pitch probabilities after keeping only the $k = 5$ largest values from calculating summation or the maximum function across the temporal axis of the CQT and applying HPSS. The correct pitches of c4 and g5 are highlighted in red. (Color figure online)

5   Experiments

In this section, we will introduce the datasets and the setup of experiments.

5.1   Evaluation Measure

All experiments use the Jaccard measure [21] for evaluation, which is defined as

$$J = 1 - \frac{|X \cap Y|}{|X \cup Y|} \tag{3}$$

for two non-empty sets X and Y. In our case, they are the candidate set X_c and target set X_t, as introduced in Sect. 3.1. A more intuitive formulation of this measure is

$$J = \frac{FP + FN}{TP + FP + FN}, \tag{4}$$

where TP is the number of true positives, FP and FN are the numbers of false positives and negatives respectively. Note that this method excludes true negatives, as most classes will be correctly labelled as negatives in our experiments.

5.2   The First Dataset: Ground Truth Search

As mentioned above, we make use of two datasets for our experiments. The first one consists of 1000 single-onset polyphonic audio mixes generated from our sample library. These examples were created the same way as individuals in our algorithm upon the first initialisation. This means that the target signal can be exactly re-created by our algorithm, giving us a good measure on whether the algorithm is able to find the global optimum reliably. In this case, the global optimum will have a fitness value of 0, as the resulting signals are identical.

For each experiment, the algorithm is run once for all 1000 targets from the dataset. At each step of the algorithm, we calculate the Jaccard errors for instrument classes (J_i), pitch classes (J_p), and joint instrument-pitch tuples (J_{ip}) for the individual with the highest fitness in that generation. The algorithm terminates prematurely if an individual with zero fitness is found to avoid unnecessary computation; otherwise, an experiment ends if the algorithm has reached its 10 000th generation.

5.3   The Second Dataset: Artificial Audio Multitracks

For our full piece approximation experiments, we use the "tiny" version of the Artificial Audio Multitracks (AAM) dataset [30]. This version of the dataset consists of 20 fully annotated pieces that were artificially created from instrument samples and then further processed in a digital audio workstation to make them sound less mechanical and more akin to human playing. Annotations for these pieces consist of the onset times, as well as the instrument-pitch tuples that are present in the audio at each onset. We use the annotated onset times to slice the piece into onset-onset windows that start at one onset and end at the subsequent onset. Each of these windows are jointly approximated as suggested in [33] by computing an individual's fitness for all windows, and then calculating the mean for only the $\phi = 5\%$ of onsets with the best fitness values, as described in Sect. 3.3.

Due to the high stochasticity of the algorithm, with random initialisation and mutations, we repeat the experiments 20 times and then average the Jaccard errors per piece across those runs. Errors are once again divided into J_i, J_p, and J_{ip} as above. Unlike the experiments in the ground truth search, we do not expect to find individuals with exactly zero fitness, because our algorithm uses a different sample library to the AAM pieces and does not apply any further processing on the generated audio signals. We run each experiment for a total of 10 000 generations and again log the Jaccard errors at the end of each generation. A deviation from the default parameters is also necessary, as the small population size of 10 is unlikely to accurately capture a wide range of onsets in the given piece. The population size μ is therefore increased to 300 for our full piece experiments.

6   Results and Discussion

In this section, we first present the results of the single onset ground truth search experiments, then the results for the full piece approximation, and finally, a short analysis of the correlation between fitness values and instrument-pitch detection errors.

6.1   Ground Truth Search

Figure 4 shows mean detection errors for the baseline algorithm across 10 000 generations. On initialisation, errors are close to 1, meaning that almost no correct instruments or pitches are found. After convergence, errors settle in around 0.18 for pitches, 0.21 for instruments, and 0.23 for the joint mean error. On closer inspection of the graphs for J_i and J_{ip}, one can see that the two graphs closely resemble each other. This resemblance indicates that the instrument and the joint detection errors often decrease together. Therefore, in these cases where the joint error decreases, it was the instrument mutation that caused the decrease, while the correct pitch was already found in the individual. This pattern is what motivates our modification to the algorithm, as it appears that the correct pitch must generally be found before finding the right instrument.

Before running experiments with the modified algorithm, we must first find a fitting cut-off value k. For that, we generate initial populations with the modified algorithm for 20 values of k and measure the pitch detection errors. The results in Fig. 5 show that larger values of k cause increased error rates and that the summation method clearly outperforms the maximum method. For all our experiments, we settled on $k = 3$ with the summation method.

Results for the modified algorithm are shown in Fig. 6. It is clear that the initial mean pitch error is drastically lower compared to the baseline algorithm, but it happens to increase in the first 80 generations. Errors after convergence are comparable to those of the baseline algorithm, however the slopes in the first few thousand generations are steeper in our modified algorithm. In cases where there is not ample time to run the algorithm for the full 10 000 generations, using our proposed method will provide better detection upon early termination.

To address the initial pitch error increase, we ran further experiments with a reduced mutation strength. The plots in Fig. 7 show the effect almost completely vanishing when setting the mutation strength to the lowest possible setting, where only a single mutation is applied to an individual. Despite the low mutation strength, final errors after convergence are not compromised. It is also necessary to note that the Jaccard error is a rather rough error metric, as it excludes many cases of correctly classified true negatives. Especially with our large set of 51 instrument labels, we consider mean errors around 0.2 to be promising for the potential of this approach.

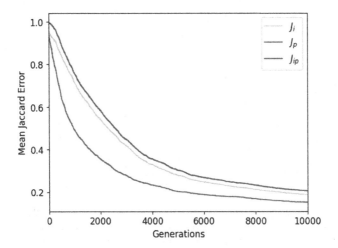

Fig. 4. Mean Jaccard errors for instrument (J_i), pitch (J_p), and joint detection (J_{ip}) across 10 000 generations of the evolutionary detection algorithm on the ground truth search dataset. The pitch detection error is the lowest, while the instrument and joint detection error closely resemble each other.

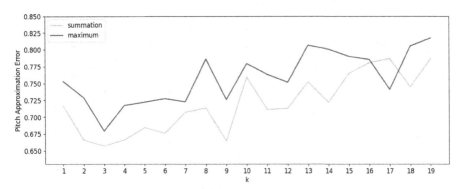

Fig. 5. Mean pitch approximation errors on the ground truth search dataset after probabilistic pitch initialisation with different values for the parameter k. The summation method outperforms the maximum method in almost all cases. Values above $k = 9$ produce increasingly higher errors.

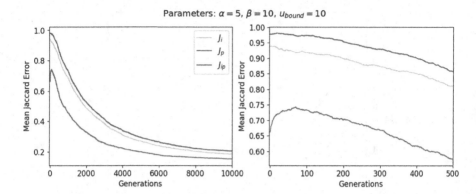

Fig. 6. Left: mean Jaccard errors for instrument (J_i), pitch (J_p), and joint detection (J_{ip}) across 10 000 generations of the evolutionary detection algorithm with probabilistic initialization on the ground truth search dataset. Right: a zoomed in view on the first 500 generations provides a closer look at the initial increase in J_p in the first 80 generations.

6.2   Artificial Audio Multitracks

Our experiments on the AAM dataset are closer to real world applications where detection tasks are often applied to entire pieces of music. As this task is significantly harder than the previous single-onset experiments, higher error rates are expected. The graphs in Fig. 8 show mean detection errors for both the baseline and the modified algorithm. We see lower initial errors for all three error classes, but surprisingly, the largest difference occurs for the instrument detection error J_i. It appears that, across an entire piece and with a much larger population, our initialisation method which only uses pitch information somehow improves the mean instrument error more than the mean pitch error. Keeping in mind the pattern noticed in the previous experiment, where pitches were generally found before instruments, this effect may be in place here too. The larger population size simply amplifies the effect to make it mostly appear in the initial population. Unlike the single-onset experiments, we can also observe a slight improvement in errors after convergence for the modified algorithm. The initial gap in J_i is strongly reduced by the time of convergence, again showing that our modification brings the largest advantage when faced with constraints in computation time. As the computation time for our pitch probabilities is negligible compared to the runtime of the algorithm itself, even its slight improvements in converged errors makes it a useful addition regardless.

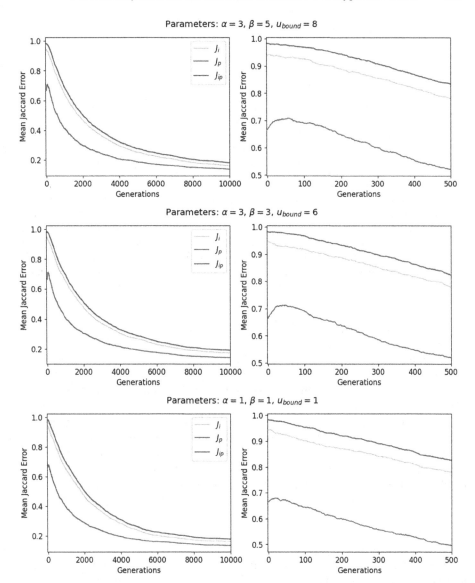

Fig. 7. Left: mean Jaccard errors for instrument (J_i), pitch (J_p), and joint detection (J_{ip}) across 10 000 generations of the evolutionary detection algorithm with probabilistic initialization on the ground truth search dataset. Mutation strength parameters introduced in Eq. 1 are varied to address the initial error increase in the first 80 generations. Right: zoomed-in views on the first 500 generations.

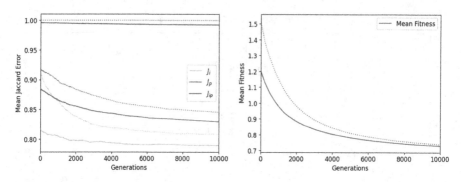

Fig. 8. Mean Jaccard errors (left) for instrument (J_i), pitch (J_p), and joint detection (J_{ip}) and mean fitness (right) across 10 000 generations of the evolutionary detection algorithm with probabilistic initialisation (solid lines) and without modification (dotted lines) on the AAM dataset. The modified algorithm produces lower mean errors than the baseline algorithm in all three error categories. Mean fitness values of the modified algorithm are lower at initialisation, but almost converge with the unmodified algorithm's fitness after 10 000 generations.

6.3   Investigating Error-Fitness Correlation

To investigate the rather high errors rates the algorithm produces on the AAM dataset, we ran a validation test in which we used a-priori knowledge from the annotated pitches and instruments to create "correct" individuals from our sample library. For each of these individuals, we ran a random search for 100 steps to find a good selection of instrument styles based on their fitness. We then calculated the fitness of these correct individuals when only compared to the onset-onset window they correctly label. This test revealed that our evolutionary approach indeed found individuals with better fitness than those with correct labels for 80.8% of onsets. The result is a high rate of false positives in both instrument and pitch detection tasks. It appears that the correlation between our magnitude-based COSH distance fitness function and the correct detection of instrument-pitch tuples is rather low, signalling a need for a more closely correlated fitness function.

Figure 8 also shows graphs for the mean fitness values in our full piece experiments, demonstrating that the fitness indeed keeps decreasing for a long time, while detection errors barely improve. One can also see that by the 10 000th generation, the mean fitness values of our modified and the baseline algorithm are close; another indication that the fitness function may indeed be a bottleneck that is hiding the true potential of our modification. Despite this, our algorithm performs far better than a random classifier (please recall a very high number of possible classes to predict), and our modification yields noticeable improvements over the baseline algorithm, as particularly demonstrated by the results in Sect. 6.1.

7   Conclusion

We presented a novel approach to initialise an algorithm for instrument and pitch detection through evolutionary synthesis. Our method uses information from the CQT spectrum to create a probability distribution from which our algorithm draws pitches for its first generation of individuals. The experiments show that this modification greatly improves mean pitch and instrument detection errors upon initialisation, but also leads to smaller errors after convergence on a full-piece detection task. The advantages of this evolutionary synthesis approach are its unsupervised nature, ability to increase or reduce the number of instrument and pitch labels without need for re-training, and its additional application in feature generation by finding spectrally close approximations of a piece that may be comprised of other instruments. Closer investigation into the detection errors made by the algorithm revealed that it indeed found individuals with good fitness values. Unfortunately, those fitness values were even better than those of individuals manually created with correct instrument-pitch tuples, suggesting a need for a modification to the current fitness function.

As an analysis on alternative single-objective fitness functions was already done in [12], we believe that a more sophisticated approach to individual fitness is required. A multi-objective fitness function that combines a wider variety of features may improve detection errors of the algorithm significantly. Among further investigations of alternative fitness functions, future work may also explore other mutation methods and extend individual's dimensions with aspects such as loudness or the addition of audio effects, e.g., reverb or compression. Other evolutionary algorithms or the integration of recombination may also be considered. For method comparison on more popular datasets, one may also reduce the size of the sample library to only the known set of instruments present in a given dataset.

In conclusion, evolutionary synthesis is a potent method for music feature calculation and joint instrument-pitch detection. Further research into potential fitness functions and more elaborate initialisation methods is required, but may lead to a promising alternative to current state-of-the-art supervised deep learning methods. To support further work on the topic, our modular code is publicly released on GitHub[1].

References

1. Bansal, M., Sircar, P.: Parametric representation of voiced speech phoneme using multicomponent am signal model. In: Proceedings of the 2018 IEEE/ACIS 17th International Conference on Computer and Information Science, ICIS, pp. 128–133 (2018)
2. Benetos, E., Dixon, S., Duan, Z., Ewert, S.: Automatic music transcription: an overview. IEEE Sig. Process. Mag. **36**(1), 20–30 (2019)

---

[1] https://github.com/jd-rub/EvoMUSART23-repro-code.

3. Bittner, R.M., Salamon, J., Tierney, M., Mauch, M., Cannam, C., Bello, J.P.: MedleyDB: a multitrack dataset for annotation-intensive MIR research. In: Proceedings of the 15th International Society for Music Information Retrieval Conference, ISMIR, pp. 155–160 (2014)
4. Brown, J.C., Houix, O., McAdams, S.: Feature dependence in the automatic identification of musical woodwind instruments. J. Acoust. Soc. Am. **109**(3), 1064–1072 (2001)
5. Brown, J.C., Puckette, M.S.: An efficient algorithm for the calculation of a constant Q transform. J. Acoust. Soc. Am. **92**(5), 2698–2701 (1992)
6. Eerola, T., Ferrer, R.: Instrument library (MUMS) revised. Music. Percept. **25**(3), 253–255 (2008)
7. Eiben, A.E., Smith, J.E.: Introduction to Evolutionary Computing. Natural Computing Series, 2nd edn. Springer, Heidelberg (2015). https://doi.org/10.1007/978-3-662-44874-8
8. Elsken, T., Metzen, J.H., Hutter, F.: Neural architecture search: a survey. J. Mach. Learn. Res. **20**, 55:1–55:21 (2019)
9. Fitzgerald, D.: Harmonic/percussive separation using median filtering. In: Proceedings of the 13th International Conference on Digital Audio Effects, DAFx, pp. 1–4 (2010)
10. Fricke, L., Vatolkin, I., Ostermann, F.: Application of neural architecture search to instrument recognition in polyphonic audio. In: Johnson, C., Rodríguez-Fernández, N., Rebelo, S.M. (eds.) EvoMUSART 2023. LNCS, vol. 13988, pp. 117–131. Springer, Cham (2023). https://doi.org/10.1007/978-3-031-29956-8_8
11. George, E.B., Smith, M.J.: Analysis-by-synthesis/overlap-add sinusoidal modeling applied to the analysis and synthesis of musical tones. J. Audio Eng. Soc. **40**(6), 497–516 (1992)
12. Ginsel, P.: Abstandsmaße zur evolutionären Klangapproximation auf Audiodaten. Master's thesis, TU Dortmund University, Department of Computer Science (2021)
13. Gong, Y., Chung, Y.A., Glass, J.: AST: audio spectrogram transformer. In: Proceedings of the 22nd Annual Conference of the International Speech Communication Association, Interspeech, pp. 571–575 (2021). https://doi.org/10.21437/Interspeech.2021-698
14. Gray, A., Markel, J.: Distance measures for speech processing. IEEE Trans. Acoust. Speech Sig. Process. **24**(5), 380–391 (1976)
15. Han, Y., Kim, J., Lee, K.: Deep convolutional neural networks for predominant instrument recognition in polyphonic music. IEEE/ACM Trans. Audio Speech Lang. Process. **25**(1), 208–221 (2017)
16. Heittola, T., Klapuri, A., Virtanen, T.: Musical instrument recognition in polyphonic audio using source-filter model for sound separation. In: Proceedings of the 10th International Society for Music Information Retrieval Conference, ISMIR, pp. 327–332 (2009)
17. Humphrey, E., Durand, S., McFee, B.: OpenMIC-2018: an open data-set for multiple instrument recognition. In: Proceedings of the 19th International Society for Music Information Retrieval Conference, ISMIR, pp. 438–444 (2018)
18. Itakura, F.: Analysis synthesis telephony based on the maximum likelihood method. In: Reports of the 6th International Congress on Acoustics, pp. C17–20 (1968)
19. Klapuri, A.: Multiple fundamental frequency estimation based on harmonicity and spectral smoothness. IEEE Trans. Speech Audio Process. **11**(6), 804–816 (2003)

20. Koutini, K., Schlüter, J., Eghbal-zadeh, H., Widmer, G.: Efficient training of audio transformers with patchout. In: Proceedings of the 23rd Annual Conference of the International Speech Communication Association, Interspeech, pp. 2753–2757 (2022). https://doi.org/10.21437/Interspeech.2022-227

21. Levandowsky, M., Winter, D.: Distance between sets. Nature **234**(5323), 34–35 (1971)

22. Li, X., Wang, K., Soraghan, J., Ren, J.: Fusion of Hilbert-Huang transform and deep convolutional neural network for predominant musical instruments recognition. In: Romero, J., Ekárt, A., Martins, T., Correia, J. (eds.) EvoMUSART 2020. LNCS, vol. 12103, pp. 80–89. Springer, Cham (2020). https://doi.org/10.1007/978-3-030-43859-3_6

23. Livshin, A., Rodet, X.: The significance of the non-harmonic "noise" versus the harmonic series for musical instrument recognition. In: Proceedings of the 7th International Conference on Music Information Retrieval, ISMIR, pp. 95–100 (2006)

24. Manilow, E., Wichern, G., Seetharaman, P., Le Roux, J.: Cutting music source separation some Slakh: a dataset to study the impact of training data quality and quantity. In: Proceedings of the IEEE Workshop on Applications of Signal Processing to Audio and Acoustics, WASPAA (2019)

25. Marques, J., Moreno, P.J.: A study of musical instrument classification using Gaussian mixture models and support vector machines. Cambridge research laboratory technical report series CRL **4**, 143 (1999)

26. Mauch, M., Dixon, S.: Approximate note transcription for the improved identification of difficult chords. In: Proceedings of the 11th International Society for Music Information Retrieval Conference, ISMIR, pp. 135–140 (2010)

27. McFee, B., et al.: librosa: audio and music signal analysis in Python. In: Proceedings of the 14th Python in Science Conference, vol. 8, pp. 18–25 (2015)

28. Müller, M., Ewert, S.: Towards timbre-invariant audio features for harmony-based music. IEEE Trans. Audio Speech Lang. Process. **18**(3), 649–662 (2010)

29. Instruments, N.: Komplete 11 Ultimate. Native Instruments North America Inc., Los Angeles (2016)

30. Ostermann, F., Vatolkin, I., Ebeling, M.: AAM: a dataset of artificial audio multitracks for diverse music information retrieval tasks. EURASIP J. Audio Speech Music Process. **2023**(1), 13 (2023). https://doi.org/10.1186/s13636-023-00278-7

31. Schmid, F., Koutini, K., Widmer, G.: Efficient large-scale audio tagging via transformer-to-CNN knowledge distillation. In: Proceedings of the 2023 IEEE International Conference on Acoustics, Speech and Signal Processing (ICASSP), pp. 1–5 (2023). https://doi.org/10.1109/ICASSP49357.2023.10096110

32. Singh, S., Wang, R., Qiu, Y.: DeepF0: end-to-end fundamental frequency estimation for music and speech signals. In: Proceedings of the IEEE International Conference on Acoustics, Speech and Signal Processing, ICASSP, pp. 61–65 (2021)

33. Vatolkin, I.: Evolutionary approximation of instrumental texture in polyphonic audio recordings. In: Proceedings of the 2020 IEEE Congress on Evolutionary Computation, CEC, pp. 1–8 (2020)

Modelling Individual Aesthetic Preferences of 3D Sculptures

Edward Easton[✉], Ulysses Bernardet, and Anikó Ekárt

Aston Centre for AI Research and Application (ACAIRA), Aston University,
Birmingham, UK
{eastonew,u.bernardet,a.ekart}@aston.ac.uk

Abstract. Aesthetic preference is a complex puzzle with many subjective aspects. This subjectivity makes it incredibly difficult to computationally model aesthetic preference for an individual. Despite this complexity, individual aesthetic preference is an important part of life, impacting a multitude of aspects, including romantic and platonic relationships, decoration, product choices and artwork. Models of aesthetic preference form the basis of automated and semi-automated Evo-Art systems. These range from looking at individual aspects to more complex models considering multiple, different criteria. Effectively modelling aesthetic preference greatly increases the potential impact of these systems. This paper presents a flexible computational model of aesthetic preference, primarily focusing on generating 3D sculptures. Through demonstrating the model using several examples, it is shown that the model is flexible enough to identify and respond to individual aesthetic preferences, handling the subjectivity at the root of aesthetic preference and providing a good base for further extension to strengthen the ability of the system to model individual aesthetic preference.

Keywords: Aesthetic judgement · 3D Art Generation · Aesthetic modelling

1 Introduction

Many theories of aesthetic appreciation exist; some consider aesthetic aspects and their values as constants. However, in most theories, these aspects are considered subjective and reliant on who is assessing the item. For example, when calculating a measure for the Global Contrast Factor [20], it was noted that during the experiment to determine the weights to apply to the contrast, there were a large number of contradictions between participants. These contradictions clearly indicate that aspects of assessing artwork are often subjective. The subjectivity has been investigated where aspects of aesthetic judgement have been split into categories of private and shared taste [14]. Other research [16,21] has investigated which factors influence aesthetic judgement and found that a wide variety of factors affect an individual's aesthetic preferences and judgement. This includes expertise with art (often determined by how often a person interacts

© The Author(s), under exclusive license to Springer Nature Switzerland AG 2024
C. Johnson et al. (Eds.): EvoMUSART 2024, LNCS 14633, pp. 130–145, 2024.
https://doi.org/10.1007/978-3-031-56992-0_9

with art), which does not easily fit within the private and shared categories as, whilst it is a subjective aspect of assessment, it will change over time. Parallels can be drawn between expertise, user fatigue [28] and novelty [2,3,31], as they are all highly dependent on time. Initially created artwork can seem exciting and attractive to the user; however, very quickly, without the ability to introduce significant levels of variety, the user becomes bored as they become more knowledgeable about the types of artwork they are looking at.

Aspects that change over time should not be categorised solely as subjective aspects, as there is a high chance that even within a short time, an individual's preferences can change. Instead, they should be treated as volatile aspects of aesthetic judgement, suggesting a further category should be introduced over private and shared to appropriately represent these aspects: Transient. The private and shared categories will be referred to as subjective and constant, respectively, to align with the naming of the transient category. A full description of each category is shown below:

1. Constant Aesthetics - Universal aspects which apply to the majority of people regardless of characteristics such as age, culture or art experience.
2. Subjective Aesthetics - The aspects which will vary due to individual taste but are relatively constant within an individual.
3. Transient Aesthetics - All aspects which vary due to individual taste, but which are highly volatile. The judgement of these aspects is very likely to change, even at short notice.

Recently, the concept of generating art has started encroaching into the general public's consciousness, thanks to systems like DALL-E [24], MidJourney and Stable Diffusion [25]. The potential power of these systems to generate images from text prompts has brought about a wave of negative press about the impact of AI [13,23,29]. Whilst these systems have had enormous success, one glaring issue is that they work from a much more generalised point, collecting vast amounts of data in order to train the system, effectively reducing the ability to cope with subjectivity. How the data is collected also becomes a problem, introducing a significant bias into the results: BAME people are routinely excluded when simple prompts are used, and women can be portrayed in misogynistic ways. In addition to this, the systems understanding of abstract concepts, something very important within artwork, is minimal. These points are illustrated in Fig. 1, showing images generated through the DALL-E system using the term "attractive". These images only showcase a narrow band of what humanity considers attractive, clearly indicating the bias introduced in these systems through the training data. It should be noted that the system can generate more inclusive images, however, as these rely on the user inputting specific prompts, once again, the emphasis is back on the human to understand the subjective ideas and guide the system rather than the system understanding them itself.

The state of these systems suggests that other approaches are required to auto-generate artwork that are more suitable to handle the subjective nature of artwork. Evolutionary Computation is a technology capable of doing so. Generating art using Evolutionary Algorithms, more commonly known as Evo-Art,

Fig. 1. DALL-E 2 generated images for the prompt "Attractive". Generated by E. Easton

has been ongoing for decades, [27]. Its significant benefit over other types of systems is that the generation can be determined at an individual level, however, it requires introducing a model of aesthetic appreciation that requires thinking about which aspects fit into aesthetic judgement. These models can refer to single measures or more complex sets of criteria. Using more complex models to represent individual aesthetic preference is not a new approach within this field, other examples exist, such as [9,17,18,26], all of which are different to the approach presented here.

To model aesthetic preference, all contributing aspects of aesthetic judgement need to be identified. However, aesthetic judgement is a multi-faceted process with innumerable aspects contributing to the assessment [10,15,22,30]. Some aspects that contribute to aesthetic judgement are relatively well-known, such as the order and complexity [1,11,12]. However, many other aspects are left in relative obscurity. Whilst identifying which aspects contribute to aesthetic judgement becomes easier with practice, a higher level of expertise in art can influence how items are judged [16,21]. However, experts are often used to help create models of aesthetic preference, which potentially introduces bias into these models, excluding non-experts. This shows that, in addition to identifying the contributing aspects to fully model the process, non-experts also need to be included, paving the way for computer systems which can generate artwork that appeals to a wider variety of people, expanding the reach, impact, and potential uses these systems can have.

This paper presents a novel approach to modelling the aesthetic preferences of individuals, which updates in line with changes in the user's preferences and allows unique sculptures to be generated for that individual. The remainder of the paper is organised as follows: Sect. 2 provides the basis for the model and details how the model is implemented. Section 3 provides several examples that utilise the model, indicating how effective the process is at quickly generating individualised sculptures. Finally, the conclusions and potential experimental designs are suggested to enable the further investigation of the model are shown in Sect. 4.

2 Creating an Aesthetic Model of Individual Preference of 3D Sculptures

The starting point of building this model is the three categories of aesthetic aspects: constant, subjective and transient. The category determines how an aspect should be updated within the aesthetic model based on user feedback. Multiple aspects which contribute to aesthetic judgement have been identified by asking numerous people to group and categorise sculptures using high-level aesthetic aspects [5]. These aspects and the category they fall within are shown in Table 1.

Table 1. Final categories of each tag

Constant	Subjective	Transient
Dynamic	Connected	Original
Curved	Bright	Sophisticated
Calm	Busy	Unrefined
Interesting	Natural	Cold
Simple	Unnatural	Unoriginal
	Ordered	Loud
	Disordered	Boring
	Complex	Warm
	Static	Friendly
	Separate	Unfriendly
	Angular	Quiet
		Drab

As indicated by these aspects, it is important to note that the model works at the tag level, allowing different aspects to be tracked at different times, and the category alone indicates how an aspect is treated within the model. The constant aspects are the most straightforward items to handle within the model; they are constantly present for the entirety of the time the user is generating new sculptures. The only changing aspect will be the value, following the rules outlined below.

The subjective aspects need to be controlled in a manner which accounts for their changeability, with different combinations utilised at different stages of the generation process. However, as subjective aspects should be similar throughout an individual's choices, the tags will only be amended based on multiple choices. The model starts by adding all subjective aspects present in the initial choices of sculptures. Then, in the second choice, any additional subjective tags present in the choices will be added. In the third set of choices, any new subjective aspects will once again be added; however, if any of the aspects from the first set of choices have not been present in the current set or second set of choices, they will be removed from the mode. This last step is repeated for subsequent choices, keeping the subjective aspects updated with the user preferences.

Finally, the transient items represent the most volatile aspects contributing to the judgement. These items are assumed to be constantly changing, and their presence in a previous selection, either the aspect or the value of the measure, is not considered enough evidence that the user wants to view these aspects within all of the sculptures they are presented. Due to this, the transient tags used within the model will just be taken from the most recent choices, this is also true for the values for the transient aspects, which will also have no persistence within the model. This approach solves three potential problems: situations where the presence of a tag is masked by others and was unintentionally selected by the user, the potential volatility of user choices and finally, the potential for the miscategorisation of the subjective and transient aspects.

The values of each aspect will be handled in two ways; for constant and subjective aspects, an average value will be taken and used across user choices. The reason for using this process is to accommodate the information learned in [6], where the overall preference of a particular aspect is influenced by the user's choices. As multiple sculptures may be selected in each round of choices, the value calculated for each set of choices is the average value of the measure across all chosen sculptures. If the aspect is not selected in a set of choices, the value is not updated to avoid influencing the model without having enough information to accurately make the change. Due to how the transient aspects have no persistence in the model, only the values from the most recent set of choices will be used. The entire process is detailed in Algorithm 1. Initially, each of the sculptures that have been chosen by an individual are iterated over to obtain a sum of each of the aesthetic value (Lines 3–9), this provides the overall details for the choices the participant has made. After the sum has been calculated, whether this is the first set of choices provided by the user is checked; if it is the first set of choices, then the average value for each of the measures is added into the model (Lines 12–15). If this is not the first set of choices, then the measures being tracked within the current model are iterated through, transient measures have any existing value removed (Line 19), constant and subjective measures, which are already part of the model, have the stored value updated to include the average from the latest set of choices (Lines 21–24). If a measure is not part of the existing choices, if it was added two or more iterations previously, it is removed from the model (Lines 26–29). One final loop through the latest set of choices is made, adding any transient or subjective aspects which are not currently being tracked as part of the model (Lines 34–40).

Algorithm 1. Algorithm for creating and updating the model of aesthetic appreciation

---

1: **procedure** UPDATEMODEL(sculptures: Sculpture[], currentIteration: int, currentModel: Dictionary)
2:    $choiceValues \leftarrow Dictionary$
3:    **for each** $sculpture \in sculptures$ **do**
4:        **for each** $measure \in sculpture.measures$ **do**
5:            **if** $notExists(choiceValues, measure.name)$ **then**
6:                $choiceValues.Add(measure.name, measure.value)$
7:            **else**
8:                $choiceValues[measure.name]+ = measure.value$
9:            **end if**
10:        **end for**
11:    **end for**
12:    **if** $currentIteration == 0$ **then**
13:        **for each** $measure \in choiceValues$ **do**
14:            $currentModel.Add(measure.name, measure.value/sculptures.count())$
15:        **end for**
16:    **else**
17:        **for each** $measure \in currentModel$ **do**
18:            **if** $measure.isTransient$ **then**
19:                $currentModel.remove(measure)$
20:            **else**
21:                **if** $choiceValues.has(measure.name)$ **then**
22:                    $measure.lastUpdatedIteration \leftarrow currentIteration$
23:                    $average \leftarrow choiceValues[measure.name]/sculptures.count()$
24:                    $measure.value \leftarrow (measure.value + average)/2$
25:                **else**
26:                    **if** $measure.lastUpdatedIteration <= (currentIteration - 2)$ **then**
27:                        **if** $measure.isSubjective$ **then**
28:                            $currentModel.remove(measure)$
29:                        **end if**
30:                    **end if**
31:                **end if**
32:            **end if**
33:        **end for**
34:        **for each** $measure \in choiceValues$ **do**
35:            **if** $measure.isTransient$ **or** $measure.isSubjective$ **then**
36:                **if not** $currentModel.has(measure.name)$ **then**
37:                    $currentModel.Add(measure.name, measure.value/sculptures.count())$
38:                **end if**
39:            **end if**
40:        **end for**
41:    **end if**
42: **end procedure**

The methods of determining which tags should be included in the model and which values to aim for allow the model to converge separately on the user's preferred tags and values.

Applying this model to the Genetic Program requires an objective function to be determined that identifies the sculptures which most closely match the user model, guiding the evolutionary search. To achieve this, the Euclidean Distance will be used, Eq. 1, which determines how well the sculpture meets the current user model. The objective function will be minimised during the running of the GP, which finds sculptures which are closer matches.

$$d(model_{user}, model_{sculpt}) = \sqrt{\sum_{i=1}^{n} (model^i_{sculpt} - model^i_{user})^2} \qquad (1)$$

3   Modelling Individual Preference

In order to demonstrate the model, three examples are presented. Calculating the initial user profile, simulating user choices and presenting the final generated sculptures after two choices from each user, all starting from the same set of sculptures. The three profiles will make different choices from the initial set of sculptures; two will make the opposite choices to one another, and the third will select half of the items from each of the other two examples. This allows the final set of sculptures to be visually compared, showcasing how the process is capable of quickly generating unique sculptures based on individual choices. The worked examples will only consider the following aspects: how calm the sculpture appears, how connected the sculpture appears and how friendly a sculpture appears, which have been formalised in [4], along with the form symmetry and Normalised Kolmogorov Complexity calculations used in [8] (Table 2).

Table 2. Evolutionary Algorithm parameters

Total Generations	10
Mutation Probability	0.7
Max tree depth	5
Initialisation Method	Ramped Half and Half
Selection Method	Tournament (k = 3)
Population Size	14
Fitness Measure	User model replication

The initial sculptures have been generated through the AGP, which uses expression trees to place geometric objects around a central axis [7,8]. The sculptures were generated using a distance search approach, which maximised a measure of how different a sculpture was from an archive of existing sculptures, calculated using a reduced version of the approach specified in [19], which calculates the difference based on multiple metrics, with the most different sculptures being added to the archive. This ensures that the range of the presented values is as wide as possible, the initial sculptures are shown in Fig. 2. Figure 3 shows the values of all formalised measures across all sculptures.

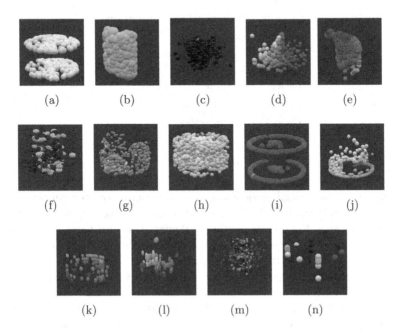

Fig. 2. Initial sculptures generated on which the user model emulation is based

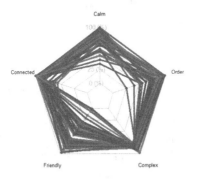

Fig. 3. The aesthetic value measurement values exhibited by the initial set of sculptures

User one selects the first seven sculptures (a–g), and from this, the initial model is calculated, shown in Table 3. User two selects the other seven sculptures (h–n) with their entire model progression shown in Table 5, and user three selects four sculptures from the first user and three from the second user (d–j), with their initial model shown in Table 6.

Table 3. Initial model for example one

Aspect	Category	Initial value
Connected	Constant	0.78
Order	Subjective	0.7
Complexity	Subjective	0.8
Calm	Subjective	0.72
Friendly	Transient	0.66

Using each of the initial models as the initial population, ten generations are run in the GP, taking the initial set of sculptures as the population, crossing over and mutating. The fitness is then calculated as the Euclidean distance, Eq. 1, between the 5-vector generated for the sculpture and the user model. The lower the distance, the better the generated solution. Several examples of the sculptures are shown in Figs. 4, 7 and 8, for Users one, two and three respectively.

This second set of sculptures is then presented to each user, who selects their favourites again. Instead of the originally specified choices, each user selects the seven sculptures that most closely match their current aesthetic model. With the new choices complete, the model is then updated and refined. The constant and subjective aspects have had different values selected and, therefore, updated within the model, and the transient values have been replaced, leading to the model being updated to the definitions shown in Table 4, along with the user's choices.

Table 4. Second model for the first example

Aspect	Category	Model value
Connected	Constant	0.66
Order	Subjective	0.68
Complexity	Subjective	0.77
Calm	Subjective	0.66
Friendly	Transient	0.75

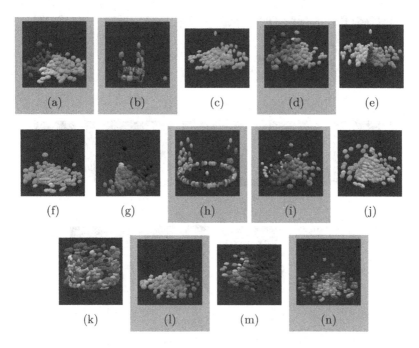

Fig. 4. First set of sculptures generated for Example One, highlighted items are the choices the user makes for the next generation

The values for the constant and subjective aspects are calculated using the averaging function, Eq. 2, which calculates the mean between the existing value for the measure and the newly selected value. The same aspects are persisted in these examples due to the limited number of formalised aspects available. As this list of aspects increases, different terms are expected to form the model at each stage. The GP runs another ten generations on the population of sculptures, optimising the distance between the newly created items and the latest user model. This leads to new items being generated and presented to the user, Fig. 5. From these items, the closest matching sculpture to the user model would be considered the best-fit for the user, shown in Fig. 6a. The same process is repeated for the remaining two examples with the model progression for user two shown in Table 5 and their generated best-fit sculpture shown in Fig. 6b. The model for user three is shown in Table 6 and their final sculpture in Fig. 6c.

$$val_{new} = (val_{exist} + val_{select})/2 \qquad (2)$$

Visually, all three best-fit sculptures generated are very different, even after only two choices and minimal algorithmic input, indicating the effectiveness of this approach for generating unique sculptures specific to an individual's preferences. The influence of the initial choices can be seen in the final sculptures presented to each user. Users one and three have similar forms, and Users two

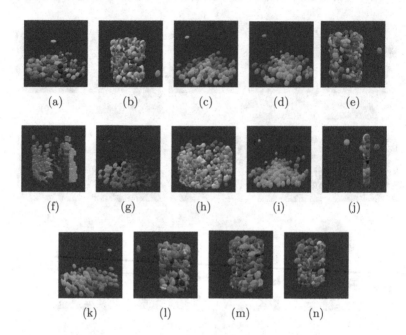

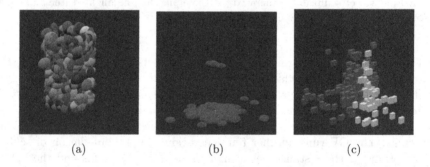

Fig. 5. Final set of sculptures generated for Example One

Fig. 6. Best sculpture generated for Examples One (a), Two (b) and Three (c)

Table 5. Model values for Example Two

Aspect	Initial value	Second value
Connected	0.93	0.97
Order	0.87	0.91
Complexity	0.9	0.88
Calm	0.72	0.8
Friendly	0.76	0.69

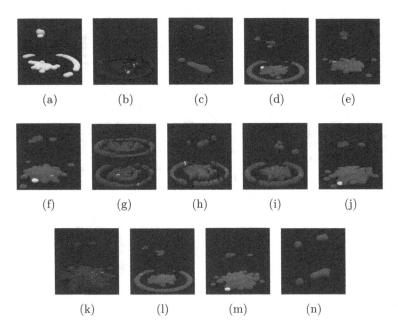

(a)   (b)   (c)   (d)   (e)

(f)   (g)   (h)   (i)   (j)

(k)   (l)   (m)   (n)

Fig. 7. Final set of sculptures presented for Example Two

Table 6. Model values for Example Three

Aspect	Initial value	Second value
Connected	0.95	0.96
Order	0.7	0.73
Complexity	0.83	0.8
Calm	0.87	0.81
Friendly	0.67	0.67

and three have similar colours, with limited similarity between users one and two. With the additional subjectivity introduced by using real humans, these similarities would be much less defined, indicating how capable the model is at representing different preferences, a very important aspect considering the subjectivity of art preference. The results from the model also indicate how well the formalised measures work in combination. For example, in Fig. 6a, the user model asks for a relatively low level of each of the aspects, which the final sculpture reflects by having elements of each but no dominant factor. This contrasts with the examples for users two and three, who both have a high level of connectivity, an aspect which is more clearly visible in their respective best sculptures.

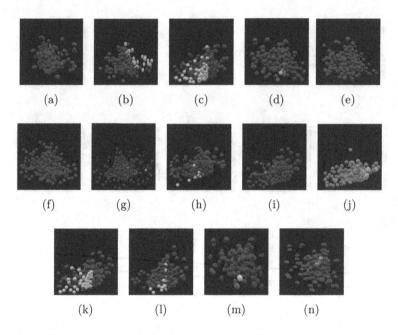

Fig. 8. Final set of sculptures presented for Example Three

4   Conclusion

This paper presents a novel approach to modelling individual aesthetic appreciation, which is suitable for generating sculptures unique to individuals, quickly identifying and generating unique sculptures based on an individual's preference.

The model and its application represent a lightweight approach to modelling individual user preferences across various aesthetic aspects. This model provides several benefits: it allows easy integration into a Genetic Program, which allows sculptures to be generated using the model, updating in real-time based on user choices and changing preferences. In addition, the feedback loop allows the model to become more refined based on more choices being made by a user, generating more personalised sculptures. The model considers each user as an individual, allowing sculptures to be quickly generated which are unique to an individual and their preferences, even if multiple users overlap in their choices. This means the user can spend less time making choices in order to generate sculptures which appeal to them, helping to keep the user engaged in the process.

As mentioned in Sect. 1, one of the current concerns with the proliferation of AI-based art generation systems is how they are trained. This is another of the benefits of the modelling approach presented, as it can generate artwork specific to a person without requiring large amounts of existing artwork to be used to train the system and generalised bias. It also allows these items to be generated without the user having to be able to explicitly identify their preferences, keeping the entire process for the user at a very high level, making it more usable, especially by people who have less expertise in art.

There are some limitations with this approach, even though individual preferences can be identified quickly within the model, accurately creating the model of the user's preferences may require the user to spend more time refining the model in order to generate the artwork; this runs the risk of hitting issues relating to user fatigue, and it is possible that the user would get bored with making choices. However, this can be negated by introducing some considerations for novelty, for example, checking whether the newly generated sculptures are substantially different from existing ones. Another potential improvement is knowing exactly how the aspects interact with one another; the presence of one aspect may prevent another from being strongly present, either due to the generation method or a conflict between how the values are calculated. This could lead to the model trying to introduce a specific aspect to the sculptures that is unnecessary and may impact the presence of other aspects. The AGP was a generation process designed to generate a wide range of visually different sculptures, however, it is not capable of generating all possible sculptures. This means that the process in its current form may not apply to all users, however, due to the modular design of the process, the generation method can easily be amended, allowing the process to be applied to other art styles and keeping it flexible.

The implementation of the presented model cannot represent a full aesthetic judgement model. However, with additional experimentation, the scope of the system detailed here can be improved in impact and reliability. One approach to determine the success of this modelling approach would be to use the art gallery VR system described in [5], where it would be possible to implement the modelling and generation process using the grouping mechanism to identify a user's preferred sculptures. User feedback and choices about preferred sculptures can determine whether the system markedly improved the quality of the sculptures generated in terms of user assessment.

Overall, the presented theoretical model meets all the stated aims by quickly generating sculptures unique to an individual's choices. The model is a theoretical approach that has shown high potential for future use from the emulation of different user choices. Even with the stated limitations, the implementation of the model has shown impressive results in handling the subjectivity of aesthetic preference. This indicates that once some of the mitigations have been applied to overcome the limitations mentioned above, this process will be further strengthened, becoming a powerful tool for automatically and semi-automatically generating unique sculptures based on a specified aesthetic judgement model.

References

1. Birkhoff, G.: Aesthetic Measure. Harvard University Press, Cambridge (1933)
2. Boden, M.A., et al.: The Creative Mind: Myths and Mechanisms. Psychology Press (2004)

3. Canaan, R., Menzel, S., Togelius, J., Nealen, A.: Towards game-based metrics for computational co-creativity. In: 2018 IEEE Conference on Computational Intelligence and Games (CIG), pp. 1–8. IEEE (2018)
4. Easton, E.: Measuring the impact of subjective and transient aesthetics in the generation and appreciation of 3D virtual artwork. Ph.D. thesis, Aston University, Birmingham, UK (2023)
5. Easton, E., Bernardet, U., Ekárt, A.: Contributors to the aesthetic judgement of 3D virtual sculptures. In: 2023 Third International Conference on Digital Creation in Arts, Media and Technology (ARTeFACTo). IEEE (2023)
6. Easton, E., Bernardet, U., Ekárt, A.: Is beauty in the age of the beholder? In: Johnson, C., Rodríguez-Fernández, N., Rebelo, S.M. (eds.) EvoMUSART 2023. LNCS, vol. 13988, pp. 84–99. Springer, Cham (2023). https://doi.org/10.1007/978-3-031-29956-8_6
7. Easton, E., Ekárt, A., Bernardet, U.: Axial generation: a concretism-inspired method for synthesizing highly varied artworks. In: Romero, J., Martins, T., Rodríguez-Fernández, N. (eds.) EvoMUSART 2021. LNCS, vol. 12693, pp. 115–130. Springer, Cham (2021). https://doi.org/10.1007/978-3-030-72914-1_8
8. Easton, E., Ekárt, A., Bernardet, U.: Axial generation: mixing colour and shapes to automatically form diverse digital sculptures. SN Comput. Sci. 3(6), 505 (2022). https://doi.org/10.1007/s42979-022-01329-0
9. Ekárt, A., Joó, A., Sharma, D., Chalakov, S.: Modelling the underlying principles of human aesthetic preference in evolutionary art. J. Math. Arts 6(2–3), 107–124 (2012)
10. Hayn-Leichsenring, G.U., Chatterjee, A.: Colliding terminological systems-Immanuel Kant and contemporary empirical aesthetics. Empir. Stud. Arts 37(2), 197–219 (2019)
11. den Heijer, E., Eiben, A.E.: Comparing aesthetic measures for evolutionary art. In: Di Chio, C., et al. (eds.) EvoApplications 2010. LNCS, vol. 6025, pp. 311–320. Springer, Heidelberg (2010). https://doi.org/10.1007/978-3-642-12242-2_32
12. Johnson, C.G., McCormack, J., Santos, I., Romero, J.: Understanding aesthetics and fitness measures in evolutionary art systems. Complexity 2019, 3495962 (2019)
13. Koidan, K.: Legal & ethical aspects of using DALL-E, midjourney, & stable diffusion (2023). https://medium.com/@katekoidan/legal-ethical-aspects-of-using-dall-e-midjourney-stable-diffusion-cc5606a76d8e
14. Leder, H., Goller, J., Rigotti, T., Forster, M.: Private and shared taste in art and face appreciation. Front. Hum. Neurosci. 10, 155 (2016)
15. Leder, H., Nadal, M.: Ten years of a model of aesthetic appreciation and aesthetic judgments: the aesthetic episode-developments and challenges in empirical aesthetics. Br. J. Psychol. 105(4), 443–464 (2014)
16. Leder, H., Tinio, P.P., Brieber, D., Kröner, T., Jacobsen, T., Rosenberg, R.: Symmetry is not a universal law of beauty. Empir. Stud. Arts 37(1), 104–114 (2019)
17. Li, Y., Hu, C., Chen, M., Hu, J.: Investigating aesthetic features to model human preference in evolutionary art. In: Machado, P., Romero, J., Carballal, A. (eds.) EvoMUSART 2012. LNCS, vol. 7247, pp. 153–164. Springer, Heidelberg (2012). https://doi.org/10.1007/978-3-642-29142-5_14
18. Johnson, C.G.: Aesthetics, artificial intelligence, and search-based art. In: Machado, P., Romero, J., Greenfield, G. (eds.) Artificial Intelligence and the Arts. CSCS, pp. 27–60. Springer, Cham (2021). https://doi.org/10.1007/978-3-030-59475-6_2
19. Mahmoudi, M., Sapiro, G.: Three-dimensional point cloud recognition via distributions of geometric distances. Graph. Models 71(1), 22–31 (2009)

20. Matkovic, K., Neumann, L., Neumann, A., Psik, T., Purgathofer, W.: Global contrast factor-a new approach to image contrast. In: Computational Aesthetics 2005, pp. 159–168 (2005)

21. Monteiro, L.C.P., do Nascimento, V.E.F., da Silva, A.C., Miranda, A.C., Souza, G.S., Ripardo, R.C.: The role of art expertise and symmetry on facial aesthetic preferences. Symmetry **14**(2), 423 (2022)

22. Pelowski, M., Markey, P.S., Lauring, J.O., Leder, H.: Visualizing the impact of art: an update and comparison of current psychological models of art experience. Front. Hum. Neurosci. **10**, 160 (2016)

23. Plunkett, L.: AI creating "art" is an ethical and copyright nightmare (2022). https://kotaku.com/ai-art-dall-e-midjourney-stable-diffusion-copyright-1849388060

24. Radford, A., et al.: Learning transferable visual models from natural language supervision. In: International Conference on Machine Learning, pp. 8748–8763. PMLR (2021)

25. Rombach, R., Blattmann, A., Lorenz, D., Esser, P., Ommer, B.: High-resolution image synthesis with latent diffusion models. In: Proceedings of the IEEE/CVF Conference on Computer Vision and Pattern Recognition, pp. 10684–10695 (2022)

26. Romero, J., Machado, P., Santos, A., Cardoso, A.: On the development of critics in evolutionary computation artists. In: Cagnoni, S., et al. (eds.) EvoWorkshops 2003. LNCS, vol. 2611, pp. 559–569. Springer, Heidelberg (2003). https://doi.org/10.1007/3-540-36605-9_51

27. Sims, K.: Artificial evolution for computer graphics. In: Proceedings of the 18th Annual Conference on Computer Graphics and Interactive Techniques, pp. 319–328. ACM (1991)

28. Takagi, H.: Interactive evolutionary computation: fusion of the capabilities of EC optimization and human evaluation. Proc. IEEE **89**(9), 1275–1296 (2001)

29. Taylor, J.: From trump Nevermind babies to deep fakes: DALL-E and the ethics of AI art (2022). https://www.theguardian.com/technology/2022/jun/19/from-trump-nevermind-babies-to-deep-fakes-dall-e-and-the-ethics-of-ai-art

30. Ventura, D., Gates, D.: Ethics as aesthetic: a computational creativity approach to ethical behavior. In: ICCC, pp. 185–191 (2018)

31. Wiggins, G.A.: A preliminary framework for description, analysis and comparison of creative systems. Knowl.-Based Syst. **19**(7), 449–458 (2006)

Adaptation and Optimization of AugmentedNet for Roman Numeral Analysis Applied to Audio Signals

Leonard Fricke[1]([✉]) [iD], Mark Gotham[2] [iD], Fabian Ostermann[1] [iD],
and Igor Vatolkin[3] [iD]

[1] Department of Computer Science, TU Dortmund University, Dortmund, Germany
{leonard.fricke,fabian.ostermann}@tu-dortmund.de
[2] Department of Computer Science, Durham University, Durham, UK
mark.r.gotham@durham.ac.uk
[3] Department of Computer Science, RWTH Aachen University, Aachen, Germany
igor.vatolkin@rwth-aachen.de

Abstract. Automatic harmonic analysis of music has recently been significantly improved by AugmentedNet, a convolutional recurrent neural network for predicting Roman numeral labels. The original network was trained on a combination of computer encodings of digital scores and human harmonic analyses thereof. Learning from these pairs, the system predicts new harmonic analyses for unseen examples. However, for much music, no score symbolic is available. For this study, we adjusted AugmentedNet for a direct application to audio signals (represented either by chromagrams or semitone spectra). We also implemented and compared further modifications to the network architecture: adding a preprocessing block designed to learn pitch spellings, increasing the network size, and adding dropout layers to avoid over-fitting. A thorough statistical analysis helped to identify the best among the proposed configurations and has shown that some of the optimization steps significantly increased the classification performance. We find that this adapted Augmented-Net can reach similar accuracy levels when faced with audio features as it achieves with the "cleaner" symbolic data on which it was originally trained.

Keywords: Musical harmony · Roman numeral analysis · Convolutional recurrent neural networks · Optimization of neural network architecture

1 Introduction

1.1 Harmonies, and Harmonic Analysis

Harmony is a core parameter in many musical repertoires. In Western music, not only is harmony almost always central, but even many of the finer details are shared across hundreds of years of ostensible very different repertoires (classical, folk, and popular musics). Certain chordal qualities are highly prevalent (major, minor, . . .) as are certain ways of organizing and conceptualizing them in relation to keys, modes, etc.[1].

Equally, there are of course many differences. Most relevantly for our purposes, in some musical repertoires and practices it is common to state explicitly in the notation what "the harmonies" are, while in other contexts, they are only implied. Jazz lead sheets provide an example where chords symbols are explicitly annotated alongside the main melodic line (and sometimes the bassline). While this serves simply as a starting point, and the reality of the resulting music may be very far away from that notated, there is at least an *explicit statement* about the chordal structure.

Baroque classical music has an arguably similar convention in the encoding of figures (e.g., '6' indicating the generic interval of a sixth) above a bass note. In both cases, the *timing* (a.k.a. harmonic rhythm, *when* chords change) and the *content* (separating chord from non-chord tone) are explicit.

"Functional harmonic analysis" expresses that same chord information (content and timing) with the addition of asserting a *key context* for each chord. Roman numeral (hereafter 'RN') analysis is one kind of this functional analysis in which chords are expressed relative to the key position.[2] So while a leadsheet may include the chord 'C Major', and no explicit statement about the key, RN analysis would express this as (for example) a tonic chord ('I') in the *key* of C Major.

RN analyses (and other forms of functional analysis) are almost never explicitly set out by the creator of the music, but rather exist as an *analytical statement*, usually by a third-party, created in response to the music, after the fact of composition. The practice emerged in the 19th century as part of a wider effort towards more systematic forms of musical analysis.

1.2 Why Do Harmonic Analysis at All?

With RN harmonic analysis not serving in a *prescriptive* role (indicating what a player should do) one might reasonably wonder why we do it at all. This question really concerns the value of musical analysis. One answer is that musical analysis exists to enrich our understanding and appreciation of music, and while this clearly aspires to *add* value, it often achieves this goal through several processes of *reduction*, which serve to reveal underlying hierarchies (of parameters including harmony, form, and more). This is simultaneously a *motivation* for doing RN analysis, and a limitation thereof: RN analysis expresses some hierarchical distinctions (between chord and non-chord tones, and 'primary' and 'secondary' triads, for instance), though it is very brittle in relation to the often ambiguous

---

[1] For a wide-ranging account of similarities and differences across tonal musics very much in the plural, and spanning from micro- to macro- temporal scales, see Tymoczko's [22] and "Tonality an Owner's Manual", (forthcoming).

[2] A basic introduction to RN analysis can be found in the chapter "Roman Numerals and SATB Chord Construction" in the Open Music Theory textbook [9].

nature of harmonic change and the different approaches of different analysts. This "inter-annotator disagreement" is important for our purposes, as it undermines the notion of (a single) ground truth, and puts a limit on how successful any automatic approach can help to be, at least while the output consists of one single, flat analysis per source.[3] In short, "100% accuracy" is not really attainable or meaningful for subjective analysis.

1.3   Types of Musical Data: Audio and/or Symbolic

So sources vary in *whether* they express harmonic information and (where applicable) *how* they do so. But musical sources and their computational encodings vary in more significant ways than that, of course, most obviously between audio (.wav, .mp3, .flac, etc.) and so-called "symbolic" (.midi, .musicXML, .abc, etc.) formats.

The choice of format has a strong bearing of our ability to gather data at the scale required for machine learning. While most (all recorded) music exists in "audio" formats, few musical traditions historically and globally have adopted print-visual notation in (symbolic) scores. Even among those musical traditions that do use notation (notably for our purposes, Western classical music), comparatively little of that repertoire is readily available in encoded machine-readable formats, under suitable licences.

The primary distinction between audio and symbolic data is the type of musical data encoded. Broadly speaking, symbolic formats serve as means of conveying *score* information (i.e., the notes and more that composers wrote down as instructions for players) while audio encodes *performance* data (sound as produced by specific musicians, in a specific room, creating compression and rarefaction in the air). In practice, because audio effectively includes both score data (*which* notes) and performance data (*how* they were played), many computational methods for dealing with audio rely on transcription to a semi-symbolic format. Moreover, in machine learning, we create vector encodings of musical data in ways that fit the model at hand. We often do so in ways that encode symbolic and audio data rather similarly to each other, making the seemingly large distinction between the two rather moot.

1.4   AugmentedNet, Data Types, and Use Cases

While there have been automatic systems for harmonic analysis for decades, recent years have seen a surge of interest in creating datasets of harmonic analyses and using that data to train machine learning systems for the task. A notable recent success was achieved with AugmentedNet [14,20] which builds on the "When in Rome" meta-corpus [10,11] of all RN analyses that *human* annotators have encoded in *computational* formats. Given the above discussion, when we observe the success of a model like AugmentedNet we should not consider

---

[3] For more on this topic see [11, §7.1].

this result strictly limited to symbolic data, but explore ways to apply it to the more abundant provision of audio data. This project explores that possibility.

To be more specific, certain core parts of AugmentedNet are self-evidently available to wider data formats, while others are not. Let us take one example of each. First, pitch spelling is not encoded in audio formats (nor even intermediate formats like MIDI). AugmentedNet pays a great deal of attention to the specific encoding of pitch spelling:[4] while C4 and B♯4 refer to the same key on the piano, they are not necessarily to be considered the same musical object, as they have very different implications for the harmonic reading. This enharmonic information is simply not available from an audio signal.

Conversely, synthetic data augmentation is highly relevant. The name "AugmentedNet" comes from the attention given by that model to data augmentation in general, and to a form of augmentation we called "texturization". While most models have explored augmentation of musical data using transposition this is very limited in scope: the absolute pitches change, but the internal intervallic relationships do not. Texturization, by contrast, serves to create additional data by making several synthetic entries from the same harmonic analysis.[5] And while this process is synthetic, it has a close relative in real musical practice where it is common for entirely *distinct pieces* to be based on the *same chord progression*. This is perhaps most clear in the very widespread use of certain highly canonical chord progressions in popular music,[6] but the practice is much older and more widespread still. For example, in the baroque era there was a fashion for creating ground bass variation movements on a single repeating chord progression and bassline.[7] In a way, AugmentedNet's texturization process emulates that compositional task.[8]

1.5   Scope of This Work

As symbolic representations are not always available for music pieces, and variations in individual performances or applied digital effects may not be visible in the score, harmonic analysis of audio signals is a very important application scenario. AugmentedNet in its original form was designed to work with symbolic inputs, therefore several modifications are required and have to be evaluated.

For this study, we have proposed four steps in re-designing AugmentedNet: (1) integration of audio input features (chromagram and semitone spectrum) instead of pitch spelling; (2) increase of the network capacity; (3) addition of dropout layers; (4) extension with a preprocessing convolutional block at the

---

[4] As do other models like [16].

[5] For more, see Figure 2 and surrounding text of [14].

[6] For some examples, see Open Music Theory's [9] intro to pop schemas.

[7] See, for instance, the many 'La Folia' compositions from that time period.

[8] In practice, we could overstate that relationship. AugmentedNet naturally focussed on the task at hand: the vector encoding and what kind of variation will be easy both to generate and provide the system with pertinent variation. This is not intended as a meaningful tool for generation.

front of the original network. The source code of the entire project is publicly available on GitHub.[9]

The results of experiments and their rigorous evaluation by means of statistical tests show that additional preprocessing layers and an increased network size lead to a significant improvement of performance, while dropout does not bring any significant advantages. Furthermore, using our best configurations, the accuracies for 13 classification tasks are quite close to the baseline fed with pitch spellings from the original symbolic score, and even outperform it for two tasks. Finally, we discuss several promising further extensions which should be investigated in future.

2   Methods and Implementation

2.1   Original AugmentedNet

The original AugmentedNet [14,20] is a convolutional recurrent neural network which utilizes multi-task learning for simultaneous prediction of multiple harmonic categories (key, chord root, bass note, harmonic rhythm, etc.). The architecture is similar to the network in [16] but uses a different setup of convolutional layers. The input pitch information is split into bass and chroma, each of them represented with a vector of 19 dimensions: 12 pitch classes $(0, 1, \ldots, 11)$, as well as 7 note names $(A, B, \ldots, G)$. Each pitch is a two-hot encoded vector where one dimension of all pitch classes and one dimension of the note names is set to one, all others to zero. The lengths of both inputs are fixed to 640 timesteps per sequence, where each step is a fixed symbolic length of a 32nd note (i.e., 8 per quarter note).[10]

The inputs are individually fed into two separate convolutional blocks. To detect different time-dependencies, the network reuses feature maps of previous layers by concatenating the output of each convolution with the previous layer. The architecture starts with more filters and smaller filter size in the initial layers to prioritize short-time dependencies. In the further convolutional layers, the filter size increases and the number of filters decreases in powers of 2. These deeper layers can detect the longer-term context. In total, AugmentedNet consists of six convolutional layers. All layers use the Rectified Linear Unit (ReLU) [18] for the activation function.

The outputs of both convolutional blocks are concatenated before passing them to the rest of the network. Two dense layers with 64 and 32 neurons are applied to the concatenated output to reduce the dimensionality before entering the recurrent part of the network which consists of two bidirectional Gated Recurrent Unit (GRU) layers [4] with 30 units each. Both layers return their output at every timestep so that the length of the timestep axis remains constant.

In order to output predictions, AugmentedNet uses a multi-task learning approach. The part of the network described so far is shared for all classification

---

[9] https://github.com/Leo1998/AugmentedNet/tree/audio.

[10] See discussion in [16] §3.2.

tasks. The second GRU layer connects to separate fully-connected dense layers to output the predictions for each of the individual tasks.

In total, we predict 13 classification tasks: Bass, Tenor, Alto, Soprano, ChordQuality, ChordRoot, Inversion, LocalKey, PitchClassSet, PrimaryDegree, RomanNumeral, SecondaryDegree, and TonicizedKey[11]. In AugmentedNet the tasks are often codenamed with their number of output classes appended (e.g., Bass35). Each of these tasks describe different components of the RN annotation which can be combined to construct the final RN label. For a detailed explanation of the individual tasks and the process of constructing RN labels, see [20].

2.2 Datasets and Features

The following experiments were executed using five datasets: Annotated Beethoven Corpus (ABC) [19], Beethoven Piano Sonatas (BPS) [3], Theme and Variation Encodings with RN (TAVERN) [6], When-in-Rome (WiR) [10,11,23], and Well-Tempered Clavier (WTC) [23]. The splitting into training, validation, and test set is presented in Table 1. It is done in the same way as in [14], however some tracks were excluded because they had shown inconsistent behavior when remapping their time position to seconds which is necessary for audio rendering.

The original datasets are sampled using 32nd notes as time steps. This leads to a fine-grained sampling, but remapping to seconds depends on the playback tempo of the track (which can also change along the same track). To make these annotations compatible with audio, the time units of the annotations (quarter length) were remapped to seconds. The annotations can then be merged with any audio feature that has a fixed time step to provide precisely labeled training data.

Table 1. Numbers of music pieces in training, validation, and test sets.

Dataset	Training	Validation	Test
ABC [19]	27	5	5
BPS [3]	12	6	4
TAVERN [6]	24	8	6
WiR [10,11,23]	42	6	8
WTC [23]	11	5	5
Total	116	30	28

In order to use AugmentedNet on audio, the dataset tracks had to be converted to audio files. Because all scores contain annotations of the notes currently

---

[11] The classification tasks went through multiple revisions during the development of AugmentedNet, we used the latest version: v1.9.1.

playing at each time step (the input features in the symbolic version), the representations were converted to MIDI files. Those were rendered to audio using FluidSynth [17]. We chose the soundfont "Salamander Grand Piano" [12] with high quality piano recordings and a sample rate of 44100 Hz.

For audio features used as input to the neural network, the NNLS Chroma plugin [15] for the tool SonicAnnotator [2] was used. It processes audio framewise and extracts a log-frequency spectrum (constant-Q) [1] with three bins per semitone. On that spectrum, it performs a tuning step so that each center bin matches exactly with one semitone. It also applies normalization to the spectrum by subtracting the moving average and dividing by the moving standard deviation. Then, the resulting log-frequency spectrum is used as input for NNLS approximate transcription which leads to a semitone-spaced representation hereinafter called semitone spectrum (with 84 dimensions). Further, this spectrum can be multiplied with chroma or bass chroma profiles, one to extract all the pitches for the chromagram and one for the bass-specific chromagram which contains only the lower frequencies. These are then mapped to the 12 pitch classes by summing the values in order to receive two 12-dimensional vectors. Finally, the resulting chroma frames are normalized using the maximum norm (L_∞ metric). The step size is set to 2048 and the window size to 16384 which are suggested as default values[12] for the NNLS Chroma plugin [15]. A sample sequence for chroma and bass chroma is shown in Fig. 1a and 1b, respectively. The full semitone spectrum for the same sequence is displayed in Fig. 1c.

2.3 Modifications

AugmentedNet was designed to work with symbolic data which differs significantly from audio data as discussed in Sect. 1.3. In the following, we will explore several modifications of the network architecture that aim to improve performance of the AugmentedNet on audio data:

- C or S: one of the two different representations (C: chromagram, S: semitone spectrum) as input to the neural network
- 2×: increasing of network capacity (doubling the number of neurons in the dense and GRU layers)
- D: adding of dropout layers for regularization
- P: more convolutional layers at the front for additional preprocessing

The bass chromagram and the chromagram extracted with NNLS Chroma [15] provide two separate 12-dimensional input vectors to be fed into the network. This approach is very similar to the original AugmentedNet, however the audio chromagrams provide pitch information less accurate than the annotated pitch spellings from the symbolic score. The second approach is to use the full semitone spectrum with 84 dimensions from the NNLS Chroma. The semitone spectrum should include more precise information about the individual pitches and notes than the 12-dimensional chroma vectors. With the semitone spectrum as input,

---

[12] With this setting, the timestep is $\frac{2048}{44100\,\text{Hz}} \approx 0.0464\,\text{s}$.

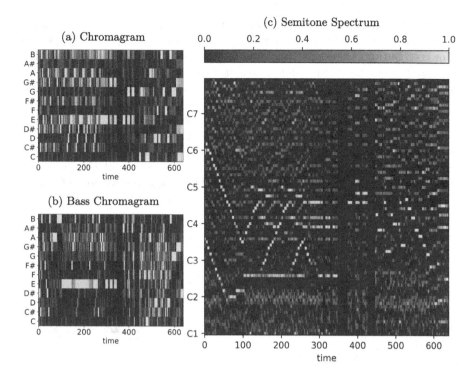

Fig. 1. An example sequence of chroma and semitone spectrum from Beethoven Piano Sonatas Opus 2: Piano Sonata No.2 in A major (bps-02-op002-no2-1). The x-axis measures time frames since the start of the track (640 frames of 2048 samples using a sampling rate of 44100 Hz, so the 640 frames here equate to an approx. 30-second segment). The y-axis shows the pitch: in (a) and (b) this is the pitch class (0–11); in (c) it is equivalent to the midi number. In all three cases the pitch is expressed in terms of one possible pitch spelling simply for ease of reading (spelling information is not present in audio data).

the network has only one input block and thus also one convolutional block. Moreover, using the whole semitone spectrum adds the additional hidden task of detecting the bass note to the prediction task.

Next, we increased the size of the dense and GRU layers of the network. The two fully-connected dense layers as well as the two bidirectional GRUs were tested with twice as many neurons as in the original AugmentedNet.

As another modification, dropout was added to the architecture to reduce overfitting [21]. The dropout layers were placed after each convolutional layer and after each dense layer, except for the final output layers. We tested dropout probabilities of 0.0, 0.25, and 0.5.

Finally, we added a preprocessing block of convolutions at the front side of the network. The idea behind this is that this block may learn a mapping from the input audio features to the pitch spellings provided by the symbolic score data that the original system had available as input information. This block is constructed with 2-dimensional convolution layers instead of the 1-dimensional time convolutions that the rest of AugmentedNet uses. Before passing the inputs to the convolutions, it reshapes the data so that the octaves of the spectrum are distinguished by the channel dimension. In case of 12-dimensional inputs, this simply adds a channel dimension of size 1 to the data because the chromagrams cannot represent different octaves. The semitone spectrum with 84 dimensions gets mapped to 7 channels which leads to a dimensionality of 12×7 for each timeframe. The block consists of three 2-dimensional convolutional layers with a kernel size of 7×5 and filters of decreasing size. Finally, the channel dimension is merged with the 12-dimensional axis to make the output of the block compatible with the following 1-dimensional convolutions. This block is set up so that it does not change the number of time-sequences. Figure 2 presents the full architecture.

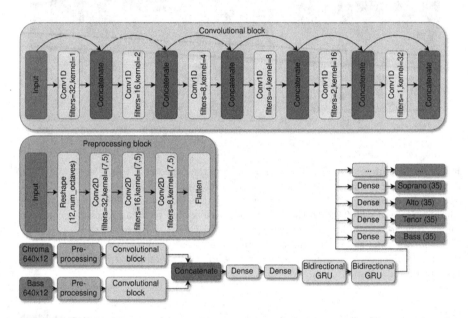

Fig. 2. The architecture of AugmentedNet with modifications for audio. This figure shows the network with NNLS Chroma inputs (two input blocks with 12 chroma bins). The preprocessing block can be added at the front of the network as an additional encoder for pitch estimation.

2.4   Training Procedure

The different network configurations were all trained for a maximum number of 100 epochs with early stopping to abort the training if there is no loss reduction after 5 successive epochs. The original AugmentedNet [14,20] did not use early stopping but we determined in preliminary experiments that it is useful for training on audio data. We calculate the sparse categorical cross-entropy loss for each of the output layers. The total loss is the mean of all output tasks. To optimize the neural networks' parameters, we used ADAM [13].[13] Every combination of the modifications from Sect. 2.3 was trained and iterated 20 times for the statistical measurement of the stability of results.

All experiments were executed on an HPC cluster on high-end GPUs (Nvidia A100) using the machine learning library Keras [5] in version 2.13.1. Each of the 20 runs took about 6 hours of computing time on these.

3 Results

In the following, we will evaluate all proposed modifications and compare them also to the baseline of the original AugmentedNet applied to symbolic inputs.

To train the baseline network, we generated pitch annotations for all tracks from the dataset. The inputs for the baseline are hence the pitch spellings from the original score represented by the encoding explained in Sect. 2.1. The rest of the network remains untouched with none of the presented modifications applied. The only difference between this setup and the network from the original publication [14] is the underlying dataset because the original one has slightly more training tracks. For a fair comparison, we used the same tracks for all experiments.

For the evaluation of results, we measure prediction accuracies of each individual task on the test dataset. Each configuration was repeated 20 times. In Fig. 3, a heatmap of the mean and the standard deviation over the 20 repetitions of each experiments accuracies is provided. The performance of the baseline is shown in the top row of that figure. For the rightmost column, the mean accuracy over all of the classification tasks is calculated for each run individually. These values are then accumulated to compute the mean value and standard deviation over all 20 runs.

Each parameter configuration is denoted by a tag consisting of multiple tokens separated by underscores. The first token is either C or S for chroma or semitone spectrum. The second token denotes the factor by which the number of neurons of the dense layers and the GRU units was multiplied (1 or 2). The third token describes the selected dropout probability (zero when no dropout is used). If the token P is present, the network was trained with the additional preprocessing block at the front of the neural network architecture.

At first glance, we may observe that the configurations with a dropout of 0.5 and a size multiplier of 1 are almost always significantly worse than all others. The configurations with a dropout of 0.5 and the increased network size are better, but all are outperformed by the configurations with a dropout of 0.25. A value of 0.25 appears to be the best, however it only achieves a slight improvement of less than 1% compared to configurations with no dropout.

In general, configurations with an increased network size (2×) always outperform their competitors with a normal network size (1×). The networks with a higher capacity show improvements of approximately 2–3% in accuracy. Also, the

---

[13] The original AugmentedNet implementation [14] initially used RMSprop, but the code was later changed to use ADAM.

	Bass35	Tenor35	Alto35	Soprano35	ChordQuality11	ChordRoot35	Inversion4	LocalKey38	PitchClassSet121	PrimaryDegree22	RomanNumeral31	SecondaryDegree22	TonicizedKey38	Mean
Baseline	0.726 ±0.006	0.697 ±0.005	0.669 ±0.007	0.701 ±0.007	0.731 ±0.008	0.762 ±0.009	0.712 ±0.007	0.726 ±0.008	0.730 ±0.008	0.652 ±0.010	0.612 ±0.010	0.879 ±0.003	0.755 ±0.006	0.719 ±0.003
C_1_0.0	0.645 ±0.003	0.608 ±0.004	0.585 ±0.005	0.632 ±0.004	0.678 ±0.004	0.721 ±0.003	0.664 ±0.005	0.715 ±0.009	0.663 ±0.004	0.619 ±0.006	0.561 ±0.005	0.877 ±0.002	0.734 ±0.006	0.669 ±0.003
C_1_0.0_P	0.646 ±0.008	0.618 ±0.007	0.594 ±0.006	0.645 ±0.004	0.696 ±0.007	0.728 ±0.004	0.666 ±0.006	0.717 ±0.011	0.676 ±0.005	0.627 ±0.008	0.572 ±0.007	0.877 ±0.003	0.735 ±0.006	0.676 ±0.004
C_1_0.25	0.629 ±0.006	0.597 ±0.007	0.575 ±0.005	0.617 ±0.005	0.655 ±0.009	0.710 ±0.005	0.645 ±0.001	0.723 ±0.010	0.649 ±0.006	0.592 ±0.009	0.532 ±0.010	0.877 ±0.000	0.728 ±0.005	0.656 ±0.004
C_1_0.25_P	0.641 ±0.009	0.620 ±0.007	0.588 ±0.006	0.635 ±0.008	0.696 ±0.008	0.727 ±0.007	0.645 ±0.002	0.726 ±0.012	0.673 ±0.008	0.615 ±0.011	0.562 ±0.012	0.876 ±0.001	0.732 ±0.008	0.672 ±0.006
C_1_0.5	0.574 ±0.014	0.548 ±0.016	0.538 ±0.011	0.587 ±0.009	0.616 ±0.022	0.698 ±0.007	0.644 ±0.000	0.721 ±0.013	0.624 ±0.010	0.568 ±0.010	0.489 ±0.021	0.876 ±0.001	0.715 ±0.009	0.631 ±0.010
C_1_0.5_P	0.588 ±0.013	0.571 ±0.012	0.554 ±0.008	0.597 ±0.007	0.631 ±0.015	0.705 ±0.006	0.644 ±0.000	0.710 ±0.015	0.635 ±0.006	0.569 ±0.010	0.494 ±0.016	0.875 ±0.001	0.712 ±0.010	0.637 ±0.007
C_2_0.0	0.654 ±0.003	0.623 ±0.004	0.599 ±0.004	0.647 ±0.004	0.697 ±0.003	0.730 ±0.003	0.681 ±0.005	0.729 ±0.009	0.676 ±0.004	0.639 ±0.004	0.587 ±0.004	0.882 ±0.003	0.745 ±0.005	0.684 ±0.002
C_2_0.0_P	0.657 ±0.007	0.628 ±0.005	0.607 ±0.008	0.659 ±0.007	0.711 ±0.005	0.739 ±0.005	0.683 ±0.006	0.727 ±0.012	0.686 ±0.005	0.649 ±0.006	0.595 ±0.006	0.882 ±0.004	0.741 ±0.005	0.690 ±0.004
C_2_0.25	0.656 ±0.007	0.627 ±0.007	0.599 ±0.006	0.647 ±0.007	0.687 ±0.008	0.729 ±0.005	0.667 ±0.008	0.740 ±0.011	0.673 ±0.006	0.629 ±0.010	0.575 ±0.012	0.879 ±0.002	0.744 ±0.006	0.681 ±0.006
C_2_0.25_P	0.666 ±0.007	0.643 ±0.006	0.612 ±0.007	0.663 ±0.005	0.713 ±0.009	0.740 ±0.003	0.678 ±0.007	0.738 ±0.013	0.689 ±0.006	0.649 ±0.006	0.596 ±0.012	0.880 ±0.003	0.745 ±0.007	0.693 ±0.006
C_2_0.5	0.640 ±0.005	0.613 ±0.005	0.585 ±0.006	0.630 ±0.006	0.680 ±0.007	0.724 ±0.005	0.646 ±0.001	0.745 ±0.010	0.662 ±0.005	0.619 ±0.007	0.561 ±0.009	0.877 ±0.001	0.741 ±0.006	0.671 ±0.004
C_2_0.5_P	0.645 ±0.007	0.623 ±0.007	0.593 ±0.007	0.638 ±0.008	0.690 ±0.008	0.731 ±0.006	0.645 ±0.001	0.738 ±0.012	0.672 ±0.006	0.622 ±0.009	0.568 ±0.011	0.877 ±0.001	0.740 ±0.006	0.676 ±0.005
S_1_0.0	0.653 ±0.006	0.622 ±0.007	0.595 ±0.006	0.634 ±0.007	0.670 ±0.005	0.716 ±0.005	0.668 ±0.006	0.721 ±0.012	0.656 ±0.005	0.612 ±0.007	0.554 ±0.006	0.877 ±0.001	0.732 ±0.006	0.670 ±0.004
S_1_0.0_P	0.682 ±0.006	0.655 ±0.007	0.627 ±0.007	0.658 ±0.006	0.693 ±0.005	0.729 ±0.006	0.685 ±0.006	0.724 ±0.009	0.674 ±0.006	0.630 ±0.010	0.574 ±0.008	0.878 ±0.002	0.736 ±0.006	0.688 ±0.005
S_1_0.25	0.645 ±0.007	0.616 ±0.007	0.591 ±0.006	0.623 ±0.007	0.652 ±0.007	0.710 ±0.004	0.646 ±0.002	0.724 ±0.011	0.646 ±0.005	0.594 ±0.007	0.532 ±0.009	0.877 ±0.000	0.728 ±0.008	0.660 ±0.005
S_1_0.25_P	0.673 ±0.011	0.653 ±0.011	0.619 ±0.009	0.647 ±0.010	0.689 ±0.012	0.723 ±0.006	0.657 ±0.007	0.724 ±0.009	0.667 ±0.008	0.614 ±0.012	0.561 ±0.016	0.877 ±0.001	0.731 ±0.006	0.680 ±0.008
S_1_0.5	0.598 ±0.018	0.575 ±0.019	0.558 ±0.012	0.592 ±0.008	0.611 ±0.018	0.697 ±0.005	0.644 ±0.000	0.726 ±0.014	0.620 ±0.008	0.564 ±0.010	0.487 ±0.019	0.876 ±0.001	0.716 ±0.008	0.636 ±0.009
S_1_0.5_P	0.603 ±0.033	0.584 ±0.034	0.565 ±0.022	0.599 ±0.012	0.628 ±0.027	0.700 ±0.005	0.644 ±0.000	0.713 ±0.016	0.628 ±0.009	0.571 ±0.012	0.496 ±0.024	0.876 ±0.001	0.709 ±0.011	0.640 ±0.014
S_2_0.0	0.667 ±0.005	0.636 ±0.007	0.612 ±0.005	0.652 ±0.005	0.690 ±0.005	0.728 ±0.004	0.685 ±0.009	0.732 ±0.004	0.670 ±0.006	0.635 ±0.006	0.580 ±0.006	0.881 ±0.003	0.741 ±0.004	0.685 ±0.003
S_2_0.0_P	0.693 ±0.005	0.665 ±0.004	0.639 ±0.004	0.673 ±0.005	0.705 ±0.006	0.738 ±0.005	0.702 ±0.005	0.740 ±0.013	0.685 ±0.005	0.650 ±0.005	0.597 ±0.008	0.885 ±0.003	0.747 ±0.006	0.702 ±0.003
S_2_0.25	0.678 ±0.006	0.651 ±0.006	0.623 ±0.008	0.660 ±0.007	0.694 ±0.007	0.732 ±0.005	0.682 ±0.007	0.744 ±0.009	0.676 ±0.004	0.635 ±0.006	0.581 ±0.007	0.879 ±0.002	0.746 ±0.004	0.691 ±0.004
S_2_0.25_P	0.699 ±0.006	0.676 ±0.005	0.645 ±0.007	0.675 ±0.006	0.708 ±0.008	0.739 ±0.005	0.699 ±0.005	0.746 ±0.010	0.686 ±0.006	0.647 ±0.007	0.597 ±0.008	0.881 ±0.003	0.747 ±0.006	0.703 ±0.005
S_2_0.5	0.665 ±0.010	0.640 ±0.009	0.611 ±0.008	0.643 ±0.008	0.683 ±0.008	0.725 ±0.005	0.648 ±0.004	0.749 ±0.010	0.666 ±0.005	0.623 ±0.008	0.567 ±0.009	0.877 ±0.001	0.743 ±0.005	0.680 ±0.006
S_2_0.5_P	0.673 ±0.009	0.651 ±0.010	0.617 ±0.008	0.644 ±0.008	0.682 ±0.011	0.725 ±0.006	0.648 ±0.004	0.737 ±0.009	0.665 ±0.007	0.618 ±0.009	0.563 ±0.012	0.877 ±0.001	0.735 ±0.006	0.680 ±0.006

Fig. 3. Heatmap of the mean accuracies of the 20 executed runs evaluated on the test dataset. The colors are normalized for each column. The rightmost column lists mean accuracies over all classification tasks.

addition of the preprocessing block to the network (P) shows improved accuracy for almost all configurations. The preprocessing block adds about 2–4% accuracy when comparing to their counterparts without preprocessing.

With respect to input features, the semitone spectrum is beneficial for performance of the predictions for classes Bass, Tenor, Alto, Soprano, Inversion, and LocalKey. Other tasks reach approximately the same classification performance when comparing semitone spectrum to chromagram.

The configuration using semitone spectrum, double size, dropout of 0.25 and preprocessing (S_2_0.25_P) is the overall best for the tasks Bass, Tenor, Alto, Soprano, RomanNumeral, and TonicizedKey. The corresponding configuration with chroma features (C_2_0.25_P) is the best one for ChordQuality, Chord-Root, and PitchClassSet. The tasks Inversion, PrimaryDegree, and Secondary-Degree are classified at best by the configuration S_2_0.0_P. LocalKey is best classified by configuration S_2_0.5. However, the difference to the configuration with a dropout of 0.25 is so small that it is hard to judge whether a dropout value of 0 makes a general benefit to these classes without further tests. Therefore, we will report and discuss results of statistical significance tests below.

Judging from the mean accuracy values in the rightmost column, the configuration with semitone spectrum, twice as many neurons, 0.25 dropout and preprocessing (S_2_0.25_P) is the overall best (the mean accuracy is equal to 0.703). However, the identical configuration without dropout reaches almost the same mean accuracy (0.702). The corresponding configuration which used NNLS chroma instead of semitone spectrum only performed 1% worse, so using the semitone spectrum only accomplished a small improvement in that case.

To further substantiate the results, we also applied one-sided Wilcoxon rank-sum tests for the evaluation of differences between all pairs of AugmentedNet configurations (except for the baseline). Observation vectors with 20 mean accuracies after experiment repetitions were compared against each other separately for all classification tasks. Figure 4 summarizes the results. A green cell in the i-th row and the j-th column means that the i-th configuration is significantly superior to the j-th configuration. Light green indicates a statistical significance of $p < 0.05$, whereas any darker green color corresponds to smaller p-values that fell below $\alpha \in \{0.01, 0.001, 0.0001\}$.

As already discussed above, we may clearly observe that the configuration C_1_0.5 has the lowest performance with all 23 other configurations showing significantly superior results (cf. column 5 in Fig. 4). C_1_0.5_P, S_1_0.5, and S_1_0.5_P also show low performances in 20 of the other cases each. The best performing configuration with the largest number of most significantly high mean accuracies is S_2_0.0_P (22 of 23 possible cases highly significant, cf. row 20 in Fig. 4), followed by S_2_0.25_P (22 cases, but slightly less significant), C_2_0.0_P, C_2_0.25_P (20 cases each) and S_1_0.0_P (19 cases).

Additionally, we compared sets of configurations that differ in only one token (also using one-sided Wilcoxon rank-sum test). We were able to show that adding the extra preprocessing layers does highly significantly improve the model's performance ($p = 5.7 \cdot 10^{-13}$). Further, the semitone spectrum should be preferred

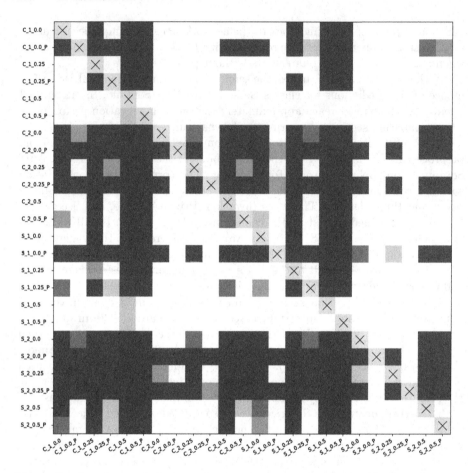

Fig. 4. Results of statistical tests of pairwise comparison of all configurations. The darker the cell color, the lower the p-value falls under significance levels of $\alpha \in \{0.05, 0.01, 0.001, 0.0001\}$. White cells report that there is no significance. (Color figure online)

over the chromagrams ($p = 0.001$). Also, increasing the network's capacity is clearly beneficial ($p < 9.0 \cdot 10^{-42}$). However, adding dropout did not bring any benefit. A smaller dropout probability is significantly more preferable in all three possible comparisons of smaller to larger probabilities of 0.0, 0.25, and 0.5 ($p < 0.001$). This reveals, that the overall best configuration (S_2_0.25_P with a mean accuracy of 0.703), which actually uses dropout, must be considered a slight outlier. The identical configuration but without dropout indeed reaches nearly the same accuracy (0.701).

In conclusion, the classification performance of all tasks gets rather close to the baseline that used the pitch spellings from the original symbolic score. From the presented experiments on audio, there is an overall gap of only about

2% to the baseline experiment. For the tasks LocalKey and SecondaryDegree, the audio networks even outperformed the baseline. However, we assume that the reason lies in the fact that the modifications we applied to the audio network's architecture would also improve the baseline. The original annotations from the baseline will still be a barrier that the predictions on audio cannot exceed, because working directly on audio should always suffer from more noisy data.

4 Conclusions

In this work, we have presented a new approach to predict Roman numeral labels directly from audio signals. We have adjusted the existing convolutional recurrent neural network AugmentedNet and tested several modifications to the network's architecture. We used the same pool of datasets as in the original AugmentedNet [14,20], but rendered them to audio data using high quality realistic piano samples. Compared to the first adaptation close to the original AugmentedNet, the performance could be significantly improved after proposed modifications, being very similar to the baseline experiment which had the pitch annotations provided by the symbolic score available, and even outperforming it for two of the classification tasks.

The preprocessing block that we added at the front of the network improved accuracy values by about 2–4%. Doubling the size of the dense layers and the GRU units also resulted in higher accuracies of about 3%. Using semitone spectrum, which contains more individual tones, instead of the NNLS chroma vectors, which just contain one entry per pitch class, showed improvements for some of the tasks. The mean of all the tasks' accuracies reached a maximum value of 0.703 for a configuration that uses the semitone spectrum as input data, twice as many neurons in the dense layers and GRU units, a dropout probability of 0.25, and adds the extra preprocessing block at the front of the network. This value is only about 2.23% lower than the mean accuracy for the baseline (0.719) which could benefit from using the original symbolic score as "clean" inputs.

Future work may further explore improvements to the network architecture, for example by applying neural architecture search [7] in order to fine-tune AugmentedNet for audio inputs. Especially the additional preprocessing block, which showed promising results in this work, can benefit from exhaustive hyperparameter tuning. Also, replacing GRU with Long Short-Term Memory cells would be a next logical step, but of course is not guaranteed to result in better performances and will be at the expense of much higher computational costs. More advanced changes may include the use of an intermediate loss term to help some layers improving in the pitch classification task, which would fully exploit the ground-truth pitch information present in the datasets. This approach would work in a similar way to that shown in [8].

Acknowledgements. The authors gratefully acknowledge the computing time provided on the Linux HPC cluster at Technical University Dortmund (LiDO3), partially funded in the course of the Large-Scale Equipment Initiative by the Deutsche Forschungsgemeinschaft (DFG, German Research Foundation) as project 271512359.

References

1. Brown, J.: Calculation of a constant Q spectral transform. J. Acoust. Soc. Am. **89**, 425–434 (1991)
2. Cannam, C., Jewell, M.O., Rhodes, C., Sandler, M., d'Inverno, M.: Linked data and you: bringing music research software into the semantic web. J. New Music Res. **39**(4), 313–325 (2010)
3. Chen, T.P., Su, L.: Functional harmony recognition of symbolic music data with multi-task recurrent neural networks. In: Proceedings of the 19th International Society for Music Information Retrieval Conference (ISMIR), pp. 90–97 (2018)
4. Cho, K., et al.: Learning phrase representations using RNN encoder-decoder for statistical machine translation. In: Proceedings of the 2014 Conference on Empirical Methods in Natural Language Processing (EMNLP) (2014)
5. Chollet, F., et al.: Keras (2015). https://keras.io. Accessed 18 Jan 2024
6. Devaney, J., Arthur, C., Condit-Schultz, N., Nisula, K.: Theme and variation encodings with roman numerals (TAVERN): a new data set for symbolic music analysis. In: Proceedings of the 16th International Society for Music Information Retrieval Conference, pp. 728–734 (2015)
7. Elsken, T., Metzen, J.H., Hutter, F.: Neural architecture search: a survey. J. Mach. Learn. Res. **20**, 55:1–55:21 (2019)
8. Georgescu, M.I., Ionescu, R.T., Verga, N.: Convolutional neural networks with intermediate loss for 3D super-resolution of CT and MRI scans. IEEE Access **8**, 49112–49124 (2020)
9. Gotham, M., et al.: Open music theory. https://viva.pressbooks.pub/openmusictheory/. Accessed 18 Jan 2024
10. Gotham, M., Jonas, P.: The openscore lieder corpus. In: Poster at the Music Encoding Conference (2021)
11. Gotham, M., Micchi, G., Nápoles-López, N., Sailor, M.: When in Rome: a meta-corpus of functional harmony. Trans. Int. Soc. Music Inf. Retrieval **6**(1), 150–166 (2023)
12. Holm, A.: Salamander grand piano soundfont. https://freepats.zenvoid.org/Piano/acoustic-grand-piano.html. Accessed 18 Jan 2024
13. Kingma, D., Ba, J.: Adam: a method for stochastic optimization. In: Proceedings of the 3rd International Conference on Learning Representations (ICLR) (2015)
14. López, N.N., Gotham, M., Fujinaga, I.: AugmentedNet: a roman numeral analysis network with synthetic training examples and additional tonal tasks. In: Proceedings of the 22nd International Society for Music Information Retrieval Conference (ISMIR), pp. 404–411 (2021)
15. Mauch, M., Dixon, S.: Approximate note transcription for the improved identification of difficult chords. In: Proceedings of the 11th International Society for Music Information Retrieval Conference (ISMIR), pp. 135–140 (2010)
16. Micchi, G., Gotham, M., Giraud, M.: Not all roads lead to Rome: pitch representation and model architecture for automatic harmonic analysis. Trans. Int. Soc. Music Inf. Retrieval **3**(1), 42–54 (2020)

17. Moebert, T., Ceresa, J.J., Weseloh, M.: Fluidsynth: a soundfont synthesizer (2023). https://www.fluidsynth.org/. Accessed 18 Jan 2024
18. Nair, V., Hinton, G.E.: Rectified linear units improve restricted Boltzmann machines. In: Proceedings of the 27th International Conference on Machine Learning (ICML), pp. 807–814 (2010)
19. Neuwirth, M., Harasim, D., Moss, F.C., Rohrmeier, M.: The annotated Beethoven corpus (ABC): a dataset of harmonic analyses of all Beethoven string quartets. Front. Digit. Human. **5**, 16 (2018)
20. Nápoles López, N.: Automatic Roman numeral analysis in symbolic music representations. Ph.D. thesis, McGill University (2022). https://escholarship.mcgill.ca/concern/theses/qr46r6307
21. Srivastava, N., Hinton, G., Krizhevsky, A., Sutskever, I., Salakhutdinov, R.: Dropout: a simple way to prevent neural networks from overfitting. J. Mach. Learn. Res. **15**(1), 1929–1958 (2014)
22. Tymoczko, D.: Geometry of Music: Harmony and Counterpoint in the Extended Common Practice. Oxford University Press, New York, Oxford (2011)
23. Tymoczko, D., Gotham, M., Cuthbert, M., Ariza, C.: The Romantext Format: a flexible and standard method for representing roman numeral analyses. In: Proceedings of the 20th International Society for Music Information Retrieval Conference (ISMIR), pp. 123–129 (2019)

Generating Smooth Mood-Dynamic Playlists with Audio Features and KNN

Shaurya Gaur[1,2]([✉])[iD] and Patrick J. Donnelly[2][iD]

[1] Radboud University, Nijmegen, The Netherlands
shaurya.gaur@ru.nl
[2] Oregon State University, Corvallis, USA
patrick.donnelly@oregonstate.edu

Abstract. Users curate music playlists for many purposes, including focus, enjoyment and therapy. Popular music streaming services generate playlists automatically which are constant in genre or mood. We propose a method to automatically create playlists dynamic in both the Arousal-Valence emotion space and the audio features of songs. Our playlist algorithm uses a two-stage approach to sequentially choose songs, employing a K-Nearest Neighbors (KNN) model to gather potential songs based on emotion and analyzing them with acoustic similarity metrics. To evaluate the effectiveness of various audio feature data, KNN parameters, and similarity metrics, we developed a testing protocol which generates playlists that traverse both Arousal-Valence and audio feature spaces. We define evaluation metrics to measure a playlist's smoothness and evenness using the Pearson correlation coefficient between dimensions and the variance of steps between songs, respectively. Our algorithm successfully creates smooth and evenly-spaced playlists that transition cohesively in both mood and genre. We explore how the choice of audio feature data, similarity metric, and KNN parameters all have an effect on playlists' smoothness and evenness across these two spaces.

Keywords: K-Nearest Neighbors · Affective Computing · Automatic Playlist Generation

1 Introduction

Listening to as few as five minutes of music while working or studying has been correlated with improved mood [17] and productivity [14,15]. In particular, listening to songs with a major key [3] and high tempo [13] has correlated with one's happiness in conversations and stress during exercise, respectively, though these relationships vary on listeners' preferences and familiarity with the songs [10] and their listening environment [15]. It has also been shown that playlists with smooth transitions between songs can be more satisfying to the ear [31].

S. Gaur—Work completed at Oregon State University.

We seek to create playlists that smoothly transition a listener from their current emotional state towards one they desire, while also gently nudging them through different styles of songs. To examine the affect of songs, we rely on Russell's circumplex model [30], which maps emotions to continuous points on a two-dimensional space by their Valence (positivity) and Arousal (intensity). This model's continuous nature makes it a popular representation of emotions and multiple studies use it to generate playlists of songs similar in affect [8,23].

There are several ways to generate automatic playlists. Some researchers approach it as a continuation problem, using audio similarity [25], collaborative filtering [23], or K-Nearest Neighbor (KNN) techniques [19] to recommend the next song for a user based on their listening history or preferred songs. Others approach this recommendation problem in two stages [36], first recalling many candidate songs and then re-scoring them to choose the single best song.

Various studies seek to build playlists that maintain the user's current mood. Some approaches integrate sensor data [16], facial images [24], or text input [22] to recognize a listener's emotion, while others use audio features [8,23] to predict the affect of recently-played songs. Cardoso et al. [5] develop a tool that allows users to make playlists of a single mood based on a seed song, or to draw a path in the Valence-Arousal plane to create a playlist with songs that lie within a specified distance to the path. However, this study does not specify an order for these songs or make an attempt to evaluate their dynamic playlists.

Recent approaches aim to generate playlists which transition between songs or artists. Some create new playlists based on existing playlist data [6,27], while others build similarity graphs of acoustic features [31] or artist popularity data[1] to guide song order. Flexer et al. [12] use spectral features to represent songs as Gaussian mixture models (GMMs) and create playlists based on their similarity to specified start and end songs. While this method creates dynamic playlists, the study only evaluates their quality through discrete genre labels, and existing music datasets do not contain enough emotion data to follow this approach [7].

We build upon previous work [9] to create an algorithm which builds playlists dynamic in both emotional and audio qualities. Our two-stage algorithm [36] utilizes a nearest-neighbors search [5,12] to choose songs sequentially, forming a linear path. We evaluate playlists by their smoothness and evenness, and demonstrate our algorithm's ability to create an emotional and audio journey in music.

2   Datasets and Preprocessing

To approach this task, we select the Deezer 2018 dataset, which contains Valence and Arousal estimations of 18,644 songs [7]. Researchers from the Deezer streaming service collected each song's affect-related tags from LastFM[2], which have been shown to correlate with the perceived emotional qualities of a song [33]. Using a dataset that mapped English words to embeddings in the Valence-Arousal space, the Deezer team aggregated and normalized these LastFM tags'

---

[1] https://musicmachinery.com/2013/01/02/boil-the-frog-2/.

[2] https://www.last.fm/.

emotion scores. This approach uses crowd-sourced tags for songs instead of manual Valence-Arousal annotations, making these estimations inherently synthetic. As a result, the Deezer dataset yields only 2,762 unique points for 18,644 songs.

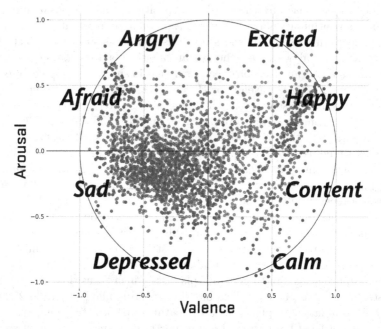

Fig. 1. Visualization of the Deezer dataset [7] in the Valence-Arousal circumplex space. Note the dense clusters of songs in the "Happy" and "Sad" regions.

Figure 1 shows the relative distribution of the dataset in these two dimensions. We observe that this distribution skews towards the region of sad songs, with smaller clusters of songs in the regions that indicate happiness and fear. Although its Valence-Arousal scores are based on synthetic labeling instead of manual human ratings, and they feature duplicate points and an uneven distribution, the Deezer dataset provides a large sample of popular songs from which to build playlists and more easily align with other datasets or streaming services.

2.1   Audio Feature Datasets

We aim to create mood-dynamic playlists that also feature smooth audio-based transitions between their first and final songs. To align with the format of the Deezer dataset, we seek acoustic features for an entire song. There are various datasets which collect audio features and other relevant metadata for large libraries of songs, such as the Spotify API[3] and the Million Song Dataset [2]. We augment the emotion data of the Deezer dataset with features from these

---

[3] https://developer.spotify.com/documentation/web-api/.

two sources. In particular, we extract as many features as possible which relate directly to either a song's audio or its cultural perception, since a song's content and societal context factor into its emotional effect on listeners [10].

Spotify Audio Features. We first extract features from two endpoints of Spotify's API: `Get Audio Features`, which returns aggregate acoustic features obtained through processing raw audio [26], and `Get Track`, which contains track metadata, such as a song's popularity (from 0 to 100) and a flag indicating explicit language. To locate a matching Spotify track ID for each Deezer song, we query the API with the song's title and artist. This successfully returned IDs for 17,755 of the Deezer dataset's 18,644 songs (95.2%). Table 1 shows the 15 features we extract from Spotify. These are primarily continuous, except for a song's key, modality, time signature, popularity, and use of explicit language.

Table 1. Statistics for 13 features from Spotify's `Get Audio Features` API endpoint, and two features from the `Get Track` endpoint, on 17,755 Deezer songs collected.

	Mean	St.Dev.	Min	Median	Max
acousticness	0.2618	0.3051	0	0.1110	0.995
danceability	0.5242	0.1590	0	0.5270	0.978
duration_ms	251336	81137	35587	239	1449973
energy	0.6222	0.2462	0	0.6510	0.999
instrumentalness	0.0844	0.2089	0	0.0002	0.983
key	5.2506	3.5622	0	5	11
liveness	0.1878	0.1557	0.0136	0.1250	1
loudness	−8.4547	3.9755	−46.2840	−7.6410	3.744
mode	0.6823	0.4656	0	1	1
speechiness	0.0605	0.0632	0	0.0387	0.954
tempo	122.0198	29.2322	0	120.1160	217.520
time_sig	3.9018	0.3680	0	4	5
valence	0.4609	0.2481	0	0.4360	0.984
popularity	34.1625	16.6780	0	33	84
explicit	0.0444	0.2059	0	0	1

Million Song Dataset. To compare the utility of two sources of audio features, we also collect from the Million Song Dataset (MSD) [2]. One of the most popular datasets in music information retrieval [1,18], the MSD includes acoustic features and metadata for 1,000,000 songs released between 1922 and 2011 from 44,745 artists [2]. The Deezer dataset includes a MSD track ID for each song, which we use to connect its Valence-Arousal points with MSD features. We first collect each song's key, mode, and time signature (as well as confidence measures for these estimations), and estimations on loudness and tempo. The MSD provides peak loudness and spectral features for each segment of the song. To align with our other datasets, we aggregate the mean, variance, minimum, maximum, and median of the segment-based values to capture the peak loudness and timbre for a whole song. In total, we collect 78 features from the MSD dataset.

2.2 Preprocessing

Since the Spotify dataset has only 15 features compared to the MSD's 78, the difference in dimensionality may have an impact, as the "curse of dimensionality" is particularly problematic in nearest-neighbors approaches [20]. To investigate this, we first apply Principal Component Analysis (PCA) on both the Spotify and MSD datasets, and the combination of the two. To maintain a consistent feature size, we reduce each dataset to 12 principal components. Table 2 describes the variance retained from the transformation on each audio feature dataset.

Table 2. Variance retained after reducing to 12 dimensions with PCA on each dataset. The Spotify, containing fewer original features, retains most of its variance.

Dataset	Features	Variance Retained
Spotify	15	96.7%
MSD	78	68.3%
All (Spotify + MSD)	93	62.7%

Because KNN approaches work best when features are on the same scale, we next transform our data such that no distance in one dimension overweighs others. To mitigate against outliers, we employ a Yeo-Johnson power transformation on each feature [35]. While this reduced their prominence, some features still retained extreme outliers, which we define as values that stray $5.5 \times IQR$ from the quartiles. To combat this, we discretize features that contained extreme outliers into 20 bins. Lastly, we scale all features to the $[-1, 1]$ range.

3   Playlist Algorithm

Given a pair of songs and a desired playlist length, our algorithm generates a playlist, such as the one in Fig. 2, in which the songs form a path between the two songs in the Valence-Arousal space and a specified audio feature space. We choose each song in a two-stage loop, calculating the trajectory from the most recent song to the chosen final song with both emotional and acoustic features.

3.1   Stage 1: K-Nearest Neighbors

We first employ a K-Nearest Neighbors (KNN) model to recall a specified K candidate points from the Deezer dataset. Given the last chosen song S_c and

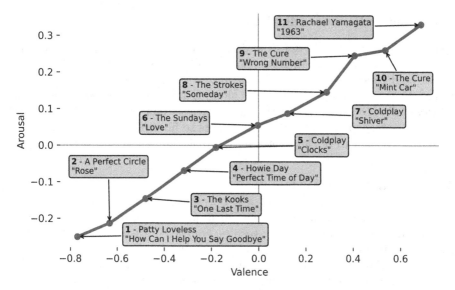

Fig. 2. A playlist generated by our algorithm which begins at sad songs (negative Valence and Arousal) and gradually transitions to happy and excited songs.

the final song S_f as points in the Valence-Arousal space, and the number of songs remaining N, we calculate a hypothetical target point S_t using Eq. (1), representing the first of N equal steps in the linear trajectory from S_c to S_f.

$$S_t = S_c + \frac{S_f - S_c}{N} \tag{1}$$

Utilizing an unsupervised `NearestNeighbors`[4] model from `scikit-learn` [28], which stores distances in a K-dimensional (KD) tree, we query the LK Valence-Arousal points closest to S_t, where L is the playlist's current length. The distance of these candidate neighbors depends on the density of the Deezer dataset's points at the target region. This query has an $\mathcal{O}[P \log(P)]$ time complexity, where $P = 2,762$ is the number of unique points in the Deezer dataset.

We use the same KD tree throughout the playlist creation process for time efficiency, since rebuilding the KD tree has an $\mathcal{O}[P \log(P)]$ time complexity [28]. However, this means that points already in our playlist may reappear as candidates, which requires us to query for LK neighbors and manually filter to find the closest K not yet chosen. This avoids repetition in the playlist. Though this method has a worst-case time complexity of $\mathcal{O}[L^2 K]$, L is likely to be far smaller than P or K, making manual filtering more efficient. We then extract the song IDs for all songs corresponding to these K points in the Deezer dataset.

---

[4] https://scikit-learn.org/stable/modules/neighbors.html.

3.2   Stage 2: Distance Metrics

Given the Deezer song IDs for candidate points, we query an audio dataset for these songs' features. We then use Eq. (1) to calculate a target audio feature vector, and choose the song which minimizes a specified distance score to the target. Evaluating these K neighbors along D audio features has a time complexity of $\mathcal{O}[KD]$, which makes this more efficient than the KNN-based method of the first stage. This two-stage loop repeats to find the following songs in the path from the first to the final song, until the playlist reaches the desired length.

4   Evaluation Metrics

We seek to measure the quality of playlists as paths in both the Valence-Arousal space, and a space of audio feature vectors. For each of these spaces, we evaluate playlists on two key priorities: (1) *Smoothness* and (2) *Evenness*.

First, a playlist should be *smooth* such that the songs fall along a linear path from the first to the final song. This ensures that the emotional and audio characteristics consistently change in the same direction. To evaluate this linear smoothness, we use the Pearson Correlation Coefficient (PCC), a method used to evaluate smoothing techniques [29] which measures how well two variables A and B fall on a straight line (see Eq. (2)). For emotional smoothness, we calculate the PCC score between Valence and Arousal to see how well our algorithm creates a line in these two dimensions. For audio feature spaces, we calculate the average PCC score of all pairs of features using a correlation matrix. A score of 1.0 implies a perfectly smooth playlist as a line from the first to the final song.

$$\rho_{AB} = \left| \frac{\text{Cov}(A,B)}{\sigma_A \sigma_B} \right| \qquad (2)$$

Second, a playlist should be *evenly spaced* such that each transition between songs should be an equivalent step in an emotional or audio feature space. Uneven playlists could startle users with a sequence of similar songs followed by one with very different characteristics. We measure evenness as the variance of a playlist's step sizes, using the Root Mean-Square Error (RMSE) to calculate the distance between each adjacent pair of songs S_i and S_{i+1} over N features, as shown in Eq. (3). In a perfectly even playlist, the size of each step from one song to the next should be the same, and the variance of their RMSEs should be zero.

$$RMSE_i = \sqrt{\frac{||S_{i+1} - S_i||_2}{N}} \qquad (3)$$

5   Experiments and Results

We examine our playlist algorithm's behavior against a wide variety of parameters. Below, we discuss our tests on the effectiveness of various distance metrics, audio datasets and K values. When we experimentally vary one parameter, we use default values for the others to ensure consistency. We use $K = 7$ for our algorithm's first stage, Euclidean distance for Stage 2, and audio features from both Spotify and the MSD. We generate 12-song playlists across all tests in this study, a choice informed by a 2014 survey which found that three datasets of playlists from popular sources contained 11.6 tracks on average [4].

Since these playlists are paths in the Deezer dataset's unevenly distributed Valence-Arousal space, we generate playlists to explore its different regions. We sample 100 random songs in each quadrant of the Valence-Arousal space (see Fig. 3), and for each permutation of two quadrants, we make playlists that start at each point in the first and end at each point in the second. Across the 12 quadrant pairs, we generate 120,000 playlists for each parameter we test.

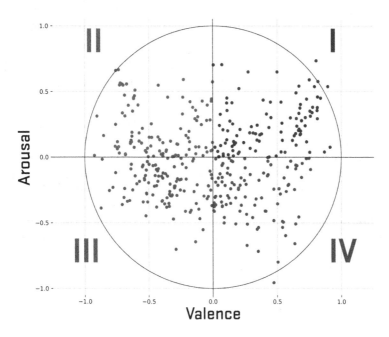

Fig. 3. Map of 100 sample songs per labeled quadrant in the Deezer 2018 dataset [7]. All playlists in the following experiments start from one region and travel to another.

5.1 Distance Metrics

We first test the effects of different common distance metrics in our algorithm's second stage, examining their smoothness and evenness across audio and emotion spaces. Previous studies in mood-dynamic playlist generation [9] only examine the smoothness of a few random playlists built on various distance metrics. Here, we examine the merits in comparing candidate songs with the Euclidean, Manhattan, Cosine, and Jaccard distances, against a simple baseline method which randomly chooses a single candidate song during the second stage.

When testing across regions, we find that the choice of distance metric for the audio features of second stage has no effect on a playlist's smoothness or evenness in the Valence-Arousal space. This indicates that all the candidates chosen in the first stage enable strong transitions emotionally, which allows us to optimize travel on audio feature spaces with no trade-off in emotional transitions.

Table 3. Audio-based scores of playlists by distance, measuring smoothness and evenness across the combined Spotify and MSD audio feature space. The Euclidean and Manhattan distances create the most smooth and even playlists.

Distance Metric	Pearson Correlation			Step-Size Variance		
	Mean	St.Dev.	Median	Mean	St.Dev.	Median
Euclidean	0.2820	0.0308	0.2789	0.0088	0.0041	0.0081
Manhattan	0.2826	0.0312	0.2795	0.0099	0.0045	0.0091
Cosine	0.2755	0.0283	0.2730	0.0234	0.0105	0.0218
Jaccard	0.2711	0.0348	0.2653	0.0243	0.0116	0.0224
Random	0.2618	0.0256	0.2597	0.0223	0.0099	0.0208

Table 3 shows that employing Euclidean and Manhattan distances in our algorithm's second stage generates the smoothest and most evenly-spaced playlists when considering audio transitions. Given that they are used as the L1 (Manhattan) and L2 (Euclidean) vector norms, it is perhaps unsurprising that they find the most direct distance to a target point in an ideal line. Using the Cosine and Jaccard distances yields smoother playlists than the baseline, but fails to improve playlist evenness. This is likely because these two scores do not consider the length of the vector between the candidate and target points—only the angle between them (Cosine) or the minima and maxima of their values (Jaccard).

5.2   Audio Datasets

Next, we compare the effectiveness of using Spotify, MSD, and both sets of audio features in Stage 2. We also explore the impact of variance and dimensionality on these three datasets by applying a 12-feature PCA to each. To understand if including audio features improves upon mood-only playlists [9], we reuse the Deezer dataset's Valence-Arousal points in Stage 2 as a baseline method.

Employing only affective features for Stage 2 leads to the smoothest and most even playlists in mood. On average, playlists that use the Deezer features for both stages achieve a mood-based PCC of around 0.9336 and step-size variance of 0.0003. The choice in audio features shows no impact on playlist smoothness or evenness in emotional transitions, as all audio datasets yield an average mood-based PCC of 0.8734 and step-size variance of 0.0010. This discrepancy between purely emotional data and audio data is to be expected: utilizing affective features means that Stage 2 chooses the song closest to a target based on emotional, instead of acoustic, similarity, yielding the best emotional transitions.

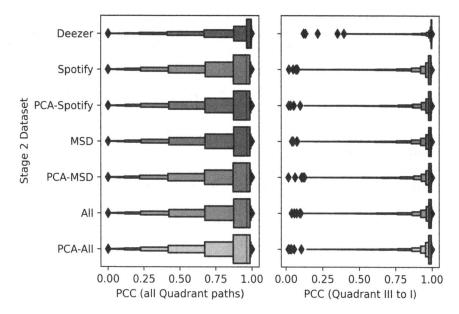

Fig. 4. Smoothness in the Valence-Arousal space of playlists by dataset. The left plot includes all playlists, and the right shows only playlists that travel from the sad quadrant (III) to the happy region (I). Most playlists in both plots have a very high PCC score, regardless of dataset (*Deezer* indicates reusing emotional features for Stage 2).

However, Fig. 4 reveals two other findings. First, the distributions of all of the box-plots indicates that the vast majority of playlists generated with audio-based features still transition smoothly in mood, and those that do not are outliers. Second, the box-plot on the right-hand side visualizes only the playlists that start in Quadrant III (sad and depressed songs) and end in Quadrant I (happy and excited songs)—a diagonal path through two dense regions of the Deezer dataset, as visualized in Fig. 1. The difference in smoothness between using emotional and acoustic features for Stage 2 is far smaller for playlists with this path than with other trajectories. The availability of Valence-Arousal data along this path may improve emotional transitions, as playlists generated from one dense region of this space to another yielded the smoothest playlists.

We evaluate transitions in the MSD and Spotify feature spaces separately in this experiment to better examine the relationships between audio feature sets. Table 4 indicates that in the 78-feature MSD space, no dataset yields significant improvements in smoothness. However, Fig. 5 visualizes that both the original 15 Spotify features and those transformed with PCA yield the smoothest playlists when evaluating against its feature space. This implies that the availability of a large number of acoustic features may in fact limit potential smoothness, because the algorithm must ensure a linear path for far more dimensions.

Table 4. Smoothness (PCC scores) and evenness (step-size variances) of playlists based on dataset, scored in the MSD audio feature space. All datasets yield similarly smooth playlists, and, besides results using the MSD alone, similarly uneven playlists.

DATASET (STAGE 2)	PEARSON CORRELATION			STEP-SIZE VARIANCE		
	Mean	St.Dev.	Median	Mean	St.Dev.	Median
Mood Only						
Deezer	0.2999	0.0171	0.2982	0.0063	0.0029	0.0058
Audio Datasets						
Spotify	0.2966	0.0174	0.2943	0.0063	0.0030	0.0058
MSD	0.2995	0.0208	0.2959	0.0031	0.0014	0.0028
All (Spotify + MSD)	0.2981	0.0203	0.2947	0.0034	0.0015	0.0031
Audio (with PCA)						
Spotify	0.2963	0.0167	0.2943	0.0062	0.0028	0.0057
MSD	0.3063	0.0197	0.3040	0.0055	0.0025	0.0051
All (Spotify + MSD)	0.3050	0.0195	0.3026	0.0055	0.0025	0.0051

Table 4 shows that only the original MSD-based feature sets generate playlists with greater evenness when evaluated in the MSD space. Applying PCA to the MSD features led to a greater playlist step-size variance, and Fig. 5 visualizes the same effect in the Spotify feature space. From this, and from the lack of improvement in smoothness, we conclude that PCA discards key information about song points and do not recommend its use on our algorithm's audio features.

However, Table 4 also indicates no significant difference between employing emotional or acoustic features in Stage 2 for audio-based smoothness or evenness. The first stage of our algorithm chooses candidate songs only based on emotion, which may only present candidates with suboptimal audio-based transitions. This set of candidates may mean that the second stage has very limited ability to improve acoustic smoothness or evenness, resulting in the low PCC scores seen in Table 4. However, the inconclusivity of our findings underscores the need for further exploration of the balance between emotional and acoustic transitions and large high-quality datasets of music annotated with emotion.

5.3 K Values

Finally, we evaluate different numbers for K, or the number of candidate points our algorithm extracts from the Deezer dataset. We compare the effectiveness of extracting $K = 3, 7, 11, 15, 19, 23, 27, 31$ songs in Stage 1 from our KNN model. In Figs. 6 and 7, we see that as K increases, we generate playlists with less smooth and even transitions in the Valence-Arousal space. However, as this change occurs, our algorithm generates playlists that are more smooth and even in acoustic transitions. While the trend in these figures appears when the playlist length is 12 songs, we observe similar results in playlists containing 5 to 20 songs.

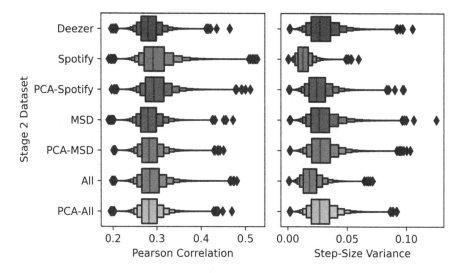

Fig. 5. Smoothness (left) and evenness (right) scores of playlists based on dataset, scored in the Spotify feature space, visualizing PCA's impact on playlist quality.

When recalling fewer candidates, we encourage more precise transitions in the Valence-Arousal space at the expense of acoustic performance. By recalling more songs, we explore more candidates in Stage 2, where songs less than ideal emotionally may lie closest to the ideal audio feature path. For example, when $K = 19$, the algorithm chooses the 19 closest points to the target in the Valence-Arousal space during Stage 1. The song at the 19th closest point will be the farthest candidate from the target point emotionally. However, in Stage 2, this song may be the closest to the target when considering audio features. If so, it will become the next song in the playlist because of this audio similarity even though its emotional transition may be suboptimal. Developers could use K to balance the performance of playlists between emotional and audio feature spaces.

However, it should be emphasized that this behavior is exacerbated by the number of points in our dataset. With only 2,762 unique points, the Deezer dataset is dwarfed by the millions of songs available on streaming services. Within a normalized range of song features at $[-1, 1]$ processed in a similar way to our approach, these songs will yield a much greater density of points in the Valence-Arousal space for our algorithm's first stage. As a result, when $K = 19$, the 19th closest point in the streaming service's data will likely be much closer to the target point than such a point in our current dataset. Therefore, the emotional transformation will be only slightly suboptimal if the 19th closest song in the Valence-Arousal space is chosen compared to the closest song. This could encourage developers to use larger values of K than we used in our experiments to maintain similar smoothness and evenness in emotion spaces.

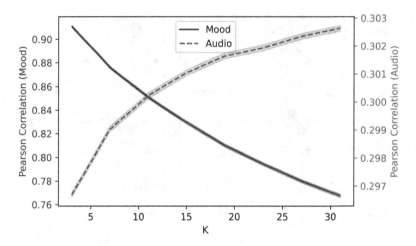

Fig. 6. Smoothness (PCC score) of 12-song playlists by K values, visualizing an inverse relationship between affective and audio smoothness as K changes.

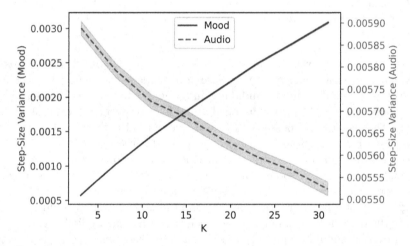

Fig. 7. Evenness (step-size variance) of 12-song playlists by K values, visualizing an inverse relationship between affective and audio smoothness as K changes.

6   Conclusions

In this work, we demonstrate an algorithm which generates playlists that gradually travel through emotion and audio feature spaces. Our approach creates a smooth and linear path through vector spaces of song features, and thus requires no data on existing playlists or users. This can enable applications to create high-quality playlists for users without potential privacy issues. We also define a testing methodology for dynamic playlists which covers many regions of a feature space and evaluates playlists on their smoothness and evenness.

We find that audio features make playlists less smooth and even in an emotional space than considering only affect alone, and that their impact on playlist travel through acoustic similarity is inconclusive. However, the balance between these two feature spaces can be tuned easily by adjusting the recall of our algorithm's first stage. Employing dimensionality reduction algorithms such as PCA fail to yield significantly smoother playlists and come at the cost of consistent steps between songs, since they lose key information regarding a song's acoustics.

Our approach is primarily limited by the quantity and quality of music emotion datasets. While the Deezer 2018 dataset [7] is the largest public repository of songs and their Valence-Arousal scores, their synthetic annotation process leads to many duplicate emotional points. Additionally, their uneven distribution in this plane may potentially yield jarring playlists when traveling through a sparse region in the dataset space. Unfortunately, datasets that contain high quality human-collected emotional annotations contain far fewer songs.

This playlist algorithm presents plenty of opportunities for future work. We create promising measures of a playlist's emotional change, but evaluating a change in mood is highly subjective. As a next step, we plan to test this algorithm's effectiveness with a sample of human listeners, with a study that (1) asks users to report their initial mood, (2) creates a playlist to guide them towards a desired state, and (3) requests them to report their mood after they listen to each song. By measuring a listener's emotion during and after the playlist listening experience, we can assess the quality of the algorithm, and of existing datasets in modeling emotional and acoustic qualities relevant for listeners.

Due to the Deezer dataset's limitations, we currently only examine data at the full song level. However, emotions often change within a song. Perhaps through examination of emotional and audio travel within a song's segments, and the transitions between the end of a song to the beginning of the next, we may create an even smoother listening experience for the user.

Researchers building on our approach could explore integrating this algorithm into a human-in-the-loop system that incorporates sensor readings [16], sentiment analysis of writing [22], or foreground computer processes [32] to gather a user's current mood and continuously recommend songs that travel towards a desired emotional state. However, for these types of systems, it is important to maintain transparency with the listener and only operate with their informed consent, since these playlists are designed to change their mood.

Several studies show that music has a positive impact in elementary school classrooms, encouraging positive and productive behavior among children [34], helping them stay relaxed [11], and even as a means for therapy for children with autism or dyslexia [21]. Our algorithm could be extended as a novel therapeutic tool, enabling the generation of dynamic playlists that begin at a listener's current mood and simultaneously consider their preferred music genres to slowly nudge them towards a desired state of excitement, relaxation or focus.

References

1. Allik, A., Thalmann, F., Sandler, M.: MusicLynx: exploring music through artist similarity graphs. In: Companion Proceedings of the The Web Conference 2018. WWW 2018, pp. 167–170. International World Wide Web Conferences Steering Committee, Republic and Canton of Geneva (2018). https://doi.org/10.1145/3184558.3186970

2. Bertin-Mahieux, T., Ellis, D.P.W., Whitman, B., Lamere, P.: The million song dataset. In: Klapuri, A., Leider, C. (eds.) Proceedings of the 12th International Conference on Music Information Retrieval. ISMIR 2011, pp. 591–596. International Society for Music Information Retrieval, Miami (2011). https://doi.org/10.5281/zenodo.1415820

3. Blood, D.J., Ferriss, S.J.: Effects of background music on anxiety, satisfaction with communication, and productivity. Psychol. Rep. **72**(1), 171–177 (1993). https://doi.org/10.2466/pr0.1993.72.1.171

4. Bonnin, G., Jannach, D.: Automated generation of music playlists: survey and experiments. ACM Comput. Surv. **47**(2), 1–35 (2014). https://doi.org/10.1145/2652481

5. Cardoso, L., Panda, R., Paiva, R.P.: MOODetector: a prototype software tool for mood-based playlist generation. In: Proceedings of the 2nd Portuguese National Symposium on Informatics. INForum 2011, Coimbra, Portugal (2011)

6. Chen, S., Moore, J.L., Turnbull, D., Joachims, T.: Playlist prediction via metric embedding. In: Proceedings of the 18th ACM SIGKDD International Conference on Knowledge Discovery and Data Mining. KDD 2012, pp. 714–722. Association for Computing Machinery, New York (2012). https://doi.org/10.1145/2339530.2339643

7. Delbouys, R., Hennequin, R., Piccoli, F., Royo-Letelier, J., Moussallam, M.: Music mood detection based on audio and lyrics with deep neural net. In: Gómez, E., Hu, X., Humphrey, E., Benetos, E. (eds.) Proceedings of the 19th International Conference on Music Information Retrieval. ISMIR 2018, pp. 370–375. International Society for Music Information Retrieval, Paris (2018). https://doi.org/10.5281/zenodo.1492427

8. Deng, J.J., Leung, C.: Emotion-based music recommendation using audio features and user playlist. In: 6th International Conference on New Trends in Information Science, Service Science and Data Mining. ISSDM 2012, pp. 796–801. Institute of Electrical and Electronics Engineers, Taipei (2012)

9. Donnelly, P., Gaur, S.: Mood Dynamic Playlist: interpolating a musical path between emotions using a KNN algorithm. In: Ahram, T., Taiar, R. (eds.) Proceedings of the 9th International Conference on Human Interaction & Emerging Technologies: Artificial Intelligence & Future Applications. IHIET-AI 2022, vol. 23. AHFE Open Access (2022). https://doi.org/10.54941/ahfe100894

10. Donohoe, R., McNeely, T.: The effect of student music choice on writing productivity. Technical report ED448472, US Department of Education (1999)

11. Feldman, S.: Music affects productivity. Manag. Rev. **80**(7), 6–7 (1991)

12. Flexer, A., Schnitzer, D., Gasser, M., Widmer, G.: Playlist generation using start and end songs. In: Bello, J.P., Chew, E., Turnbull, D. (eds.) Proceedings of the 9th International Conference on Music Information Retrieval. ISMIR 2008, pp. 173–178. International Society of Music Information Retrieval, Philadelphia (2008). https://doi.org/10.5281/zenodo.1418272

13. Flint, M.: The effects of music on physical productivity. Ph.D. thesis, The Ohio State University (2010)
14. Fox, J., Embrey, E.: Music – an aid to productivity. Appl. Ergon. **3**(4), 202–205 (1972). https://doi.org/10.1016/0003-6870(72)90101-9
15. Kellogg, R.T.: Writing habits and productivity in technical writing. Technical report ED249502, US Department of Education (1982)
16. Kim, H.G., Kim, G.Y., Kim, J.Y.: Music recommendation system using human activity recognition from accelerometer data. Trans. Consum. Electron. **65**(3), 349–358 (2019). https://doi.org/10.1109/TCE.2019.2924177
17. Lesiuk, T.: The effect of music listening on work performance. Psychol. Music **33**(2), 173–191 (2005). https://doi.org/10.1177/0305735605050650
18. Liebman, E., Saar-Tsechansky, M., Stone, P.: DJ-MC: a reinforcement-learning agent for music playlist recommendation. In: Proceedings of the 2015 International Conference on Autonomous Agents and Multiagent Systems. AAMAS 2015, pp. 591–599. International Foundation for Autonomous Agents and Multiagent Systems, Istanbul (2015). https://doi.org/10.5555/2772879.2772954
19. Ludewig, M., Kamehkhosh, I., Landia, N., Jannach, D.: Effective nearest-neighbor music recommendations. In: Proceedings of the ACM Recommender Systems Challenge 2018. RecSys Challenge 2018. Association for Computing Machinery, New York (2018). https://doi.org/10.1145/3267471.3267474
20. Marimont, R.B., Sharpiro, M.B.: Nearest neighbour searches and the curse of dimensionality. IMA J. Appl. Math. **24**(1), 59–70 (1979). https://doi.org/10.1093/imamat/24.1.59
21. Mayer-Benarous, H., Benarous, X., Vonthron, F., Cohen, D.: Music therapy for children with autistic spectrum disorder and/or other neurodevelopmental disorders: a systematic review. Front. Psychiatry **12**, 643234 (2021). https://doi.org/10.3389/fpsyt.2021.643234
22. Meyers, O.C.: A mood-based music classification and exploration system. Master's thesis, Massachusetts Institute of Technology, Cambridge, USA (2007)
23. Moscato, V., Picariello, A., Sperlí, G.: An emotional recommender system for music. IEEE Intell. Syst. **36**(5), 57–68 (2021). https://doi.org/10.1109/MIS.2020.3026000
24. Nathan, K.S., Arun, M., Kannan, M.S.: EMOSIC - an emotion based music player for Android. In: Proceedings of the 17th IEEE International Symposium on Signal Processing and Information Technology. ISSPIT 2017, pp. 371–276. Institute of Electrical and Electronics Engineers, Bilbao (2017). https://doi.org/10.1109/ISSPIT.2017.8388671
25. Pampalk, E., Pohle, T., Widmer, G.: Dynamic playlist generation based on skipping behavior. In: Proceedings of the 6th International Conference on Music Information Retrieval. ISMIR 2005, pp. 634–637. International Society of Music Information Retrieval, Victoria (2005). https://doi.org/10.5281/zenodo.1414932
26. Panda, R., Redinho, H., Gonçalves, C., Malheiro, R., Paiva, R.P.: How does the Spotify API compare to the music emotion recognition state-of-the-art? In: Mauro, D.A., Spagnol, S., Valle, A. (eds.) Proceedings of the 18th Sound and Music Computing Conference. SMC 2021. Axea sas/SMC Network, Virtual, pp. 238–245 (2021). https://doi.org/10.5281/zenodo.5045100
27. Pauws, S., Verhaegh, W., Vossen, M.: Music playlist generation by adapted simulated annealing. Inf. Sci. **178**(3), 647–662 (2008). https://doi.org/10.1016/j.ins.2007.08.019
28. Pedregosa, F., et al.: Scikit-learn: machine learning in Python. J. Mach. Learn. Res. **12**, 2825–2830 (2011). https://doi.org/10.48550/arXiv.1201.0490

29. Rosen, P., Quadri, G.J.: LineSmooth: an analytical framework for evaluating the effectiveness of smoothing techniques on line charts. IEEE Trans. Vis. Comput. Graph. **27**(2), 1536–1546 (2021). https://doi.org/10.1109/TVCG.2020.3030421
30. Russell, J.A.: A circumplex model of affect. J. Pers. Soc. Psychol. **39**(6), 1161–1178 (1980). https://doi.org/10.1037/h0077714
31. Sakurai, K., Togo, R., Ogawa, T., Haseyama, M.: Music playlist generation based on graph exploration using reinforcement learning. In: Proceedings of the IEEE 3rd Global Conference on Life Sciences and Technologies. LifeTech 2021, pp. 53–54. Institute of Electrical and Electronics Engineers, Nara (2021). https://doi.org/10.1109/LifeTech52111.2021.9391870
32. Sen, A., Popat, D., Shah, H., Kuwor, P., Johri, E.: Music playlist generation using facial expression analysis and task extraction. Annales Universitatis Mariae Curie-Sklodowska, sectio AI - Informatica **16**(2), 1–6 (2017). https://doi.org/10.17951/ai.2016.16.2.1
33. Song, Y., Dixon, S., Pearce, M.T., Halpern, A.R.: Perceived and induced emotion responses to popular music: categorical and dimensional models. Music. Percept. **33**(4), 472–492 (2016). https://doi.org/10.1525/mp.2016.33.4.472
34. White, K.N.: The effects of background music in the classroom on the productivity, motivation, and behavior of fourth grade students. Master's thesis, Columbia College, Columbia, SC, USA (2007)
35. Yeo, I.K., Johnson, R.A.: A new family of power transformations to improve normality or symmetry. Biometrika **87**(4), 954–959 (2000). https://doi.org/10.1093/biomet/87.4.954
36. Zamani, H., Schedl, M., Lamere, P., Chen, C.W.: An analysis of approaches taken in the ACM RecSys Challenge 2018 for automatic music playlist continuation. ACM Trans. Intell. Syst. Technol. **10**(5), 1–21 (2019). https://doi.org/10.1145/3344257

Pruning Worlds into Stories: Affective Interactions as Fitness Function

Pablo Gervás[(✉)] and Gonzalo Méndez

Facultad de Informática, Universidad Complutense de Madrid, 28040 Madrid, Spain
{pgervas,gmendez}@ucm.es
http://nil.fdi.ucm.es

Abstract. An important challenge when trying to find a story to tell about some set of events that has already happened is to identify the elements in that set of events that will make a story that moves the intended audience. One possible criterion is to consider events that involve significant changes in the emotional relations between the characters involved. The present paper explores a computational model of this particular approach to the task of storytelling. An evolutionary solution is used to explore the logs of an agent-based social simulation, using metrics on the evolution of affinity between characters as fitness function, to identify sequences of events that might be good candidates for moving stories.

Keywords: story generation · affective interactions · evolutionary approach

1 Introduction

The ability to tell stories originates as means for telling a target audience about something that has happened. But an important restriction operates: the telling of the story must require significantly less time than the original events took to happen. So building the story involves an effort to select from the set of events observed the optimal subset that conveys the "meat" of the story. In most cases, additional restrictions are applied to ensure that the telling of the story achieves specific purposes. A very common purpose is to ensure that the story conveyed have some emotional impact on the audience.

The present paper explores a computational model of the task of building a story from an observed set of events, with the particular purpose to maximise the emotional impact on the audience. The set of observed events to consider is taken to be the log of an agent-based social simulation in which agents engage in interactions intended to lead to romantic entanglements between them. A set

This paper has been partially funded by the projects CANTOR: Automated Composition of Personal Narratives as an aid for Occupational Therapy based on Reminescence, Grant. No. PID2019-108927RB-I00 (Spanish Ministry of Science and Innovation) and the ADARVE (Análisis de Datos de Realidad Virtual para Emergencias Radiológicas) Project funded by the Spanish Consejo de Seguridad Nuclear (CSN).

C. Johnson et al. (Eds.): EvoMUSART 2024, LNCS 14633, pp. 179–193, 2024.
https://doi.org/10.1007/978-3-031-56992-0_12

of metrics is designed to compute quantitative measures related to the affective relationships between the agents. These metrics are employed to drive the fitness function for an evolutionary search that identifies subsets of the original set of events that maximise intensity, contrast and evolution in the affective relationships between the agents featured.

2 Related Work

Three topics are considered relevant for this paper: hypotheses linking stories with the evolution of affect between characters, computational approaches to constructing stories from observed sets of events, and evolutionary solutions for building stories.

2.1 Stories and Emotion

The importance of the affective meaning of a story–in contrast to other genres such as essays–is emphasised by Egan [3]. In his analysis of plot, Egan states that in constructing a story "events have, nevertheless, been selected and juxtaposed in order to achieve a particular emotional response and assert a particular sense of causality".

An example of a computational construction of stories that exemplifies this approach is the Mexica system [17], in which emotional information about relations between characters is used as driving force in construction of stories. Mexica operates over a representation that associates to each event details on both preconditions and effects in terms of affinities between characters. Its construction algorithm is designed to build stories that show significant evolution of the affinities between characters through the course of the story.

The idea that the evolution of particular values through the course of the story is a valuable feature in stories has also been defended by Weiland [23]. Weiland attributes particular importance to the evolution of particular characters, in what he defines as *character arcs*. These represent the evolution of the character through the story.

A similar concept is considered by [18], who use machine learning techniques over a corpus of 1,327 stories from Project Gutenberg's fiction collection to identify a set of six core *emotional arcs* that recur over the plots in the corpus. They postulate that these emotional arcs constitute "essential building blocks of complex emotional trajectories." The six emotional arcs in question are:

- *Rags to riches* (affect rises from start to end)
- *Tragedy*, or *Riches to rags* (affect falls from start to end)
- *Man in a hole* (affect falls then rises)
- *Icarus* (affect rises then falls)
- *Cinderella* (affect rises then falls then rises again)
- *Oedipus* (affect falls then rises then falls again)

They further study a corpus of movies in search of correlations between presence of these emotional arcs and interest demonstrated on the movie–in terms of downloads over time–and they conclude that Icarus (affect rises then falls), Oedipus (affect falls then rises then falls again), and two sequential Man in a hole arcs (affect falls then rises then falls then rises again), are the three most successful emotional arcs.

A more recent study explores the construction of distributed representations of character networks in stories using neural networks to create representations in terms of fixed-length vectors [12]. This solution provides a very powerful means of representing character networks that allows comparability across representations of stories of different lengths.

2.2 From Worlds to Stories: Computational Approaches

There are two different research lines focused on obtaining stories from representations of worlds: approaches based on selecting events from worlds to obtain stories of particular types, and approaches based on building narrative discourses to describe subsets of events from a given world.

The significant growth of realistic simulations of worlds arising from the rise of videogames led to efforts aimed at sifting through the material generated by such simulations in search for stories. This approach has been described as *curating storyworlds* and as *story sifting* [20]. Story sifting involves processing significant volumes of events produced by a simulated storyworld to identify subsets that match particular specifications of types of stories. There is often some type of request by a user that establishes which type of story is desired. Initial approaches to the task required that the user specify their request in some kind of technical language [1,9,10,20]. Specific languages have been proposed to describe the *story sifting patterns* that may be used in these searches [9]. More recent approaches take advantage of neural representations and allow the user to establish a curve that describes the story arc they desire, and use complex algorithms to compare this curve to those obtained from potential selections from the search space [13]. When storyworld are large,[1] sifting in search of specific patterns may still yield large volumes of stories. To address this problem, statistical criteria to identify stories that appear less frequently has been proposed as ad additional filter [11].

There is a research line that focuses on the task from the specific point of view of how to convert a set of events from a storyworld–which may have taken place over a range of locations and time periods–into a linear sequence of facts, known as *narrative discourse* in such a way that: (1) all the relevant events are told, (2) the relative order in which they appear in the discourse allows the reader to perceive easily any relations between them, and (3) particular effects on the reader may be accomplished. The task is described as *composition of narrative discourse* [4]. Efforts in this research line also operate over an input

---

[1] Searches through outputs of the Bad News game required processing 140 years of simulation of a unique procedurally generated small American town [21].

that is some kind of log of a complete set of events, and select from those events a subset to include in the discourse. However, in this case, the selection is very specifically tied to the construction of the discourse, and it may involve complex decisions to modify either the granularity at which events are described or the relative order in which they are told. One particular interesting example informs the selection of events to include in the final discourse by considering potential matches between the resulting discourse and a set of possible plot schemas [5]. In this case, the plot schemas operate in a similar way to the story sifting patterns described above.

2.3 Evolutionary Solutions for Building Stories

Evolutionary solutions for building stories often combine a genetic representation that describes a combination of basic story-building units, a set of evolutionary operators over that genetic representation, and a fitness function constructed in terms of metrics of the individuals in the population interpreted as stories. Existing approaches differ in terms of what they consider basic story-building units and in terms of how they define their fitness functions. Instances of basic story-building units considered are: partially ordered graphs of events associated with particular entities [15], instantiations of plot-relevant units of abstraction [6], plot templates [8], or even full story drafts built by an automatic story generator [22]. Criteria used to develop fitness functions include: metrics on story coherence and story interest [15], causal relations between events and certain characters being involved in related events [6], metrics on semantic consistency of the story [8], or knowledge-based heuristics [22].

Of particular interest is the solution for generating small quests for games presented in [14] which evolves a set of plans–each representing a possible quest–using as fitness function the degree of matching between the tensions in the story and a target curve of evolving tensions provided as input.

More recent approaches have considered the importance of avoiding problems of non-synonymous redundancy and low locality [19] when devising genetic representations for stories [7].

3 Finding the Most Emotional Chains of Events in Relationships

Four aspects need to be considered: the agent-based simulation used as source, a genetic representation for stories in this context, metrics for measuring the affective impact of stories, and the evolutionary process for searching for affective stories.

3.1 The Charade Agent-Based Simulator

Charade [16] is a multi-agent simulation designed to express relations and interactions between characters in a storyworld based on the existing affinities between the characters, and to model the evolution of these affinities through a given period of time. The characters may have different kinds of relationships between them (mates, friends, foes or indifferent), which may change in accordance to the evolution of their affinity level, as a result of the interactions (or lack of them) that take place between characters. The result is a log of interactions and evolutions of affinity levels which are subsequently used to generate episodes within a narrative [2].

An example of a fragment of a log for the Charade sytem is shown in Table 1. The simulation is run with 15 agents who do not all know each other. Each agent may or may not have a partner, a small set of friends (between 2 and 4) and may or may not have any enemies (1 or 2 at the start). Currently, spatial relations between the agents are not considered at all, nor are there any elements that can be interacted with. Interactions are driven by affinities between characters, and also act upon them. Probability of interaction is highest for partners, lower for friends, and lowest for enemies. Acceptance of proposals raises affinity between the characters, rejections and inactivity lower it. Runs are stopped when the logs reaches 2,500 events.

Table 1. A example of a fragment of the log generated by the Charade multi-agent simulation system.

```
Megan PROPOSE friend_have_lunch Meredith
Lester PROPOSE friend_chat Robert
Suzette PROPOSE friend_chat Silvy
Betty PROPOSE friend_weekend_out Clark
Meredith PROPOSE mate_watch_tv Lester
Clark REJECT-PROPOSAL friend_weekend_out Betty
Lester REJECT-PROPOSAL mate_watch_tv Meredith
Meredith ACCEPT-PROPOSAL friend_have_lunch Megan
Lester affinity with Meredith 87
Violet PROPOSE friend_chat Megan
Clark affinity with Betty 67
Robert REJECT-PROPOSAL friend_chat Lester
Meredith affinity with Megan 72
Silvy ACCEPT-PROPOSAL friend_chat Suzette
Robert affinity with Lester 72
Betty affinity with Clark 50
(...)
```

The set of events in the log is read into a conceptual representation to allow further processing.

The simulation operates as a distributed system, with each agent responding to the general evolution of the simulation both in terms of actions performed—whether proactively proposing interactions to other agents or reacting to proposals received–and in terms of affective responses–changes in affinity towards other agents. The sequence of events as it appears in the log lists facts in the chronological order in which they have been added to the simulation. From a narrative point of view, the relations between facts corresponding to a particular interaction–proposal, reaction to the proposal, impact of the interaction on the affinities between the characters involved– constitute the basic units of story.

In the pre-processing stage, the log is parsed to group together the sets of facts relate to a particular interaction into a single unit of this type.

An example of the parse of a Charade log into a set of single narrative units is shown in Table 2.

Table 2. A example of a fragment of the parse of a Charade log into a set of plot projections.

```
PLOT-PROJECTION 0
    ProposeActivity {activity=friend_weekend_out, proposee=Clark, proposer=Betty}
PLOT-PROJECTION 1
    ProposedActivityAccepted    {activity=friend_weekend_out, proposee=Clark, proposer=Betty}
    AffinityChange {triggerer=Clark, perceiver=Betty, impact=76}
    AffinityChange {triggerer=Betty, perceiver=Clark, impact=51-->54}
PLOT-PROJECTION 2
    ProposeActivity {activity=mate_go_to_cinema, proposee=Mary, proposer=Clark}
PLOT-PROJECTION 3
    ProposedActivityRejected    {activity=mate_go_to_cinema, proposee=Mary, proposer=Clark}
    AffinityChange {triggerer=Mary, perceiver=Clark, impact=95}
    AffinityChange {triggerer=Clark, perceiver=Mary, impact=84}
    (...)
```

Affinities between two agents A and B are directed, so what A feels for B may differ from what B feels for A. They are represented on a scale between 0 and 100, with 0 representing strong dislike and 100 representing passionate love. The Charade system considers a classification of relations between agents in terms of the affinities between them:

foe affinity between 0 and 40
neutral affinity between 40 and 60
friend affinity between 60 and 80
mate affinity between 80 and 100

The type of relation that holds between two agents determines the subset of activities that they may consider together.

The behaviour of agents is informed by the affinities between them, and the reactions of agents alter the affinities between them. Although in principle all actions by the agents might trigger a change in affinity, in the current set up only the reactions to a proposal produce changes in affinity.

3.2   Genetic Representation of Stories

A story in our approach is considered to be a selection of a subset of the single narrative units that have been constructed from the log. Each individual in our populations will be one such story. Because the events are ordered chronologically in the log of the simulation, a relative ordering can be established between all the single narrative units by assigning to each of them the chronological time of the events in the single narrative unit that constitutes a reaction to the corresponding proposal. The existence of this relative order makes it possible to represent each potential story as a numerical vector that encodes which of the single narrative units is to be included in the story. At each position, a value of 1 indicates that the corresponding single narrative unit is to be included in the story, and a value of 0 that it is not.

3.3   Metrics for Affective Interactions

To establish quantitative metrics on the affective interactions featured in a story, we first construct a set of affective threads. Each *affective thread* is a numerical vector of the same length as the story, and which is made up of the numerical values of the affinity that a character A holds at each point of the story for another character B. It is therefore a numerical vector, as long as the story, of values between 0 and 100.

For a given story, the sequence of affinities between each character and the others is compiled, so that metrics can be computed for each affective thread.

The following features are considered relevant for measuring the evolution of affect through the duration of the story, as featured in a specific affective thread:

- the maximum distance of affinities shown from the neutral value (50)
- whether the affinities change polarity–from positive to negative or negative to positive–at some point
- whether the affinities change into an interval that defines a different type of relation between the characters

The following metrics are defined in terms of these features, by establishing different types of constraints over the values of the metrics for the complete set of affective threads in a story:

- highest ratio between number of affinity changes and thread length
- average distance from neutral affinity
- percentage of threads that transition across meaningful relation types at least once
- percentage of threads that change polarity at least once

3.4   Evolutionary Optimisation of Subset of Told Events

The evolutionary process operates over a given log for the Charade system. An initial population of story drafts of size N is built by assigning the value 1 to N positions chosen at random of the genetic vector that represents a draft. The system is initialised with proportional sets of story drafts for a number of given lengths.

Mutation is carried out by deciding at random whether to activate or deactivate a gene, and then selecting at random a gene of the corresponding type to change its state.

Crossover is carried out by selecting at random a position in the genetic vector, dividing each of the participating vectors at that point, and interchanging the corresponding constituent parts.

Both mutation and crossover operators can change the length of the story drafts they operate on, so some means may be required to control the size of the drafts in the population in case the metrics for affective interactions favour drafts either longer or shorter length.

The basic score assigned to the affective quality of a story–a number between 0 and 100–is established as the average value of the four metrics presented in Sect. 3.3. More complex combinations may be considered as further work.

A fitness function is defined that weighs the basic score with two additional scores designed to control the type of output obtained: (1) a score on story length that assigns 100 to drafts within a given range of length, and 0 otherwise, and (2) a score on character variety that assigns 100 to drafts with number of characters within a desired range of values, and 0 otherwise.

Selection of individuals to pass onto the next generation is done applying accumulated fitness to ensure broad enough coverage of the search space.

3.5 Template Based Rendering of Events

In order to make the outputs of the system easier to understand, the set of events in a particular drafts has been rendered into text using simple templates to express the corresponding events as simple text. This is not intended to be the final form of expression of system outputs when employed for any practical purpose.

Consideration of state of the art techniques [24] for rendering as fluent prose the conceptual representations of story that the system produces is considered beyond the scope of the present paper, as it would not involve evolutionary techniques.

4   Discussion

The results of the proposed system are presented and the relation of the proposed approach with previous work is discussed.

For basic testing of the evolutionary procedure, the parameter on character variety is set to a specific value of two characters in a story–to test the system's ability for identifying interesting developments in the relationship between two characters–and the upper bound on draft length is set to 20.

Table 3 shows an example of system output expressed in terms of the internal representation format, together with the corresponding story rendered as text automatically by the system using basic templates for each of the actions involved. The evolutionary solution was run for 30 generations, with a population size of 20 individuals. The final score for this draft is 61/100.

One can see over this example the main features of the system in operation. Figure 1 shows the evolution of affinities between the characters in the example

Table 3. Example of story draft obtained by story sifting from a Charade log followed by the automated template-based rendering of the story. For ease of understanding the sequence of values for each of the characters is shown below, followed by the scores for the various metrics.

PA-0-ProposeActivity-95	mate_dinner_with_candles / Megan / Tony
PA-0-ProposedActivityAccepted-98	mate_dinner_with_candles / Megan / Tony
PR-0-ProposeActivity-106	mate_dinner_with_candles / Tony / Megan
PR-0-ProposedActivityRejected-112	mate_dinner_with_candles / Tony / Megan
PA-1-ProposeActivity-377	friend_serious_talk / Megan / Tony
PA-1-ProposedActivityAccepted-378	friend_serious_talk / Megan / Tony
PA-2-ProposeActivity-409	friend_help / Tony / Megan
PA-2-ProposedActivityAccepted-410	friend_help / Tony / Megan
PR-1-ProposeActivity-801	friend_have_coffe / Tony / Megan
PR-1-ProposedActivityRejected-802	friend_have_coffe / Tony / Megan
PA-3-ProposeActivity-872	friend_serious_talk / Megan / Tony
PA-3-ProposedActivityAccepted-873	friend_serious_talk / Megan / Tony
PA-4-ProposeActivity-912	friend_chat / Tony / Megan
PA-4-ProposedActivityAccepted-913	friend_chat / Tony / Megan

Tony proposes to Megan to dinner with candles as mates. Megan accepts Tony's invitation to dinner with candles as mates. Megan proposes to Tony to dinner with candles as mates. Tony rejects Megan's invitation to dinner with candles as mates. Tony proposes to Megan to serious talk as friends. Megan accepts Tony's invitation to serious talk as friends. Megan proposes to Tony to help as friends. Tony accepts Megan's invitation to help as friends. Megan proposes to Tony to have coffe as friends. Tony rejects Megan's invitation to have coffe as friends. Tony proposes to Megan to serious talk as friends. Megan accepts Tony's invitation to serious talk as friends. Megan proposes to Tony to chat as friends. Tony accepts Megan's invitation to chat as friends.

Perceiver	Target												
Tony	Megan	85	85	81	81	74	74	78	78	72	72	74	74	76
Megan	Tony	96	96	96	96	82	82	80	80	80	80	82	82	78

Overall	Polarity Change	Distance to Neutral	Relation Change	Rate of Change
61	0	40	100	42

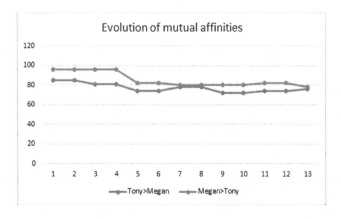

Fig. 1. Plot of the evolution of affinities between Megan and Tony. Vertical axis shows value of affinity, horizontal axis shows point in the discourse sequence for the draft.

in Table 3. An initial close relation progressively sours as Tony rejects Megan's proposals (at times 4 and 10), and in spite of a slight recovery when he accepts her offer of help (at time 8). The metrics show that there has been no polarity change (*Polarity change = 0*: in spite of the deterioration of their relationship, their affinities to one another remain positive throughout), that we are seeing a story that involves a certain passion (*Distance from neutral affinity=40*: at least at the beginning the affinities between them are high), that the relatioship between them is evolving (*Relation change=100*: Tony's affinity for Megan drops below the threshold of 80 at time point 5, making her more of a friend than a mate according to system rules), and affinities do change through the story (*Rate of change=40*).

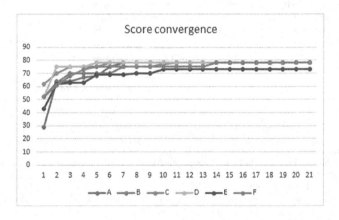

Fig. 2. Plot of convergence of the score during six different runs of the evolutionary process on the same input log from the simulation. Vertical axis shows maximum score for the population, horizontal axis shows number of generations considered to that point.

Figure 2 plots the maximum score at each generation for six different runs of the system on the same log. The graph shows several features of the system at work. The different values of the maximum score at the first generation is a result of the initial construction at random. The progressive increase of the scores shows the evolutionary algorithm at work. The rate of growth is slightly different across runs, consistent with the random nature of the evolutionary operators. Although there is a certain tendency to converge to the highest possible score at around the sixth generation, in some cases the curve does not altogether flatten out until the fourteenth generation. In view of this analysis we have considered that a value of 20 generations is adequate to ensure convergence to the maximum possible scores.

The procedure also needs to be tested on stories of more than two characters. The metric on character variety can be configured to assign maximum score to stories with a broader range of values for the number of characters that take part in the story. This should allow the construction of stories of a higher number of characters. However, the logs being used as input did not include situations that imply elaborate three-way affective interactions of the kind picked out by the relevant metrics. The logs do include instances of subsets of events involving more than two characters that read reasonably well as a story. One such example is shown in Table 4.

Table 4. Example of story draft obtained by configuring the character variety metric to assign maximum score to stories with between 2 and 4 characters.

Ray proposes to Violet to weekend together as mates. Violet accepts Ray's invitation to weekend together as mates. Violet proposes to Ray to sleep together as mates. Ray accepts Violet's invitation to sleep together as mates. Violet proposes to Drew to weekend out as friends. Drew rejects Violet's invitation to weekend out as friends. Violet proposes to Drew to serious talk as friends. Drew accepts Violet's invitation to serious talk as friends. Violet proposes to Ray to invite dinner as mates. Ray rejects Violet's invitation to invite dinner as mates

---

This example shows how the story of the romantic relationship between Violet and Ray interweaves with the friendship between Violet and Drew. The difficulty for the proposed system to produce stories like this is that such a story does not involve the kind of significant changes in affinity between the characters that the metrics are designed to detect. The relationships between the three character in this story do not change significantly. Violet and Ray remains mates and Violet and Drew remain friends throughout. Furthermore, there is no indication in the story of what the affinities are between Ray and Drew and Drew and Ray. This makes this kind of story impossible to identify by the proposed procedure.

The difficulty that the system has in finding stories with more than two characters that exhibit the desired features is an indication that the proposed metrics, while valuable for selecting interesting interactions between two particular characters, may be too restrictive if they are to be applied to the complete set of possible affective threads arising from a story. Although stories that include information about all the possible directional affinities between all the characters involved would indeed be the densest possible instances of affective interaction between members of a collective, they would read out as exhaustive enumeration. Because of that, they would very likely give an impression of lack of focus. The stories that people generally like may mention a broad range of characters, but they tend to focus on a smaller subset–the protagonist and their main relations– rather than exhaustively cover them all. The metrics that we have proposed will need to be adapted to consider this aspect of stories, that they are not well suited to reflect in their current form.

Overall, the impression is that the stories are not very interesting. This is mostly due to the fact that the logs that are currently being used as input do not constitute very interesting simulations. The quality of the stories that can be sifted out of a story log is constrained by the interest of the events already present in the log in question. An obvious solution to this problem is to find more interesting simulations to process. This will be considered as future work. An alternative approach that might provide more informative data on the relative merit of the story sifting algorithm itself would be to develop an additional set of metrics capable of measuring the relative interest of the events in the whole log. That might then be used as a baseline in the sense that stories sifted from the log cannot add interest other than by intelligent selection.

In relation to the previous work cited in Sect. 2, the type of affinities considered in this system are very similar to those used by the Mexica system [17], in the sense that they are defined by a numerical value assigned to the directional affective connections between two characters. The affective threads proposed here as means of interpreting the evolution of the affinities between characters over a story can provide a basic computational approximation to concepts like character arcs [23]. The affective threads also have similarities with the emotional arcs proposed by Reagan [18]. However, emotional arcs differ in that they are not restricted to a specific pair of characters, but rather to the evolution of the overall affinity over the storyline. The features computed in the present paper over affective threads may be used to compute emotional arcs over stories.

The solution proposed here provides a computational procedure for sifting through simulation logs [20] in search for subsets of events that involve a set of characters undergoing significant transitions in their affinities to one another. It might also be employed as an ancillary procedure to identify which subset of a storywold might be more fruitfully selected to inform a process of composition of narrative discourse [4].

In constrast to the evolutionary solutions reviewed in Sect. 2.3 for building stories, the system presented in this paper does not actually build a story by combining independent fragments, but rather selects them from the set of events present in a given log. Nevertheless, this approach differs slightly from story sifting solutions described in Sect. 2.2 in the sense that, whereas story sifting solutions attempt to select existing units of action–say scenes in which a set of characters interact with a certain unity of space and time–the present solution considers sequences of events that appear in the log and involve the same set of characters, but which may actually have taken place at completely separate moments of time within the simulation. In that sense, this procedure does build a story by combining elements from a simulation log, with those elements possible being taken out of context.

5   Conclusions

The evolutionary approach to the task of sorting from the full set of events produced by a simulation those that have a potential for being an interesting story has shown a certain ability to identify subsets that exhibit desirable properties in terms of how the affinities between the characters involved develop. The interpretation of the sequence of the affective interactions in a given story in terms of a set of affective threads, the quantitative interpretation of these threads via a set of features, and the development of a set of metrics based on those features provide valuable tools for the development of fitness functions that can inform evolutionary procedures.

The proposed system allows identification of reasonable fragments of the life of the characters that show significant affective engagement between them. The fact that the resulting stories are not especially interesting is more a question of the type of simulation employed as input. We will consider exploring other possible simulations as input in future work.

Some of the potential shortcomings of the proposed procedure have been identified and discussed. One important issue is the assumption that significant affective interactions need to hold between all the characters involved. This assumption presents a requirement so strong that the system comes up empty when confronted with input logs that are not dense enough in affective interactions. Solutions to this problem will be developed as future work.

As future work, we want to consider the development of more elaborate metrics that identify occurrence of emotional arcs [18] over the affective threads identified for a story.

References

1. Behrooz, M., Swanson, R., Jhala, A.: Remember that time? Telling interesting stories from past interactions. In: Schoenau-Fog, H., Bruni, L.E., Louchart, S., Baceviciute, S. (eds.) ICIDS 2015. LNCS, vol. 9445, pp. 93–104. Springer, Cham (2015). https://doi.org/10.1007/978-3-319-27036-4_9
2. Concepción, E., Gervás, P., Méndez, G.: Ines: a reconstruction of the Charade storytelling system using the Afanasyev framework. In: Proceedings of the Ninth International Conference on Computational Creativity, Salamanca, Spain, pp. 48–55 (2018)
3. Egan, K.: What is a plot? New Literary Hist. 9(3), 455–473 (1978)
4. Gervás, P.: Composing narrative discourse for stories of many characters: a case study over a chess game. Literary Linguist. Comput. 29(4), 511–531 (2014)
5. Gervás, P.: Storifying observed events: could i dress this up as a story? In: 5th AISB Symposium on Computational Creativity. AISB, AISB, University of Liverpool (2018)
6. Gervás, P.: Evolutionary stitching of plot units with character threads. In: De Stefano, C., Fontanella, F., Vanneschi, L. (eds.) WIVACE 2022. CCIS, vol. 1780, pp. 254–265. Springer, Cham (2022). https://doi.org/10.1007/978-3-031-31183-3_21
7. Gervás, P.: Improving efficiency and coherence in evolutionary story generation. In: 14th International Conference on Computational Creativity, Waterloo, Ontario, Canada (2023)

8. Gervás, P., Concepción, E., Méndez, G.: Evolutionary construction of stories that combine several plot lines. In: Martins, T., Rodríguez-Fernández, N., Rebelo, S.M. (eds.) EvoMUSART 2022. LNCS, vol. 13221, pp. 68–83. Springer, Cham (2022). https://doi.org/10.1007/978-3-031-03789-4_5

9. Kreminski, M., Dickinson, M., Mateas, M.: Winnow: a domain-specific language for incremental story sifting. In: Proceedings of the AAAI Conference on Artificial Intelligence and Interactive Digital Entertainment, vol. 17, no. 1, pp. 156–163 (2021)

10. Kreminski, M., Dickinson, M., Wardrip-Fruin, N.: Felt: a simple story sifter. In: Cardona-Rivera, R.E., Sullivan, A., Young, R.M. (eds.) ICIDS 2019. LNCS, vol. 11869, pp. 267–281. Springer, Cham (2019). https://doi.org/10.1007/978-3-030-33894-7_27

11. Kreminski, M., Dickinson, M., Wardrip-Fruin, N., Mateas, M.: Select the unexpected: a statistical heuristic for story sifting. In: Vosmeer, M., Holloway-Attaway, L. (eds.) ICIDS 2022. LNCS, vol. 13762, pp. 292–308. Springer, Cham (2022). https://doi.org/10.1007/978-3-031-22298-6_18

12. Lee, O.J., Jung, J.J.: Story embedding: learning distributed representations of stories based on character networks. Artif. Intell. **281**, 103235 (2020)

13. Leong, W., Porteous, J., Thangarajah, J.: Automated story sifting using story arcs. In: Proceedings of the 21st International Conference on Autonomous Agents and Multiagent Systems. AAMAS 2022, pp. 1669–1671. International Foundation for Autonomous Agents and Multiagent Systems, Richland (2022)

14. de Lima, E.S., Feijó, B., Furtado, A.L.: Procedural generation of quests for games using genetic algorithms and automated planning. In: 18th Brazilian Symposium on Computer Games and Digital Entertainment. SBGames 2019, Rio de Janeiro, Brazil, 28–31 October 2019, pp. 144–153. IEEE (2019). https://doi.org/10.1109/SBGames.2019.00028

15. McIntyre, N., Lapata, M.: Plot induction and evolutionary search for story generation. In: Proceedings of the 48th Annual Meeting of the Association for Computational Linguistics, pp. 1562–1572. Association for Computational Linguistics, Uppsala (2010)

16. Méndez, G., Gervás, P., León, C.: On the use of character affinities for story plot generation. In: Kunifuji, S., Papadopoulos, G.A., Skulimowski, A.M.J., Kacprzyk, J. (eds.) Knowledge, Information and Creativity Support Systems. AISC, vol. 416, pp. 211–225. Springer, Cham (2016). https://doi.org/10.1007/978-3-319-27478-2_15

17. Pérez ý Pérez, R., Sharples, M.: MEXICA: a computer model of a cognitive account of creative writing. J. Exp. Theor. Artif. Intell. **13**(2), 119–139 (2001)

18. Reagan, A.J., Mitchell, L., Kiley, D., Danforth, C.M., Dodds, P.S.: The emotional arcs of stories are dominated by six basic shapes. EPJ Data Sci. **5**(1), 1–12 (2016)

19. Rothlauf, F.: Representations for Genetic and Evolutionary Algorithms. Springer, Heidelberg (2006). https://doi.org/10.1007/3-540-32444-5

20. Ryan, J.: Curating simulated storyworlds. Ph.D. thesis, University of California Santa Cruz, CA, USA (2018)

21. Samuel, B., Summerville, A., Ryan, J., England, L.: A quantified analysis of *bad news* for story sifting interfaces. In: Mitchell, A., Vosmeer, M. (eds.) ICIDS 2021. LNCS, vol. 13138, pp. 142–156. Springer, Cham (2021). https://doi.org/10.1007/978-3-030-92300-6_13

22. de Silva Garza, A.G., y Pérez, R.P.: Towards evolutionary story generation. In: Colton, S., Ventura, D., Lavrac, N., Cook, M. (eds.) Proceedings of the Fifth Inter-

national Conference on Computational Creativity. ICCC 2014, Ljubljana, Slovenia, 10–13 June 2014, pp. 332–335. computationalcreativity.net (2014)

23. Weiland, K.: Creating Character Arcs: The Masterful Author's Guide to Uniting Story Structure, Plot, and Character Development. Helping Writers Become Authors. PenForASword Publishing (2016). https://books.google.es/books?id=bMRgvgAACAAJ

24. Xie, Z.: Neural text generation: a practical guide. arXiv preprint arXiv:1711.09534 (2017)

Collaborative Interactive Evolution of Art in the Latent Space of Deep Generative Models

Ole Hall$^{(\boxtimes)}$ and Anil Yaman

Vrije Universiteit Amsterdam, 1081 HV Amsterdam, The Netherlands
ole.moritz.hall@gmail.com, a.yaman@vu.nl

Abstract. Generative Adversarial Networks (GANs) have shown great success in generating high quality images and are thus used as one of the main approaches to generate art images. However, usually the image generation process involves sampling from the latent space of the learned art representations, allowing little control over the output. In this work, we first employ GANs that are trained to produce creative images using an architecture known as Creative Adversarial Networks (CANs), then, we employ an evolutionary approach to navigate within the latent space of the models to discover images. We use automatic aesthetic and collaborative interactive human evaluation metrics to assess the generated images. In the human interactive evaluation case, we propose a collaborative evaluation based on the assessments of several participants. Furthermore, we also experiment with an intelligent mutation operator that aims to improve the quality of the images through local search based on an aesthetic measure. We evaluate the effectiveness of this approach by comparing the results produced by the automatic and collaborative interactive evolution. The results show that the proposed approach can generate highly attractive art images when the evolution is guided by collaborative human feedback.

Keywords: Generative Adversarial Networks · Latent Variable Evolution · Interactive Evolutionary Computation · Collaborative Art

1 Introduction

Artificial intelligence (AI) based approaches to art generation have increased their popularity and received a great deal of attention due to available tools such as Artbreeder [34] and Nvidia Canvas [24]. A debate about the value of such art has developed at the latest since the auction of an AI generated artwork at the renowned auction house Christie's [7]. Frequent critics doubt the creativity and novelty that can be generated by AI, while other voices postulate AI mainly as a potent tool of modern artists [6].

Supplementary Information The online version contains supplementary material available at https://doi.org/10.1007/978-3-031-56992-0_13.

Deep generative models such as Generative Adversarial Networks (GANs) [14] play a major role in AI based art generation approaches. The basic idea of a GAN is to generate images that cannot be distinguished from real images. With this setup, however, doubts have been raised if artefacts generated in this way are actually creative or merely attempting to emulate the training material [10]. On the other hand, the generation process offers a creative space, as the fake images can be generated by sampling from the latent space of learned art representations, providing infinite possibilities of potential images [6].

Due to this feature, GANs open up many possibilities for human-AI interaction in co-creative processes [6,15]. For example, they can inspire designers [30,43], and artists may spend hours generating different random images to find attractive or inspiring outcomes [35]. A more promising approach to discover art images is to apply (meta-)heuristic search algorithms such as evolutionary computing (EC) [9]. Evolutionary computing has been employed to discover the design space in GANs in different fields of application (e.g. [4,28,40]), but rarely in the field of art generation [29,34].

In this paper, we employ EC to navigate within the latent space of GANs trained on art images. Firstly, we use a specific GAN architecture, the Creative Adversarial Network (CAN) [10], to introduce novel art images. CANs can achieve this because the images generated imitate the real art distribution while deviating from established art styles through a modified loss function that penalises simple categorisation into an existing art style. Then, the vector representations of latent space variables are used to encode individuals in EC. We employ evolutionary operators such as crossover and mutation to generate new individuals from the existing ones. To determine the quality of art images (i.e. fitness in EC), we use two metrics: (1) automatic aesthetic, and (2) collaborative interactive human evaluation.

Automatic aesthetic evaluation is based on Neural Image Assessment (NIMA) [38], a Convolutional Neural Network trained on an annotated dataset for aesthetic visual analysis (AVA) [23]. We also used the automatic aesthetic evaluation metric as an intelligent mutation operator that aims to perform a local search [29] to improve the quality of the images and accelerate the evolutionary process similar to the approaches used in memetic algorithms [9].

Collaborative interactive human evaluation is based on the idea of interactive evolutionary computation (IEC) [37] where human evaluations replace the fitness evaluation step in EC. The IEC is particularly suitable in cases such as art generation where the measure of the quality is subjective [27]. However, in contrast to the classic IEC setup, here, we consider collaborative evaluations from several participants to account for their subjectivity.

To prevent the generation of very similar images especially within the same generation, we introduce diversity preservation mechanisms. If two similar images are detected, one is replaced by a new randomly generated image. This also intends to accelerate the exploration of latent space, as well as to avoid user fatigue from evaluating many similar images [9].

Our results show that this methodology is generally suitable for exploring the latent space of possible art images in a GAN, and it is able to create increasingly attractive images. It was shown that both the automatic as well as the collaborative evolution achieved an increasing fitness over the evolutionary process. In addition, we tested whether the local search leads to improvements beyond a random level in the eyes of human participants, but this was not confirmed. Finally, we investigated whether the results obtained from both the automatic aesthetic and collaborative interactive evolution are in fact perceived as more attractive than randomly generated images. The findings indicate that human guidance is crucial for the evolution of art in order to achieve images that are perceived as more attractive.

2   Related Work

2.1   Generative Adversarial Networks

Introduced by Goodfellow et al. [14], GANs aim to generate artefacts that are indistinguishable from real images. For this purpose, two models are in a competing relationship, allowing for an unsupervised learning approach: the generator is trained to generate fake images that are indistinguishable from real ones, while the discriminator is trained to make the distinction by comparing the generated images with real ones. The training is therefore comparable to a min-max game for two networks. Both the generator and discriminator are designed as deep neural networks. Through mutual feedback, they constantly improve each other, eventually leading the generator to generate artefacts that are difficult to distinguish from real ones. These artefacts can be created from underlying latent variables by random sampling.

Despite impressive visual results, training GANs is considered difficult and partly unstable, and they were initially only able to generate low resolutions [18]. Therefore, many extensions and variants have been developed since then ([17], for an overview). A structural improvement for working with images was provided by Deep Convolutional Generative Adversarial Networks (DCGANs) [25], which form the basis of most of today's variants [17].

Another interesting extension is the CAN [10], which is building up on the DCGAN. Elgammal et al. argue that the classical architecture is merely emulating the training material, but that the reference to and influence by other artists' works is natural. They propose a new architecture, referring to a psychology-based theory from Martindale [21] that says that creative processes always try to evoke arousal, which can for example be achieved through stylistic breaks. At the same time, however, creative work does not want to let this arousal become too great in order to avoid negative reactions. In their architecture, the goal is to generate creative art by finding a balance between mimicking the real art distribution, while deviating from established styles. This should increase the novelty and creativity of the generated images, which was also supported by different experiments with human participants [10]. With regard to a collaborative interactive evolution of art images, this could be advantageous in that the participants are less likely to be guided in their evaluations by well-known representatives of different art styles. Therefore, the architecture of the CAN is used for the generative model in this work. It is noteworthy that the CAN generates rather abstract art images. To obtain more representational art images, the generative model can be replaced.

2.2 Latent Space Exploration

While theoretically every latent vector in a well-trained GAN leads to an output considered by the discriminator as an element of the target group, there is still a great variation between them. This diversity of possible outputs motivates the interest in exploring the latent space and opens up a creative space for human-AI interaction in co-creative processes. In their framework, Grabe et al. [15] make a distinction between four different interaction patterns in human-AI interaction with GANs such as curation, exploration, conditioning, and evolution.

By curation, they mean the most straightforward method, according to which control over the output is achieved through the selection of training data and the subsequent manual selection of generated artefacts. However, brute-forcing new artefacts and manual cherry-picking provides only a low level of control and allows only for random search [28], even if it is actually used this way in the everyday life of some artists [35]. At the same time, the model choice can also be understood as part of curation, for example the choice of the CAN architecture.

Although all four types could be subsumed under the umbrella term exploration, Grabe et al. use it to describe the iterative adaptation of generated artefacts, for example by moving a slider. One possibility is the interpolation between different images, for example CREA.blender [13] allows the exploration of space between different images. Another possibility is the conscious adaptation along either the latent vectors directly or semantic attributes represented in them. Examples of this are found in the alteration of facial features [33,42] as well as in adding, deleting, or altering aspects in images [2,3].

Conditioning refers to methods that allow people to determine desired aspects in advance, often referred to as conditional GAN. Examples are drawings or contour images that predefine the later output, which was used in the domain of fashion design [41] and in the creation of landscape paintings [24]. In addition, instructions can also be given in text form [26,43].

Latent Variable Evolution. Another form of control is offered by evolutionary computing. Grabe et al. [15] refer under evolution only to explicitly interactive setups since in IEC, human feedback directly guides the process of computer-driven exploration. However, already the definition of a target of the evolutionary process as well as the specification of an automatically evaluating fitness function provide a certain degree of control.

Evolutionary computing have already been used several times in combination with GANs, both with automatic and interactive fitness evaluation metrics. The terms Latent Variable Evolution [5,31,40] and, in an interactive setup, Deep Interactive Evolution [4] were coined. The first steps in this direction were taken by Bontrager et al. [5], who applied an evolutionary search to find a latent vector in a GAN trained on real fingerprints that matches as many subjects as possible. Bontrager et al. [4] were also the first to propose an interactive setup for developing faces and shoes towards a target image, thereby demonstrating that GANs can work as a compact and robust genotype-to-phenotype mapping. Again in the field of fashion design, an interactive setup using a conditional GAN with contour images was employed [41]. Zaltron et al. [42] chose an interactive setup with the additional possibility of fine-tuning the faces obtained during the process with the help of sliders. Also in [28] images were developed either by human feedback or by similarity to a target image. It was further shown that genetic algorithms are able to generate diverse sets of latent variables [11], which was also used to augment sparse training datasets [12]. In addition, game levels were developed both automatically [40] and interactively [31]. In a slightly different approach, Roziere et al. [29] used a (1+1) evolution strategy to find local optima in the neighbourhood of generated artefacts including artworks using automatic image quality evaluation metrics.

2.3   Evolutionary Art

Interactive Fitness Evaluation. Evolutionary art can look back on a long history ([27], for an overview). From the beginning, the subjectivity of art has been a major challenge, since the definition of an objective fitness function is not trivial [20,22]. For this reason, in various creative domains such as games, music, and image generation often an IEC approach has been chosen [37]. One problem with IEC, however, is that humans can only evaluate a relatively small number of candidates in a reasonable amount of time and get tired after only a few generations, which is known as user fatigue. Therefore, numerous attempts have already been made to reduce it [37].

One solution to these problems can be provided by crowdsourcing based interactive evolution. In image generation, this approach has been applied by Picbreeder [32], where people can further evolve the evolved images of others online. This approach is also utilised in Artbreeder [34], a creative platform for the creation of GAN-based images, which was inspired by Picbreeder.

The IEC approach proposed in this work involves a collaborative approach differing from those already described in that the fitness is formed based on the ratings of several people. Even though only a small number of the same participants evaluated the images in this study, this approach seems well suited in such a subjective domain.

In general, it opens up interesting possibilities for taking different scores into account and measure the subjectivity. Expanding this approach to different users at different times could also help overcome the problem of user fatigue. An example of this is the Electric Sheep Project [8], ongoing since 1999, in which the fitness is determined by many different users around the world.

Automatic Fitness Evaluation. Given the problems in IEC, there has been plenty of research on automatic aesthetic fitness evaluation metrics [20,22]. Using deep learning methods, great progress has been made for assessing images [38]. The availability of large annotated datasets for aesthetic visual analysis, such as AVA [23], which contains over 250,000 images rated by hobby photographers, have also contributed to these advances.

Even though most datasets consist of photographs, it has already been shown that automatic evaluation metrics can improve artworks for human viewers, at least by using a technical image quality assessment [29]. Despite their less successful results in this way, we chose an aesthetic visual metric, since we assume that art does not only function through its technical quality. The aesthetic quality of the images is determined using Neural Image Assessment (NIMA) [38], which is based on an InceptionResNet-v2 [36] image classifier architecture and is trained on AVA [23].

3   Methods

3.1   Creative Adversarial Network

In this work, the generator part of a GAN is used as genotype-to-phenotype mapping. As explained in Sect. 2.1, the CAN [10] architecture is chosen for this. It aims to generate creative art that mimic the real art distribution, but at the same time deviate from established art styles. To achieve this, in addition to the classification of true and fake images, the discriminator has the further objective of assigning the true images to a certain art style. The generator, on the other hand, still has the goal of ensuring that the generated images are not recognised as fake, but in addition the discriminator should find it as difficult as possible to classify them into a particular art style. Further information and a block diagram illustrating this setup can be found in the original work [10].

Technical Details. The design essentially follows the CAN [10] architecture, which in turn is based on the DCGAN [25] architecture. This consists of a series of strided convolutions for the discriminator and fractional strided convolutions for the generator. Each convolution is followed by a batch normalisation, except in the generator output layer and in the discriminator input layer. The activation function used in the discriminator is Leaky ReLu and in the generator ReLu, only in the generator output Tanh. A special feature of the CAN architecture is that the last strided convolution in the discriminator is followed by two heads. The first determines the probability of coming from the real image distribution using a fully connected layer. The second determines the probability of classification into the different art styles by means of three fully connected layers.

Since we showed the images on screens in 16:9 format, the original square format was converted accordingly. For this purpose, in the second, third, and fourth fractional strided convolution in the generator, the kernel sizes are adjusted from $(4, 4)$ to $(2, 4)$. The latent vectors z that are put under evolutionary control after training are of length 100 and drawn from a standard normal distribution. The exact architecture is thus:

Generator:
$z \in \mathbb{R}^{100} \rightarrow 4 \times 4 \times 1024 \rightarrow 8 \times 6 \times 1024 \rightarrow 16 \times 10 \times 512 \rightarrow 32 \times 18 \times 256 \rightarrow$
$64 \times 36 \times 128 \rightarrow 128 \times 72 \times 64 \rightarrow 256 \times 144 \times 3$ (final resolution)

Discriminator:
$256 \times 144 \times 3 \rightarrow 128 \times 72 \times 32 \rightarrow 64 \times 36 \times 64 \rightarrow 32 \times 18 \times 128 \rightarrow 16 \times 9 \times 256 \rightarrow 8 \times 4 \times 512 \rightarrow 4 \times 2 \times 512$
head 1: $4 \times 2 \times 512 = 4096 \rightarrow 1$
head 2: $4 \times 2 \times 512 = 4096 \rightarrow 1024 \rightarrow 512 \rightarrow K$ (number of art styles)

We used the publicly available WikiArts dataset [39] as training data. It consists of 81,444 artworks from 27 different art styles. The images were normalised and resized to the appropriate resolution. Since this resolution is too low for display, we upsampled the images generated by a factor of eight for human evaluation, resulting in a resolution of 2048×1152. The Laplacian Pyramid Super-Resolution Network (LapSRN) [19] was used for this purpose. The framework is available for experimentation[1].

3.2 Collaborative Interactive Evolution

After training, the generator is able to generate images from every possible latent vector z that follow the distribution of the training images but are difficult to classify into a specific art style. The first generation in the evolution consists of images resulting from randomly generated latent vectors. To not overwhelm the users, IEC classically uses small population sizes and a low number of generations [4,22]. In this work, we choose a population size of 15 and evolve it over 25 generations. In the following, the individual stages of the evolutionary algorithm are outlined. An overview of the algorithm is provided in Fig. 1.

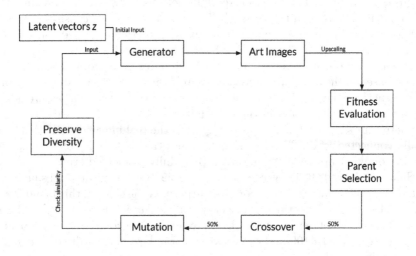

Fig. 1. Block diagram of the evolutionary algorithm.

---

[1] https://github.com/OMHall/CollaborativeArt.

Fitness Evaluation. In the collaborative interactive evolution, five participants rated each image independently on a scale of 1 to 10. For this purpose, a questionnaire with the upsampled images was sent to all participants every generation. They were advised to rate the images independently according to how much they like them, rather than comparing them. The average of these ratings then form the fitness of an image. An excerpt of the questionnaire can be found in the supplementary material in Figure 10.

In the automatic evolution, fitness is evaluated using an automatic aesthetic evaluation by NIMA [38], as outlined in Sect. 2.3. Starting from the same initial population and using the same algorithm otherwise, this allows for a comparison of the results.

Parent Selection. Due to the collaborative evaluation, no direct selection of parents can be performed as is often the case in IEC [4,9,41]. Instead, stochastic universal sampling (SUS) [1] is used for parent selection, whereby 15 parents are selected. The next generation is formed from these through crossover and mutation.

Crossover. Crossover is applied to two randomly chosen parents in 50% of the cases. As in other studies on IEC in combination with GANs, uniform crossover is chosen [4,41], whereby the probability of exchanging individual genes is set to 25%. The rationale behind this is that crossover should lead to interesting new images on the one hand, but on the other hand the selected parents should also be recognisable in order to reduce user frustration due to the loss of good solutions [9].

Mutation. In order to explore the latent space quickly, the mutation probability is also set to 50%, as in [4]. A local search is implemented as mutation, which should both accelerate exploration and further increase the quality of the images in order to keep user fatigue low. This can be understood as intelligent mutation in the sense of a hybrid or memetic evolutionary algorithm [9].

The local search is implemented following Roziere et al. [29], who search for the best possible image in the neighbourhood of a latent vector z. For this, a $(1 + 1)$ evolution strategy [9] is used. In this strategy, there is initially one parent image, from which one offspring is created through mutation of the underlying latent vector z. Based on the results of [29], 1/length of $z = 1/100$ is chosen as individual mutation rate. If a gene is mutated, the mutation comprises the addition of a random number drawn from a standard normal distribution. Survivor selection then takes place on the basis of deterministic elitist replacement, meaning that only the better evaluated image is kept. The quality of the images resulting as phenotypes is automatically evaluated using NIMA [38]. In trade-off between quality increase and preservation of diversity, each local search spans 100 generations, since too many generations tend to result in less preserved diversity [29].

(a)   (b)   (c)   (d)

Fig. 2. Illustration of the *preserve diversity* metric. While the examples (a) and (b) would be judged as too similar, the examples (c) and (d) exhibit sufficient differences.

Preservation of Diversity. Moreover, a metric is introduced that aims to keep the diversity in the population high. On the one hand, this has the goal of exploring the latent space quickly, since in IEC due to user fatigue usually only a small part of the search space is considered. On the other hand, this is intended to avoid user frustration, which can emerge after rating similar images over several generations and a feeling of being in a "blind alley" [9].

The metric chosen is the sum of the absolute distances between individual genes, which is not allowed to fall below a certain threshold. This threshold is set to 25, and examples of the effects of this metric can be found in Fig. 2. After crossover and mutation, this metric is applied to the entire population. Given that two images are too similar, one is replaced by a random immigrant, i.e. an image consisting of a completely new random latent vector. Random immigrants are a way to explicitly increase diversity [16], and have also been used for this purpose in previous work on IEC in combination with GANs [4,41]. In order to increase the quality of the random immigrants, they are first undergoing a local search over 100 generations, as implemented as mutation.

3.3 Evaluation

The final evaluation by human participants of the results obtained with this framework consists of three parts. Each part is composed of pairwise comparisons, where two images are placed next to each other and participants have to decide which one they prefer. In the first part, 20 random images are compared before and after performing the local search to test its effect. In the second and third part, the results of the automatic and collaborative evolution are compared with random images as in the first generation. For this purpose, 10 images each were selected from the hall of fame, the overall best images during the evolutionary process, and randomly combined with the comparison images. Throughout the questionnaire, the order of the images is balanced to avoid position bias. An exemplary excerpt can be found in the supplementary material in Figure 11.

A total of $N = 31$ participants were recruited for the evaluation. Among them were 16 men, 13 women, and one person each who indicated a different gender or did not want to indicate their gender. The participants were between 20 and 87 years old ($M = 31.8, SD = 13.9$). In the evaluation, none of the participants involved in the collaborative evolution took part.

4   Results

4.1   Local Search

Figure 3(a) shows the fitness improvement of 20 random images that underwent local search for 100 generations as it was applied as mutation. These 20 images were also used to evaluate whether local search resulted in quality improvement for human participants.

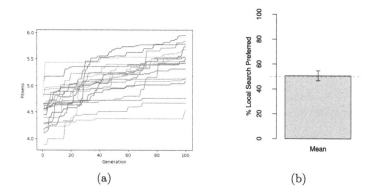

(a)   (b)

Fig. 3. (a) Fitness improvement of 20 random images over 100 generations of local search according to the automatic evaluation metric. (b) Proportion of participants preferring the local search results over the original image, averaged over all 20 images.

Figure 3(b) illustrates the proportion at which the local search was preferred over the original image, averaged over all 20 images. The mean value of 51% shows that the results are preferred only in slightly more than half of the cases, which corresponds to random level. Looking at the individual comparisons, besides many close decisions, local search led to clear improvements in some cases, but also to clear deteriorations in others. The most successful and least successful local searches are exemplified in Fig. 4. Overall, it seems that sharper contours and more intense colours were perceived as improvements, while colour or composition changes received less approval.

(a)   (b)   (c)   (d)

Fig. 4. Images (b) and (d) result from images (a) and (c) using the local search. (b) was preferred in 77% over (a), while (d) was preferred only in 10% over (c).

4.2   Automatic Evolution

The evolution based on the automatic aesthetic evaluation metric resulted in increasing fitness through the generations. This trend can be seen in Fig. 5, which displays results averaged over five runs. The fitness increased particularly at the beginning and seemed to reach a plateau at the end. This trend is presumably supported by the fact that the mutation used also evolved the individual images with respect to the same aesthetic evaluation metric.

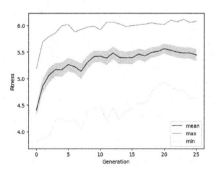

Fig. 5. Mean fitness through generations of the automatic evolution, with the shaded area representing the standard error. The results are averaged over five runs.

All images of the first and last generation are shown in the supplementary material in Figure 12, a visualisation of the entire evolution can be found online[2]. A selection of the best images overall is displayed in Fig. 6. It is apparent that the automatic evolution evolved in the direction of rather blurred images with less clear shapes. With the exception of the image in Fig. 6(d), which has clearer structures and could be reminiscent of a landscape, diffuse contours predominate. Partly, a sky might be recognisable.

Evaluation. The evaluation regarding the attractiveness of the images obtained in the automatic evolution compared to random images as in the first generation showed that human participants preferred the obtained images on average in 49% of the cases, which corresponds to random level. All individual comparisons and the resulting mean can be found in the supplementary material in Figure 13(a). Only two comparisons were clearly in favour of the automatic evolution results, while a majority of comparisons tended towards the random images.

4.3   Collaborative Evolution

The collaborative interactive evolution resulted in increasing fitness through the generations, too. This trend can be seen in Fig. 7(a). In contrast to the automatic

---

[2] Visualisation of the automatic evolution: https://youtu.be/JCRx3Ih_0hA.

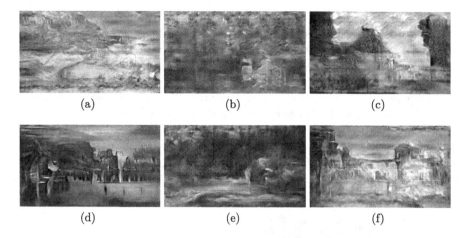

(a)   (b)   (c)

(d)   (e)   (f)

Fig. 6. A selection of the overall best images of the automatic evolution.

evolution, this increase was rather linear and kept rising until the end. Figure 7(b) shows how the automatic aesthetic evaluation metric would have assessed the fitness of the collaborative interactive evolution. Interestingly, similar to the actual automatic evolution in Fig. 5, the fitness is strongly increasing at the beginning and rather stagnant afterwards, which may also be due to the effect of the local search. Compared to the collaborative interactive evolution, there are similarities such as the local minima at generation 20, but the trends also reveal many differences.

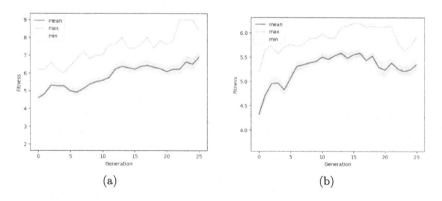

(a)   (b)

Fig. 7. (a) Mean fitness through generations of the collaborative interactive evolution. (b) The collaborative interactive evolution assessed by the automatic aesthetic evaluation metric. The shaded areas represent the standard error.

A total of 17 random immigrants were introduced into the population during the evolutionary process. Only in generation two, two immigrants were inserted, otherwise at most one. Interestingly, the immigrants were rated above average in 14 out of 17 cases and were on average 0.67 above the generation means. This could indicate that novelty was perceived as positive by human participants throughout the evolutionary process.

The collaborative interactive approach further revealed how subjectively art images are perceived. The ratings differed significantly in some cases, with ranges across the entire scale (1–10) and standard deviations of up to $SD = 3.38$. For some images, however, the ratings were also highly similar. Figure 8 shows in (a) the image with the widest range in ratings and in (b) the image with the closest agreement in ratings. In fact, the image was the only one rated equally by all participants with an 8.

   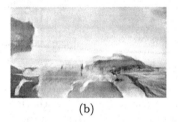

(a)   (b)

Fig. 8. (a) The image with the widest range in ratings (1–10). (b) The image with the closest agreement (all 8).

As with the automatic evolution, all images of the first and last generation are shown in the supplementary material in Fig. 12, and a visualisation of the entire evolution can be found online[3]. A selection of the best images overall is displayed in Fig. 9. In contrast to the automatic evolution, sharper contours prevail. The images are less blurry and the colours appear more diverse, also the images seem to differ more from each other. Many of the images evoke landscape-like associations or are reminiscent of abstract art.

Evaluation. The evaluation regarding the attractiveness of the images obtained in the collaborative interactive evolution compared to random images as in the first generation showed that human participants preferred the obtained images on average in 60% of the cases. Averaged across all ten comparisons and all assessments, an exact binomial test revealed that the results of the collaborative evolution were significantly preferred over random images ($T = 185, p < .001$). All individual comparisons and the resulting mean can be found in the supplementary material in Figure 13(b). Some decisions were close and do not point clearly in one direction considering the standard errors. However, with descriptively eight decisions in favour of the results of the collaborative evolution and in at least three cases a strong preference, a clear tendency is recognisable.

---

[3] Visualisation of the collaborative evolution: https://youtu.be/rG_pLiX_UFo.

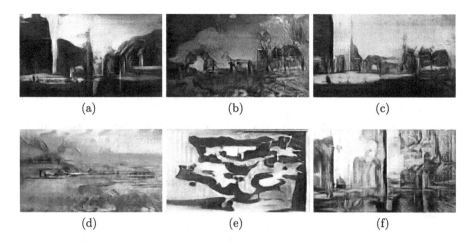

(a)   (b)   (c)

(d)   (e)   (f)

Fig. 9. A selection of the overall best images of the collaborative evolution.

5   Conclusion

In this work, we introduced a novel framework to explore the latent space of possible art images in a GAN using evolutionary computing. It was shown that the developed framework led to an increase in quality over the evolutionary process, using both a collaborative interactive and an automatic aesthetic evaluation metric. The evolutionary algorithm was hybrid, since a local search based on an automatic evaluation metric was incorporated as an intelligent mutation. However, the evaluation by human participants revealed that the local search did not lead to improvements beyond a random level. Furthermore, only the results of the collaborative interactive evolution, but not of the automatic aesthetic evolution were found to be significantly more attractive than randomly generated images. This highlights that automatic aesthetic evaluation of art is challenging and emphasises the importance of human guidance in the evolution of art.

It was demonstrated that the use of the generator part of a GAN as genotype-to-phenotype mapping offers a promising approach for the evolution of art. Throughout all generations, diverse images were generated that can be considered as art images, and the generated images increased in their attractiveness over time. While the already high quality of the randomly generated images makes it more difficult to create significantly more attractive ones in the evolutionary process, it benefits the ongoing interest of users and the prevention of user fatigue. Further, it allows techniques such as random immigrants to be used without leading to substantial fitness loss, and is thus also conducive to accelerated exploration of latent space. The introduced collaborative interactive approach indicated its usefulness due to the subjectivity of art and the positive outcomes, and opens up a multitude of future research possibilities.

References

1. Baker, J.E.: Reducing bias and inefficiency in the selection algorithm. In: Proceedings of the Second International Conference on Genetic Algorithms, pp. 14–21 (1987)
2. Bau, D., et al.: Semantic photo manipulation with a generative image prior. ACM Trans. Graph. **38**, 1–11 (2019)
3. Bau, D., et al.: GAN dissection: visualizing and understanding generative adversarial networks. In: Proceedings of the International Conference on Learning Representations (ICLR), p. 19 (2019)
4. Bontrager, P., Lin, W., Togelius, J., Risi, S.: Deep interactive evolution. In: Liapis, A., Romero Cardalda, J.J., Ekárt, A. (eds.) EvoMUSART 2018. LNCS, vol. 10783, pp. 267–282. Springer, Cham (2018). https://doi.org/10.1007/978-3-319-77583-8_18
5. Bontrager, P., Roy, A., Togelius, J., Memon, N., Ross, A.: DeepMasterPrints: generating masterprints for dictionary attacks via latent variable evolution. In: 2018 IEEE 9th International Conference on Biometrics Theory, Applications and Systems (BTAS), pp. 1–9 (2018)
6. Cetinic, E., She, J.: Understanding and creating art with AI: review and outlook. ACM Trans. Multimedia Comput. Commun. Appl. (TOMM) **18**(2), 1–22 (2022)
7. Cohn, G.: AI art at Christie's sells for $432,500 (2018). https://www.nytimes.com/2018/10/25/arts/design/ai-art-sold-christies.html
8. Draves, S.: The electric sheep screen-saver: a case study in aesthetic evolution. In: Rothlauf, F., Branke, J., Cagnoni, S., Corne, D.W., Drechsler, R., Jin, Y., Machado, P., Marchiori, E., Romero, J., Smith, G.D., Squillero, G. (eds.) EvoWorkshops 2005. LNCS, vol. 3449, pp. 458–467. Springer, Heidelberg (2005). https://doi.org/10.1007/978-3-540-32003-6_46
9. Eiben, A.E., Smith, J.E.: Introduction to Evolutionary Computing. Springer, Berlin (2003). https://doi.org/10.1007/978-3-662-05094-1
10. Elgammal, A., Liu, B., Elhoseiny, M., Mazzone, M.: CAN: creative adversarial networks, generating "art" by learning about styles and deviating from style norms. arXiv:1706.07068 (2017)
11. Fernandes, P., Correia, J., Machado, P.: Evolutionary latent space exploration of generative adversarial networks. In: Castillo, P.A., Jiménez Laredo, J.L., Fernández de Vega, F. (eds.) EvoApplications 2020. LNCS, vol. 12104, pp. 595–609. Springer, Cham (2020). https://doi.org/10.1007/978-3-030-43722-0_38
12. Fernandes, P., Correia, J., Machado, P.: Towards latent space exploration for classifier improvement. In: 24th European Conference on Artificial Intelligence (ECAI 2020) - ADGN 2020: First Workshop on Applied Deep Generative Networks (2020)
13. Gajdacz, M., et al.: CREA.blender: a GAN based casual creator for creativity assessment. In: Proceedings of the International Conference on Computational Creativity. ICCC, p. 5 (2021)
14. Goodfellow, I., et al.: Generative adversarial nets. In: Advances in Neural Information Processing Systems, vol. 27 (2014)
15. Grabe, I., Duque, M., Risi, S., Zhu, J.: Towards a framework for human-AI interaction patterns in co-creative GAN applications. In: Joint Proceedings of the ACM IUI Workshops (2022)
16. Grefenstette, J.J.: Genetic algorithms for changing environments. In: Proceedings of Parallel Problem Solving from Nature, pp. 137–144 (1992)

17. Gui, J., Sun, Z., Wen, Y., Tao, D., Ye, J.: A review on generative adversarial networks: algorithms, theory, and applications. IEEE Trans. Knowl. Data Eng. **35**(4), 3313–3332 (2023)
18. Karras, T., Aila, T., Laine, S., Lehtinen, J.: Progressive growing of GANs for improved quality, stability, and variation. In: International Conference on Learning Representations (2018)
19. Lai, W.S., Huang, J.B., Ahuja, N., Yang, M.H.: Deep Laplacian pyramid networks for fast and accurate super-resolution. In: IEEE Conference on Computer Vision and Pattern Recognition (2017)
20. Machado, P., Romero, J., Manaris, B.: Experiments in computational aesthetics: an iterative approach to stylistic change in evolutionary art. In: Romero, J., Machado, P. (eds.) The Art of Artificial Evolution: A Handbook on Evolutionary Art and Music. NCS, pp. 381–415. Springer, Heidelberg (2008). https://doi.org/10.1007/978-3-540-72877-1_18
21. Martindale, C.: The Clockwork Muse: The Predictability of Artistic Change. Basic Books (1990)
22. McCormack, J.: Facing the future: evolutionary possibilities for human-machine creativity. In: Romero, J., Machado, P. (eds.) The Art of Artificial Evolution: A Handbook on Evolutionary Art and Music. NCS, pp. 417–451. Springer, Heidelberg (2008). https://doi.org/10.1007/978-3-540-72877-1_19
23. Murray, N., Marchesotti, L., Perronnin, F.: AVA: a large-scale database for aesthetic visual analysis. In: IEEE Conference on Computer Vision and Pattern Recognition, pp. 2408–2415 (2012)
24. Park, T., Liu, M.Y., Wang, T.C., Zhu, J.Y.: Semantic image synthesis with spatially-adaptive normalization. In: Proceedings of the IEEE/CVF Conference on Computer Vision and Pattern Recognition, pp. 2337–2346 (2019)
25. Radford, A., Metz, L., Chintala, S.: Unsupervised representation learning with deep convolutional generative adversarial networks. arXiv:1511.06434 (2015)
26. Reed, S., Akata, Z., Yan, X., Logeswaran, L., Schiele, B., Lee, H.: Generative adversarial text to image synthesis. In: ICML (2016)
27. Romero, J.J., Machado, P. (eds.): The Art of Artificial Evolution: A Handbook on Evolutionary Art and Music. Springer, Heidelberg (2008). https://doi.org/10.1007/978-3-540-72877-1
28. Rozière, B., Riviere, M., Teytaud, O., Rapin, J., LeCun, Y., Couprie, C.: Inspirational adversarial image generation. IEEE Trans. Image Process. **30**, 4036–4045 (2021)
29. Roziere, B., et al.: EvolGAN: evolutionary generative adversarial networks. In: Proceedings of the Asian Conference on Computer Vision (2020)
30. Sbai, O., Elhoseiny, M., Bordes, A., LeCun, Y., Couprie, C.: DeSIGN: design inspiration from generative networks. In: Leal-Taixé, L., Roth, S. (eds.) ECCV 2018. LNCS, vol. 11131, pp. 37–44. Springer, Cham (2019). https://doi.org/10.1007/978-3-030-11015-4_5
31. Schrum, J., Gutierrez, J., Volz, V., Liu, J., Lucas, S., Risi, S.: Interactive evolution and exploration within latent level-design space of generative adversarial networks. In: Proceedings of the 2020 Genetic and Evolutionary Computation Conference, pp. 148–156 (2020)
32. Secretan, J., Beato, N., Ambrosio, D.B.D., Rodriguez, A., Campbell, A., Stanley, K.O.: Picbreeder: evolving pictures collaboratively online. In: Proceedings of the SIGCHI Conference on Human Factors in Computing Systems, pp. 1759–1768 (2008)

33. Shen, Y., Gu, J., Tang, X., Zhou, B.: Interpreting the latent space of GANs for semantic face editing. In: IEEE/CVF Conference on Computer Vision and Pattern Recognition (CVPR), pp. 9240–9249 (2020)
34. Simon, J.: Artbreeder (2018). https://www.artbreeder.com
35. Spratt, E.L.: Creation, curation, and classification: Mario Klingemann and Emily L. Spratt in conversation. XRDS: Crossroads ACM Mag. Stud. 24(3), 34–43 (2018)
36. Szegedy, C., Vanhoucke, V., Ioffe, S., Shlens, J., Wojna, Z.: Rethinking the inception architecture for computer vision. In: Proceedings of the IEEE Conference on Computer Vision and Pattern Recognition, pp. 2818–2826 (2016)
37. Takagi, H.: Interactive evolutionary computation: fusion of the capabilities of EC optimization and human evaluation. Proc. IEEE 89(9), 1275–1296 (2001)
38. Talebi, H., Milanfar, P.: NIMA: neural image assessment. IEEE Trans. Image Process. 27(8), 3998–4011 (2018)
39. Tan, W.R., Chan, C.S., Aguirre, H., Tanaka, K.: Improved ArtGAN for conditional synthesis of natural image and artwork. IEEE Trans. Image Process. 28(1), 394–409 (2019)
40. Volz, V., Schrum, J., Liu, J., Lucas, S.M., Smith, A., Risi, S.: Evolving Mario levels in the latent space of a deep convolutional generative adversarial network. In: Proceedings of the Genetic and Evolutionary Computation Conference, p. 221–228 (2018)
41. Xin, C., Arakawa, K.: Object design system by interactive evolutionary computation using GAN with contour images. In: Zimmermann, A., Howlett, R.J., Jain, L.C., Schmidt, R. (eds.) KES-HCIS 2021. SIST, vol. 244, pp. 66–75. Springer, Singapore (2021). https://doi.org/10.1007/978-981-16-3264-8_7
42. Zaltron, N., Zurlo, L., Risi, S.: CG-GAN: an interactive evolutionary GAN-based approach for facial composite generation. In: Proceedings of the AAAI Conference on Artificial Intelligence, pp. 2544–2551 (2020)
43. Zhu, S., Fidler, S., Urtasun, R., Lin, D., Loy, C.C.: Be your own prada: fashion synthesis with structural coherence. In: International Conference on Computer Vision. ICCV (2017)

Towards Sound Innovation Engines Using Pattern-Producing Networks and Audio Graphs

Björn Þór Jónsson[1,2](✉) [ID], Çağrı Erdem[2] [ID], Stefano Fasciani[3] [ID], and Kyrre Glette[1,2] [ID]

[1] RITMO Centre for Interdisciplinary Studies in Rhythm, Time and Motion, University of Oslo, Oslo, Norway
{bthj,kyrrehg}@uio.no
[2] Department of Informatics, University of Oslo, Oslo, Norway
cagrie@uio.no
[3] Department of Musicology, University of Oslo, Oslo, Norway
stefanof@uio.no
https://www.uio.no/ritmo

Abstract. This study draws on the challenges that composers and sound designers face in creating and refining new tools to achieve their musical goals. Utilising evolutionary processes to promote diversity and foster serendipitous discoveries, we propose to automate the search through uncharted sonic spaces for sound discovery. We argue that such diversity promoting algorithms can bridge a technological gap between the theoretical realisation and practical accessibility of sounds. Specifically, in this paper we describe a system for generative sound synthesis using a combination of Quality Diversity (QD) algorithms and a supervised discriminative model, inspired by the Innovation Engine algorithm. The study explores different configurations of the generative system and investigates the interplay between the chosen sound synthesis approach and the discriminative model. The results indicate that a combination of Compositional Pattern Producing Network (CPPN) + Digital Signal Processing (DSP) graphs coupled with Multi-dimensional Archive of Phenotypic Elites (MAP-Elites) and a deep learning classifier can generate a substantial variety of synthetic sounds. The study concludes by presenting the generated sound objects through an online explorer and as rendered sound files. Furthermore, in the context of music composition, we present an experimental application that showcases the creative potential of our discovered sounds.

Keywords: Sound Synthesis · Quality Diversity Search · Innovation Engines

Supported by the Research Council of Norway through its Centres of Excellence scheme, project number 262762.

C. Johnson et al. (Eds.): EvoMUSART 2024, LNCS 14633, pp. 211–227, 2024.
https://doi.org/10.1007/978-3-031-56992-0_14

1   Introduction

Either you know what sound you're looking for, or you don't know what sound you're looking for. In the latter case, inquiry, or prompting, is impossible. To discover new sounds, you must recognize them when you have found them. But if you can do that, you must have known them already. Transferring such a paraphrasing [30] of Meno's Paradox to the domain of novel sound design can be a way of establishing the usefulness of serendipitous sonic discoveries, where a new sound may not have been explicitly sought after but immediately recognised when heard. With all sound admissible as material for making music and all sounds theoretically possible with digital synthesis, there is still much more to explore considering the entirety of the sonic domain [38]. Composers and sound designers often need to create and refine new tools in order to achieve their musical goals. This endeavour may be hindered by a lack of technical expertise. Our proposed approach leverages evolutionary processes to generate novel sounds, thereby facilitating the creative journey and overcoming the technical barriers that may limit composers and sound designers in expanding their sonic repertoire.

We work towards an approach to automate navigation through previously unexplored sonic territories. As such, while entirely novel, the discovered sounds can be perceived as appealing and seemingly recognisable to the listener despite their unprecedented nature. Such investigations have been carried out interactively in the visual domain [34], demonstrating the usefulness of abandoning specific objectives, or at least switching goals as stepping stones are found while traversing paths to interesting discoveries. These findings provided a basis for proposing the Novelty Search algorithm [19] and later other variants, forming a family of Quality Diversity (QD) search algorithms [3,20,25,32]. These QD algorithms combine the open-endedness of Novelty Search with competition between solutions in their own "niche", resulting in diverse and high-performing (quality) solutions. Overall, QD algorithms serve as effective tools for illuminating high-quality solutions within a domain and are powerful search algorithms in their own right. This is due in part to their ability to exploit behavioral diversity and stepping stones during the search process, which can lead to discovering a variety of valuable solutions [7,29]. To drive automated exploration with such diversity-promoting algorithms, the Innovation Engine algorithm abstracts the process of human curiosity, replacing human judgement with a discriminative model that identifies interesting ideas [26,28]. Innovation Engines integrate two key components: Evolutionary Algorithms (EAs), such as those from the family of QD, capable of generating and gathering various novel outputs; and a model capable of distinguishing that novelty and evaluating its quality, such as Deep Neural Network (DNNs), creating niches and competition within them, thus providing selection pressure to guide QD search. The ultimate goal of this architecture is to continuously generate interesting and innovative creations in any given field.

Compositional Pattern Producing Networks (CPPNs) [35] are a foundation of the explorations leading to the Novelty Search and Innovation Engine algorithms. The networks abstract unfolding development in evolutionary processes, which build a phenotype over time. This is done by using any variety of canonical functions at each node, based on the idea that the order in which the networks compose functions can provide that abstraction. This can be compared with the process of timbral development, where musical expression depends on changes and nuances over time. The use of patterns produced by CPPNs as sources of sound- and control signals for sound synthesis has been explored in a novelty seeking Interactive Evolutionary Computation (IEC) [37] configuration, which was inspired by previous work on the generation of visual artefacts [16]. The representation of temporal unfolding provided by CPPNs has been combined with the evolution of Digital Signal Processing (DSP) graphs during several iterations of investigation, detailed in [14]. This resulted in a distinct approach to sound synthesis, where any combination of the two graphs, depicted in Fig. 1, can be rendered at any duration, revealing the sub-patterns encoded by CPPNs over varying periods of time.

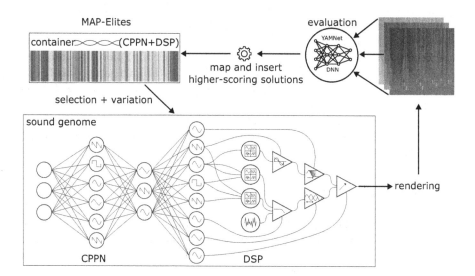

Fig. 1. The QD algorithm MAP-Elites uses the pre-trained YAMNet DNN classifier to define cells in a container and the performance of an evaluation candidate across those cells to determine placement and replacement in that archive. The genome of each evaluation candidate is rendered to a waveform, which is supplied to the classifier. Inputs to the CPPN are discussed in Sect. 2.

Given the more diverse application of sonic artefacts as material in creative processes, we argue that there may be an even higher incentive to investigate the Innovation Engine algorithm's applicability in the sound domain. Furthermore, whereas humans can evaluate images in a split second, evaluation of sounds requires more time. There is a minimal duration threshold for perceiving salient features of sonic objects [24] and we typically perceive them holistically as meaningful units in the 0.5 to 5 s range [10]. Experiments with interactive evolution of sounds [16] revealed how fatigue can set quickly in when potentially listening to a long series of taxing sounds. This further limits the ability of humans to provide sufficient quantity of selection to have a significant effect on evolution.

Automating the discovery of new sounds is the goal of this study. We achieve this by applying the Innovation Engine algorithm to the sound synthesis approach developed in previous research on interactive novelty discovery. By using the proposed technique for sound synthesis, the system does not need to be trained beforehand as the evolutionary method starts from networks with no hidden nodes and progressively evolves primitive individuals by adding nodes and connections with the NeuroEvolution of Augmenting Topologies (NEAT) algorithm [36]. In our initial experiments, we use a signal from a pre-trained discriminative model to guide QD search, without human feedback in the evolutionary loop. Investigating this setup is intended to pave the way for further explorations of unbounded discovery of interesting sounds.

Our contributions include researching the application of a special type of Innovation Engine in the sound domain with a distinct approach to sound synthesis within an EA. Furthermore, we examine different configurations of our generative system and study how our sound synthesis method interacts with the discriminative model. We also offer a web-based interface to explore the outcomes of our evolutionary processes through our Innovation Engine setup. Lastly, we showcase audio artefacts rendered from the solutions discovered during the QD runs. Experimental results, in the form of historical data from evolution runs, elite maps and genomes from each point in time, and sounds rendered from those genomes at final iterations, along with the source code to replicate the results, are available in the dataset accompanying this article [15].

2   Approach and Experimental Setup

To start evaluating the applicability of the Innovation Engine algorithm in the domain of sounds, we combine a sound synthesis technique with a supervised discriminative model. The foundation of our sound-generating system relies on using the patterned outputs from CPPNs as the raw materials for sound and control signals. These signals can be utilised in their original form or further shaped through a DSP graph. Such a design choice enables the evolutionary state to begin from a blank slate, established with random initialization of the CPPN and DSP graph counterparts. This avoids dataset constraints that might limit the potential for discovery of novel sounds. The genome evolved by the evolutionary (QD) processes is composed of the CPPN and DSP networks and the evolvable connections between them. Details of this genome configuration are discussed and diagrammed in [14]. Figure 1 illustrates the data flow of our experimental setup and shows how the genome fits within the data pipeline.

Behavioural Descriptor. To guide the QD search, we chose the Yet Another Mobile Network (YAMNet) DNN classifier to define our search space. The confidence scores output by the classifier for each class are used as selection signals for the QD algorithm, as discussed in Sect. 3.1. While this pre-trained network may limit our exploration, it was adopted in an effort to replicate a setup from previous evaluations of the Innovation Engine algorithm in the visual domain. That classifier is trained on AudioSet [9], which can be considered as a sonic sibling of the DNN classifiers trained on the ImageNet dataset [5]. YAMNet outputs 521 scores from a logistic (softmax) layer, corresponding to AudioSet classes. The classifier's output is intended "as a stand-alone audio event classifier that provides a reasonable baseline across a wide variety of audio events."[1]. Our approach to sound generation can be somewhat likened to a unique type of sound synthesiser, which is not crafted with the intention of mimicking natural sounds or creating textures that easily fit into well-known categories. Many modern generative models excel at such tasks [1], building on their prior training, but we considered the varied signal provided by this model as a good starting point for driving the EA towards diversity. We also considered it interesting to mirror the overall setup from experiments [18,26,28] that inspire our sonic investigations.

Periodic Signal Composition. One factor potentially influencing the search space is our choice of CPPN activation functions and node types in the DSP graph. CPPNs have commonly been used to compose Gaussian, sigmoid, and periodic functions, such as in [34,35]. In our case, the pattern-producing network can only compose periodic functions, commonly used as oscillators in a variety of sound synthesis techniques: sine, square, triangle, and sawtooth. The node types in the DSP graph are the same as in [33], in addition to custom nodes, which were added to the repertoire in an effort to widen the search space. Those additional nodes are a wavetable and a specialised additive synthesis node, where multiple audio signals are sourced from the CPPN to fill a table in the former and represent partials or harmonics in the latter. The wavetable is traversed according to a control signal, also sourced from the CPPN, in a manner similar to vector synthesis. The partials in the additive synthesis node can be slightly inharmonic, according to a mutable parameter to each.

The duration of sounds rendered from each genome is defined by a linear ramp of values from -1 to 1 supplied to one CPPN input, while the pitch is controlled by the rate of a periodic (sine) signal at another input. Velocity is intended to simulate stimuli of different intensities when interacting with physical instruments, which is achieved by scaling the sine wave input by a velocity factor. The inputs are sampled at the same rate as the sampling rate of the audio graph.

QD Algorithm. For the diversity-promoting algorithm, we chose Multidimensional Archive of Phenotypic Elites (MAP-Elites) [25]. Our experiments are based on a bespoke implementation of that algorithm, with the common addition of biasing it away from exploring niches that produce fewer innovations. This is achieved by assigning each niche a decrementing counter, representing a *curiosity score* as defined in [3] with constants set as in [18]. The counters start

---

[1] YAMNet audio event classifier: https://tfhub.dev/google/yamnet.

at a fixed value of 10, impacting the probability of that niche being selected for reproduction. The classification outputs of the discriminative model define the cells of the behaviour space which the QD algorithm explores, where the performance at each niche is determined by the confidence values for each class. During our main runs of QD search, evaluations were performed in batches of 32.

Parameter Search. Considering the temporal dynamics of sounds, and that the underlying pattern generator of our sound synthesis engine (CPPN) encodes sub-patterns that reveal over time, we performed preliminary experiments classifying sounds rendered at a different duration for each evaluation. One of the configurations involved 112 evaluations of each sound genome, rendering it to sounds of 4 durations, 7 pitch variations and 4 amplitudes. To explore other parameters of the QD search, such as mutation rates and their balance between the CPPN and DSP genome counterparts, as well as graph and node addition or deletion rates, we conducted a manual parameter search. Due to the computationally intensive nature of the task, these runs were based on a limited selection of parameter values. A comprehensive collection of plots from those runs can be found in the dataset accompanying this paper [15]. We found that evaluating sounds with a duration of half a second frequently led to the emergence of successful sound variants. Therefore, we decided to use this specific duration for assessing sounds in the QD runs of our primary experiments. Runs with node- and connection addition rates of 10% and corresponding deletion rates of 6% resulted in the best performance during our parameter search. As such, we ran with that as our baseline configuration along with equal probability of mutating each genome counterpart.

3   Results

For our main experiment, we ran 10 independent runs of the MAP-Elites algorithm, with the rates discussed in Sect. 2 and behaviour evaluated by YAMNet on 0.5-s sounds. Each run lasted for 300 thousand iterations, with a batch or generation size of 32. At the start of each run, 50 seed iterations were performed, which differ from the rest of the iterations in that each individual is initialised from scratch rather than mutating a randomly selected elite occupying any of the cells.

3.1   Sound Generation - and QD Algorithm Variants

Aside from the set of evolution runs using our basic configuration described above, we performed two additional sets of runs. In one set, we altered the sound generation, and in the other set, we modified the progression of the QD algorithm.

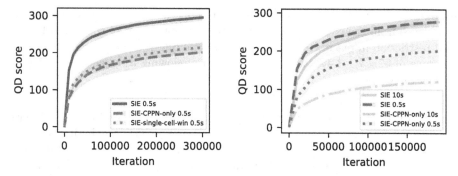

Fig. 2. On the *left* scores are plotted for the baseline configuration (Sect. 2), evaluating sounds of 0.5 s duration, along with variants where sounds are only rendered from CPPN mutations and where only one cell can be won at a time by each candidate elite. Data for each variant comes from 10 runs. The plot on the *right* compares the performance achieved when evaluating sounds rendered at two different durations—0.5 s and 10 s—from the baseline and CPPN-only run configurations, each independently executed 5 times.

Signal Processing Graph. To investigate the impact of merging CPPNs with DSP graphs, we set up evolutionary runs in two distinct configurations: one in which an evolved CPPN functioned solely as the audio signal source, providing a single output, and another where the CPPN was paired with an evolving DSP graph, allowing it to offer a multitude of audio and control signals, from up to 18 outputs.

In our experiments, we quantify the QD algorithm performance by calculating the QD-score [31, 32]. This score is determined by summarising the confidence levels of the elites across the various classes delineated by YAMNet. When comparing the results from these runs, we observe in Fig. 2 that the phenotypes (i.e., sound objects) produced from the genomes where CPPNs and DSP graphs were co-evolved achieved the highest overall QD-score. Through informal listening sessions conducted by the authors, it was observed that the sounds rendered from runs where the evolution of DSP graphs was allowed alongside CPPNs exhibited a higher degree of subjective aesthetic appeal. This phenomenon could potentially be attributed to the prevalence of classical synthesizer sounds, to which our ears have grown accustomed. In this context, the DSP graph can be seen as functioning akin to a modular synthesizer patch, rendering us less inclined to perceive the raw output generated by CPPNs as inherently pleasing. The rendered sounds can be auditioned in an online explorer (Sect. 3.7) or accompanying dataset [15].

Behaviour Space Coverage. The default behaviour of our MAP-Elites implementation allows each evaluated individual to win all cells where it performs better or where there is a vacancy, so it reaches full coverage from the first seed. To examine the effect of gradually covering the map of cells by allowing each candidate to potentially win only one cell, the one where it receives the highest confidence from the classifier, we performed an identical set of runs except with that restriction in place. Runs where at most one cell at a time is won reached a coverage of 57.4% ± 3.4%, with their QD-score following a trajectory similar to that of full coverage CPPN-only runs, as depicted in Fig. 2, *left*.

Elite Populations. Figure 3 (*left*) shows that the range of iterations where the current elites are found at the end of each run is sharply delimited around iterations 150K to 250K of the CPPN-only runs, while the CPPN+DSP runs continue to discover new elites more gradually throughout the latter half of the runs.

The set of unique elites at the end of CPPN-only runs is smaller than when co-evolving the DSP graphs, as plotted in Fig. 3 (*right*). Instead of distinguishing between individuals by their ID, where the differences could be only slight changes in e.g. connection weights, this plot is based on distinction between unique combinations of CPPN and DSP node and connection counts.

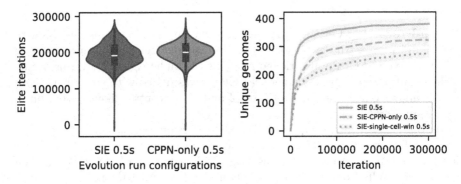

Fig. 3. *Left*: Distribution of iteration numbers at which the current class elite was discovered. *Right*: Count of unique individuals, as it evolves through iterations of the evolution runs.

3.2 Genome Complexity

The composition of audio graph nodes and CPPN activation functions can be seen in Fig. 4, where the prominence of the custom audiograph nodes (wavetable and additive synthesis, Fig. 4, *bottom*) suggest that implementing other known techniques from the history of sound synthesis may be worthwhile. The distribution of CPPN activation function types is quite uniform in all variants of our runs (Fig. 4, *top*). It's also interesting to observe in the left plot of Fig. 6 that the CPPN-only runs resulted in more complex function compositions, likely to compensate for the lack of a co-evolving DSP graph. This increased CPPN complexity resulted in longer rendering times and thus increased durations of the evolution runs, as that part of the genome is more computationally expensive, with potentially many network activations required for each sample, as discussed in [14].

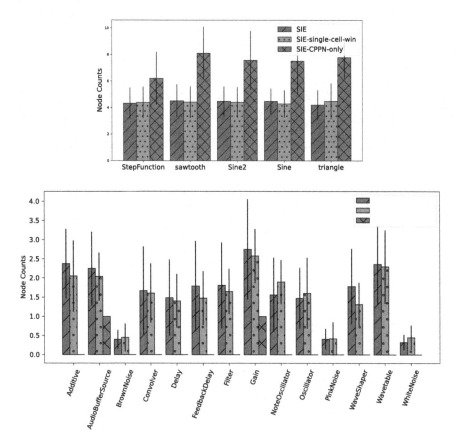

Fig. 4. Composition of CPPN activation functions (*top*) and DSP graph node types (*bottom*), from the different evolution run variants. It can be observed in the DSP chart that the CPPN-only variant does not evolve a DSP graph and only includes a source node for receiving the pattern-signal from the single CPPN output, and a gain node for passing it through to the output.

3.3 Performance Against Pre-trained Reward Signals

The YAMNet classifier chosen in this iteration of our investigations assigned high scores to the sounds generated by our system across most classes, as can be seen in Fig. 5. There we can see again how the co-evolution of CPPNs with DSP graphs achieves higher scores overall. The figure also reveals how the synthesiser struggles in the range of classes between 214 and 276, which classify musical genres, rather than distinct sounds or instruments, such as "Pop music", "Rhythm and blues", "Flamenco", etc. This is reasonable as the system is expected to generate sounds useful in the process of creating e.g. music, rather than entire musical compositions. Nonetheless it can be interesting to observe what the system came up with for those low-confidence classes, such as "Theme music": a filter can be set in the online explorer (Sect. 3.7) to audition classes containing the phrase "music" while scrubbing through the runs with a slider.

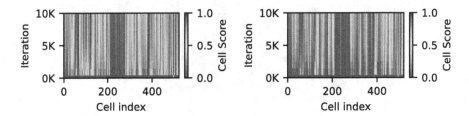

Fig. 5. Confidence scores declared by the YAMNet DNN, pre-trained on AudioSet classes (x-axis), averaged from the first 100 thousand iterations of 10 runs. Results from a set of runs where both CPPN and DSP genome counterparts are evolved can be seen on the *left* while the *right* map shows results from a set of runs restricted to evolution of the CPPN part of the genome, without evolving signal processing nodes.

3.4   Evolutionary Stepping-Stones

To assess how evolution leveraged the diversity promoted by our classifier, we conducted two measurements that explored the stepping stones across various classes. One has been called *goal switching* and defined as "the number of times during a run that a new class champion was the offspring of a champion of another class" in [26,28]. From our runs we measured a mean of 21.7 ± 3.6 goal switches, 63.2% of the 34.3 ± 4.5 mean new champions per class. This can be compared to the 17.9% goal switches reported in [26]. Another way of measuring how the evolutionary paths flow though the stepping stones laid out by the classifier is to trace through the phylogenetic tree leading to each elite and then count how often its parent comes from a class different from the one it occupies. Counting from the current elites of each class at the end of the evolution runs, we found a mean of 44.9 ± 14.7 such occurrences. In lieu of a visual phylogenetic presentation, the *generation* slider of the evolution runs explorer (Sect. 3.7) can dynamically reveal how elites for each class come from different, often unrelated classes during the course of evolution.

3.5   Abandoning Diversity

Growth of genome complexity seems to have stayed within reasonable limits, even when CPPNs were left alone to the task of performing against the classifier (Fig. 6). An exception to this is when we experimented with abandoning

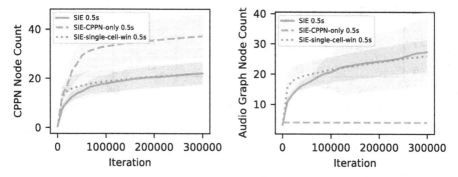

Fig. 6. Genome complexity over the course of 10 QD runs for each variant. CPPN node counts are plotted on the *left* and DSP graph node counts can be seen in the plot on the *right* (a flat line for the CPPN-only variant indicates that the DSP graph is not evolved.

diversity and adopting single objectives. Though the benefit of diversity has been demonstrated [26,28], we investigated how a similar experiment fares in the sound domain. To that end, we selected 10 classes[2] as single objectives of separate runs and compared the performance and genome complexity with the performance from the QD runs on those same classes.

Interestingly, although the performance in single objective runs is higher on average than in multi-class runs, as shown in the *first* plot in Fig. 7, the difference is accompanied by a higher level of deviation and much higher genome complexity. The *second* and *third* plots in Fig. 7 indicate that the CPPN and DSP node counts in genomes from single objective runs are significantly higher than those of genomes from the same set of classes in QD runs. The computational effort required for the complex genomes evolved during the single class runs limited our iteration count to 50 thousand, 1/6th of the iterations performed for the baseline QD runs. The unexpected result of higher performance from the single-class runs may be attributed to the narrow set of chosen classes; this experiment could benefit from further investigation.

3.6 Temporal Pattern Revelation and Classifier Characteristics

Although half a second sounds were the most prevalent renditions of successful individuals in our manual parameters search (Sect. 2), comparing sets of runs with two large variations in the duration of the evaluated phenotypes was interesting. We chose to compare runs evaluating half a second renditions of the evolved genomes with a set of runs evaluating ten-second renditions. One motivation for the choice of the longer duration, is that "YAMNet is trained on

---

[2] Single-class runs were performed on the classes Aircraft, Banjo, Beatboxing, Boom, Choir, Dubstep, Fusillade, Mandolin, Synthetic singing and Whistling.

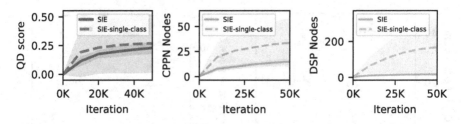

Fig. 7. The *first* plot shows performance scores from single-objective vs. multi-objective QD runs, averaged from a set of randomly selected classes. The *second* and *third* plots show how genome complexity developed during single- and multi-objective QD runs, in terms of CPPN and DSP graph node counts.

1,574,587 10-s YouTube soundtrack excerpts from within ... AudioSet"[3]. While CPPN-only runs achieved less overall confidence when rendering 0.5 s sounds for evaluation by the classifier, as can be seen on the *left* of Fig. 2, we hypothesised that allowing the classifier to sample in more detail the patterns developed by the CPPNs, when processing more frames over a longer duration, would result in higher confidence. The opposite turned out to be the case, where CPPN-only runs, rendering 10 s sounds for evaluation achieved a lower QD score than corresponding runs rendering 0.5 s sounds. Perhaps the lack of DSP becomes more significant in the evaluation of longer duration sounds. Duration has little effect when DSP graphs evolve alongside the CPPNs, as the *right* plot in Fig. 2 shows.

3.7 Access to Sound Objects and Their Application

We have facilitated open access to the generated artifacts through different means. Those include an evolution runs explorer[4], depicted in Fig. 8a. Final elites from all runs have also been rendered to (128563) WAV files, which have been included in the accompanying dataset [15]. The sound objects in the pre-rendered files reflect the render-settings used to evaluate the corresponding genome that became an elite. The online explorer (See footnote 4) provides greater flexibility as it dynamically renders sounds with the default settings, but the interface also enables users to modify these settings. This modification can potentially reveal other intriguing sonic behaviors from the same genome.

As part of our investigation into the applicability of the discovered artefacts for creating other art, we loaded subsets of them into the experimental sampler AudioStellar [8] and used that software to drive evolutionary sequences through the phenotypes. A playlist of live-stream recordings showcasing evolutionary sequences using sounds discovered by QD runs is accessible online[5]. These compositions are largely automated, with human input limited to initial settings

---

[3] YAMNet release announcement: https://groups.google.com/g/audioset-users/c/U71MxTdHqkU.

[4] Evolution runs explorer: https://synth.is/exploring-evoruns.

[5] Playlist with evolutionary sequences through sounds discovered by QD runs: https://youtube.com/playlist?list=PLSYAaR-xYhEXk0czfHYKJSWmZ8vG35xEN.

like evolutionary sequencing rates and fundamental sound effects. Nonetheless, they demonstrate the potential of the discovered sound objects to inspire creative endeavors. It is thought-provoking to consider if a human, given the same dataset, could craft more aesthetically pleasing arrangements with these sonic artefacts. We encourage the reader to obtain a copy of the files and engage in such experimentation [15].

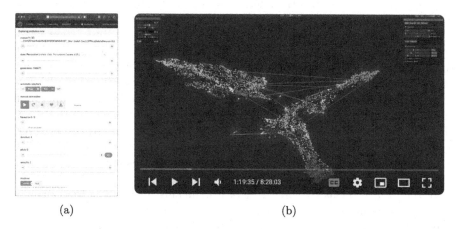

(a)   (b)

Fig. 8. (a) Evolution runs explorer, where it is possible to scrub through evolution runs, their classes and generations. The sound properties duration, pitch and velocity can be changed and favourites can be collected. (b) Live streams (recorded) of automated, evolutionary sequences through sounds rendered from the evolutionary runs discussed in this paper, as one way of experiencing and qualitatively evaluating the generated artefacts. The sequencing is performed by the experimental sampler AudioStellar.

4   Conclusion and Future Work

Applying the combination of a diversity-promoting algorithm with selection pressure from a classifier reward signal to the search for sounds has been demonstrated to be a viable approach by the results discussed in this paper. Furthermore, the distinct approach to sound synthesis employed in this work has achieved high confidence from a DNN based classifier in most classes. High-scoring sounds are, in many cases, not the most realistic representatives of their class, especially when considering non-musical instrument classes, which can be attributed to how DNNs are easily fooled [27]. Other recently proposed classification approaches may be more robust and could be worth investigating [11,13], but classification robustness may not be the most sought-after quality in a creative system. With a focus on the diversity-promoting attribute of the selection pressure applied in this investigation, the diverse and innovative sound objects generated suggest that further explorations may be based on this system. The intent would be to broaden the range of potential discoveries within the sonic domain.

Adopting YAMNet as a classifier for sounds, to provide selection pressure for a QD algorithm, was a step towards investigating a simple version of the Innovation Engine algorithm in the domain of sounds. Further explorations may include expanding the behaviour space to search beyond predefined classes. This can be done by combining the feature extraction ability of a DNN, such as the one employed in this work, with dimensionality reduction (DR), as has been done in the visual domain in [22]. Extracting features with Variational Auto Encoders (VAEs) [17], and applying a clustering algorithm in the resulting latent space to define niches, as stepping stones during QD search, is another approach [23] worth exploring further in the domain of sounds. While VAE require a training set, limiting the behaviour space to explore, periodically retraining a DR algorithm on discovered sound objects could enable autonomous and unsupervised discovery of the space of sounds which the generative system is able to render, without prior training, as proposed in [2,12]. Human intuition can also be leveraged to derive semantically meaningful diversity in the search space, as studied in [6], which can be especially important when generating sonic material leading to interesting discoveries according to individual aesthetics.

In the broadest sense, the concept of instruments has evolved from being a mere means to an end to a starting point for a journey into the unknown [21, p. 49]. The evolutionary system explored here is not intended as an instrument for serving requests from preconceived ideas but rather as a tool for discovering interesting sound objects that can steer the creative journey. The sound artefacts generated by our system, as discussed in this paper, are intended to facilitate or inspire the creation of further sonic art. This is different from the visual artefacts produced by many generative systems, which are often seen as standalone pieces without further utility. Instead of a top-down approach—where the end goals and characteristics of the desired sound are pre-defined—our method encourages a bottom-up process of exploration. This reflects the evolutionary path of human development, where cognitive skills have been shaped by the very tools that humans have uncovered. This echoes the saying, "the tool writes the toolmaker as much as the toolmaker writes the tool" ([4] as cited in [21, p. 5]). An instrument that promotes such exploratory discovery can enable us to continue on our path of evolution by developing human abilities through technology.

References

1. Choi, K., et al.: Foley sound synthesis at the DCASE 2023 challenge. https://doi. org/10.48550/arXiv.2304.12521. arXiv:2304.12521 (2023)
2. Cully, A.: Autonomous skill discovery with quality-diversity and unsupervised descriptors. In: Proceedings of the Genetic and Evolutionary Computation Conference, pp. 81–89. ACM, Prague Czech Republic, July 2019. https://doi.org/10. 1145/3321707.3321804
3. Cully, A., Demiris, Y.: Quality and diversity optimization: a unifying modular framework. IEEE Trans. Evol. Comput. **22**(2), 245–259 (2018). https://doi.org/ 10.1109/TEVC.2017.2704781

4. Davis, W.: Replications: Archaeology, Art History, Psychoanalysis. Pennsylvania State University Press, University Park. ISBN 0271015233
5. Deng, J., Dong, W., Socher, R., Li, L.J., Li, K., Fei-Fei, L.: ImageNet: a large-scale hierarchical image database. In: 2009 IEEE Conference on Computer Vision and Pattern Recognition, pp. 248–255, June 2009. https://doi.org/10.1109/CVPR.2009.5206848. ISSN 1063-6919
6. Ding, L., Zhang, J., Clune, J., Spector, L., Lehman, J.: Quality diversity through human feedback, October 2023. https://doi.org/10.48550/arXiv.2310.12103. arXiv:2310.12103 [cs]
7. Gaier, A., Asteroth, A., Mouret, J.B.: Are quality diversity algorithms better at generating stepping stones than objective-based search? In: GECCO 2019 Companion - Proceedings of the 2019 Genetic and Evolutionary Computation Conference Companion, pp. 115–116 (2019). https://doi.org/10.1145/3319619.3321897
8. Garber, L., Ciccola, T., Amusategui, J.: AudioStellar, an open source corpus-based musical instrument for latent sound structure discovery and sonic experimentation. In: Proceedings of the International Computer Music Conference, pp. 62–67 (2021)
9. Gemmeke, J.F., et al.: Audio set: an ontology and human-labeled dataset for audio events. In: Proceeding of IEEE ICASSP 2017, New Orleans, LA (2017). https://doi.org/10.1109/ICASSP.2017.7952261
10. Godøy, R.I.: Chunking sound for musical analysis. In: Ystad, S., Kronland-Martinet, R., Jensen, K. (eds.) CMMR 2008. LNCS, vol. 5493, pp. 67–80. Springer, Heidelberg (2009). https://doi.org/10.1007/978-3-642-02518-1_4
11. Gong, Y., Lai, C.I.J., Chung, Y.A., Glass, J.: SSAST: self-supervised audio spectrogram transformer, February 2022. https://doi.org/10.48550/arXiv.2110.09784. arXiv:2110.09784 [cs, eess]
12. Grillotti, L., Cully, A.: Unsupervised behavior discovery with quality-diversity optimization. IEEE Trans. Evol. Comput. **26**(6), 1539–1552 (2022). https://doi.org/10.1109/TEVC.2022.3159855
13. Huang, P.Y., et al.: Masked autoencoders that listen. In: NeurIPS (2022). https://doi.org/10.48550/arXiv.2207.06405
14. Jónsson, B.T., Erdem, C., Glette, K.: A system for sonic explorations with evolutionary algorithms. J. Audio Eng. Soc. **72**(4), (2024). https://doi.org/10.17743/jaes.2022.0137
15. Jónsson, B.T., Glette, K., Erdem, C., Fasciani, S.: Supporting data for: towards sound innovation engines using pattern-producing networks and audio graphs (2024). https://doi.org/10.18710/BAX9N5
16. Jónsson, B.T., Hoover, A.K., Risi, S.: Interactively evolving compositional sound synthesis networks. In: Proceedings of the 2015 Annual Conference on Genetic and Evolutionary Computation, GECCO 2015, New York, NY, USA, pp. 321–328. Association for Computing Machinery, July 2015. https://doi.org/10.1145/2739480.2754796
17. Kingma, D.P., Welling, M.: Auto-encoding variational Bayes. In: Bengio, Y., LeCun, Y. (eds.) 2nd International Conference on Learning Representations, ICLR 2014, 14–16 April 2014, Banff, AB, Canada. Conference Track Proceedings (2014). https://doi.org/10.48550/arXiv.1312.6114
18. Lehman, J., Risi, S., Clune, J.: Creative generation of 3D objects with deep learning and innovation engines. In: Proceedings of the Seventh International Conference on Computational Creativity: ICCC 2016, Paris, France, pp. 180–187, June 2016
19. Lehman, J., Stanley, K.O.: Abandoning objectives: evolution through the search for novelty alone. Evol. Comput. **19**(2), 189–223 (2011). https://doi.org/10.1162/EVCO_a_00025

20. Lehman, J., Stanley, K.O.: Evolving a diversity of creatures through novelty search and local competition. In: Genetic and Evolutionary Computation Conference, GECCO 2011 (GECCO), pp. 211–218 (2011). https://doi.org/10.1145/2001576. 2001606. ISBN 9781450305570

21. Magnusson, T.: Sonic Writing: Technologies of Material, Symbolic and Signal Inscriptions. Bloomsbury Academic, New York (2019)

22. McCormack, J., Cruz Gambardella, C.: Quality-diversity for aesthetic evolution. In: Martins, T., Rodríguez-Fernández, N., Rebelo, S.M. (eds.) EvoMUSART 2022. LNCS, vol. 13221, pp. 369–384. Springer, Cham (2022). https://doi.org/10.1007/978-3-031-03789-4_24

23. McCormack, J., Gambardella, C.C., Krol, S.J.: Creative discovery using QD search, May 2023. https://doi.org/10.48550/arXiv.2305.04462. arXiv:2305.04462 [cs]

24. Moore, B.C.: Hearing. Handbook of Perception and Cognition, 2nd edn. Academic Press, San Diego (1995). ISBN 0125056265

25. Mouret, J.B., Clune, J.: Illuminating search spaces by mapping elites, April 2015. https://doi.org/10.48550/arXiv.1504.04909. arXiv:1504.04909 [cs, q-bio]

26. Nguyen, A., Yosinski, J., Clune, J.: Understanding innovation engines: automated creativity and improved stochastic optimization via deep learning. Evol. Comput. **24**(3), 545–572 (2016). https://doi.org/10.1162/EVCO_a_00189

27. Nguyen, A., Yosinski, J., Clune, J.: Deep neural networks are easily fooled: high confidence predictions for unrecognizable images. arXiv, April 2015. https://doi.org/10.48550/arXiv.1412.1897. arXiv:1412.1897 [cs]

28. Nguyen, A.M., Yosinski, J., Clune, J.: Innovation engines: automated creativity and improved stochastic optimization via deep learning. In: Proceedings of the 2015 Annual Conference on Genetic and Evolutionary Computation, GECCO 2015, New York, NY, USA, pp. 959–966. Association for Computing Machinery, July 2015. https://doi.org/10.1145/2739480.2754703

29. Nordmoen, J., Veenstra, F., Ellefsen, K.O., Glette, K.: MAP-Elites enables powerful stepping stones and diversity for modular robotics. Front. Robot. AI **8**, 56 (2021). https://doi.org/10.3389/frobt.2021.639173

30. Noë, A.: The Entanglement: How Art and Philosophy Make Us What We Are. Princeton University Press, Princeton (2023). ISBN 9780691188812

31. Pugh, J.K., Soros, L.B., Szerlip, P.A., Stanley, K.O.: Confronting the challenge of quality diversity. In: Proceedings of the 2015 Annual Conference on Genetic and Evolutionary Computation, GECCO 2015, New York, NY, USA, pp. 967–974. Association for Computing Machinery, July 2015. https://doi.org/10.1145/2739480.2754664

32. Pugh, J.K., Soros, L.B., Stanley, K.O.: Quality diversity: a new frontier for evolutionary computation. Front. Robot. AI **3**, 40 (2016). https://doi.org/10.3389/frobt.2016.00040

33. Rice, D.: GenSynth: collaboratively evolving novel synthetic musical instruments. Master's thesis, The University of Oklahoma, May 2015. https://doi.org/10.13140/RG.2.1.4691.6001

34. Secretan, J., et al.: Picbreeder: a case study in collaborative evolutionary exploration of design space. Evol. Comput. **19**(3), 373–403 (2011). https://doi.org/10.1162/EVCO_a_00030

35. Stanley, K.O.: Compositional pattern producing networks: a novel abstraction of development. Genet. Program Evolvable Mach. **8**(2), 131–162 (2007). https://doi.org/10.1007/s10710-007-9028-8

36. Stanley, K.O., Miikkulainen, R.: Evolving neural networks through augmenting topologies. Evol. Comput. **10**(2), 99–127 (2002). https://doi.org/10.1162/106365602320169811
37. Takagi, H.: Interactive evolutionary computation: fusion of the capabilities of EC optimization and human evaluation. Proc. IEEE **89**(9), 1275–1296 (2001). https://doi.org/10.1109/5.949485
38. Wyse, L.: Free music and the discipline of sound. Organ. Sound **8**(3), 237–247 (2003). https://doi.org/10.1017/S1355771803000219

Co-creative Orchestration of *Angeles* with Layer Scores and Orchestration Plans

Francesco Maccarini[1]([✉])[iD], Mael Oudin[2][iD], Mathieu Giraud[1][iD], and Florence Levé[1,3][iD]

[1] Univ. Lille, CNRS, Centrale Lille, UMR 9189 CRIStAL, F-59000 Lille, France
francesco@algomus.fr
[2] Department of Music Research, Schulich School of Music of McGill University, Montreal, Canada
[3] MIS, Université de Picardie Jules Verne, F-80000 Amiens, France

Abstract. Orchestration is the process of creating music for a group of instruments, combining, blending, and contrasting their sounds to produce a unique orchestral texture. In this research/creation project, our team was commissioned to create an AI/human orchestration of two movements of *Angeles*, a piano composition by Gissel Velarde. The project turned out as a perfect case study for computational creativity in music and orchestration, where the role of the model is between *AI as a colleague* and *AI as a tool*. Our main contribution is a preliminary framework for computer assisted orchestration. By modeling a *layer score* and an *orchestration plan* in the orchestration process, we implement a simple Markov model that selects possible instrumentations for each score segment. Personalization of the AI and AI/human interaction occur through human segmentation of the score at two stages of the process (layer score, orchestral segments with loudness profile), through instrumentation presets, and finally through selection of the final orchestral plan and through the actual orchestration. We detail the research aspects of this co-creative project and analyze the roles of the actors involved in the creation of the final piece: the Music Information Retrieval (MIR) researchers, the orchestrators, and the algorithms.

Keywords: Orchestration · Computational Creativity · Co-creativity · Music Generation

1 Introduction

1.1 Orchestration and Creativity

Orchestration. Orchestration involves composing or arranging music for a large group of instruments, mixing or contrasting their sounds to create an orchestral texture. Orchestration goes well beyond distributing the voices among the instruments. Formalization and teaching of instrumentation and orchestration has improved over time, with various treatises covering topics such as the musical capabilities of each individual instrument, and their combination in order to shape the *sound of the orchestra*, and to render specific perceptual effects [2,8,23,37,47,55,57].

We can characterize orchestral music and orchestration through *musical texture*. In Western music, there are several commonly discussed types of texture, including *monophony, polyphony, heterophony*, and *homophony* [18]. In his definition of musical texture, Huron proposes to analyze the sound material according to the number of elements happening simultaneously *(density)* and their homogeneity or heterogeneity *(diversity)* [30]. Benward and Saker [7] describe texture by identifying different parts (or *layers*), which have different *roles*, such as melodies or (static, harmonic, or rhythmic) support. Complex textures can be created through the varying timbers of the large number of orchestral instruments and their combinations [47].

This richness of possible instrument combinations makes the analysis and generation of orchestral music particularly challenging for computational musicology [9,15,19,49,50,52]. Perceptual studies focused on how instruments combination affects timbral response, blending qualities, and on the use of timbral consonance and dissonance to create sounds [36,60,62]. Open corpora have been created to help research in orchestration techniques, automation, and perception. Crestel et al. published the Projective Orchestral Database (POD) linking piano and orchestral scores [15]. We proposed with Le a bar-to-bar textural analysis of 24 movements of symphonies [38]. The Orchestration Analysis & Research Database (OrchARD, `orchard.actor-project.org`, data not published) rather targets on specific auditory effects [44].

Computational Creativity and Music Generation. The spread of systems for AI-based music generation studies has contributed to drive a growing interest in machine creativity [48]. Many questions emerge regarding the definition, evaluation, and uses of *creative* machines [33,34]. Esling and Devis suggest considering generative AI algorithms as creativity-enhancing tools [20]. This proposal finds its place in Lubart's classifications of modes of human-machine interaction in a co-creative process [41]. Among them, the *computer as a colleague* mode implies a direct involvement of algorithms in the creation of the final output, rather than a mere assisting role *(algorithm as a tool)* for the creator. Similar categories are found in Kantosalo and Jordanous's description of the roles of AI in creative processes. They divide them in the categories of *co-creative colleague* and *creativity support tool* [35].

Some studies focused on human creativity in music [56], and on the way it can be enhanced with Machine Learning (ML) models [43] through modes of interaction between the artist and the model [5,24]. Difficulties emerge in the evaluation of the creativity of such systems [4]. The impact of a good interface for *steering* AI on the ability of users to express musical ideas and *"own"* the resulting creation has also been highlighted [40]. Co-creative systems in music generation [10,11,21,27,31] can be divided into two categories: for live performance improvisation [5,22,54,65], and for composition and production [1,6,58,61]. The AI song contest, a competition specifically focused on co-creation in songwriting, inspired different uses of AI and discussions on them [28]. Some of us have been part of such a team, involving a composer the right from the beginning of the system design, resulting in personalized AI models [16].

Machine Learning, Creativity, and Orchestration. In the realm of orchestral music, specific ML models were proposed for tasks related to arrangement [29,64], instrumentation [17], orchestration [12,14,26,53], and generation of orchestral music [39]. Beethoven X experiment was an AI-assisted composition of orchestral music that aimed to be a plausible Beethoven symphony [25], obtained by the collaboration between the composer Walter Werzowa, musicologists, and computational methods including generative ML models. No code nor data are publicly available, but their co-creative methodology is interesting, divided into *continuation* (expanding melodic lines and themes, from the original melodic material from Beethoven's own sketches), *harmonization* (composition of accompaniments parts for the melodic ideas, including, for example, homophony, counterpoint, and fugue), *transition* (orchestral/polyphonic inpainting, to connect different ideas), and *orchestration* (organizing across the available instruments and instrumental families of the orchestra).

1.2   Goal and Contents

We carried out this project to orchestrate two piano pieces by Gissel Velarde, focusing on *co-creativity*, i.e. putting human beings and computer models in the same loop to make music. In order to get an outcome of high quality scores, ready to be distributed to the orchestra, in a limited time frame, we focused on modeling the art of orchestration as a formal process, and on the possible ways of interaction between a human orchestrator and computational algorithms. We explicitly searched for a balance between having *simple but high-level conceptual data* that can be handled by AI and *patterns* that can be explained so that the musician can interact with them. Following Benward and Saker [7], we characterize orchestration in the "classical style" as an overlay of different layers, which have the roles of melodies or (harmonic/rhythmic) accompaniment, played by the different instruments of the orchestra and their combinations.

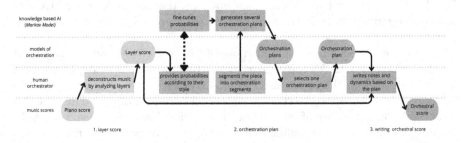

Fig. 1. The workflow of the co-creative orchestration builds on a *layer score* and on an *orchestration plan*.

Our workflow of orchestration is the following (Fig. 1):

- The first step is an analysis: the goal is to extract from the piano score a **layer score**, that is an abstract version of the piece analyzing *musical texture* (Fig. 2 and Sect. 2). Voices are represented together with their role: melody, harmonic accompaniment, or rhythmic accompaniment. At this stage we do not provide any information about the instrumentation.
- The second step is to **build an orchestration plan**: We assign each layer to an instrument or a group of instruments, taking care of the balance between the different timbres. We have developed a Markov model for this stage, that uses probabilities of finding instruments together, of instrument sequences, and instrumental density, building from instrumentation presets as well as from a segmentation of the score (Sect. 3).
- The last step is to **write the actual instrumental parts**, following the orchestration plan and taking care of the peculiarities of range and dynamics of each instrument (Sect. 4).

For this experiment, the first and third step were done by human orchestrators, who are two of us: Mael Oudin (professional orchestrator, PO), and Mathieu Giraud (amateur orchestrator, AO). They both are also researchers for the project. The focus on human/AI co-creativity is thus here in the second step but also in the very decision of modeling the layer score and the orchestration plan to enable these three steps. We neither claim that this three-steps process fully models the art of orchestration, nor that it represents an optimal process, but rather that it is a *plausible workflow* to orchestrate a piece, which could be followed by human orchestrators alone, but that has the advantage to allow interaction between humans and AI.

The rest of the article details those three steps (Sects. 2, 3, 4) and the results of this orchestration model, analyzing the roles of the actors – MIR researchers, orchestrators, and computational models. We conclude by discussing the challenges encountered, our positioning in computational creativity research, and perspectives (Sect. 5).

2   Modeling a Layer Score

2.1   Behind Orchestral Music: Analyses, Sketches

Any music, any score, may be seen as a rendering of high-level *musical ideas*, that may be intermediate steps when composing or improvising music, or serve analytical purposes. Describing and modeling these ideas is a challenge for (digital) musicology and music analysis.

Analytical Concepts. As usual in music analysis, an orchestral score can be studied at different levels. Taking an orchestral score as a "neutral level" [51], the analyst can examine orchestration techniques. Focusing on *textural/instrumental* aspects, music can be split into a number of parts, or layers, with a predominantly

(a) **Full Orchestral Score.** 13 tracks/instruments (7 in this extract)
Transcribed by ClassicMan on musescore.com
Textural labels from [38]. *Front:* Vl1, Vla, Cb; *Rhythm:* Vl2, Vc; *Harmony:* Cl, Hrn

(b) **Layer Score.** 3 layers

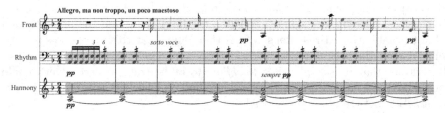

(c) **Piano reduction.** Arr. by Franz Liszt, ed. Breitkopf & Härtel

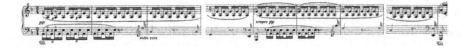

Fig. 2. First eight measures of the first movement of Symphony #9 by L. van Beethoven, op. 125. (a) The orchestral score can be decomposed into three layers: one with an *harmonic* role (red, dark), one with a *rhythmic* role (blue, light), and a third one with a *front* role, that is nevertheless difficult to categorize as a melody (yellow, very light). (b) The layer score contains one part for each of these three layers, $\ell^{harmony}$, ℓ^{rhythm}, and ℓ^{front}. (c) In the piano reduction, the harmonic and rhythmic layer are blended together and rendered with a *pianistic* texture, which is different from the orchestral version. The left hand alternates between that harmonic-rhythmic layer, and the front layer. (Color figure online)

melodic, rhythmic, or harmonic role, or a mix of them [7]. In orchestral music functional layers are highlighted by combinations of instrumental timbres [38]. At a higher level, orchestral effects arise from auditory grouping processes [44].

Composer/Orchestrator Sketches. Were these layers and effects present in the composer's – or the orchestrator's – mind? Composers may have some sketches (in their mind and/or onto paper) of these parts and then complete those sketches into more elaborated music [59,66]. The study of Beethoven's sketches have revealed precise musical ideas such as patterns, themes, and sometimes orchestration sketches [32]. The options available to composers for orchestral music encourage them to explore *textural* spaces [63], but it is unclear to what extent they can model complex organization of layers in their mind.

Concurrent musical parts and layers can thus preexist the orchestral score, which can be heard as an expansion of these initial materials. Some MIR studies on orchestration already used this idea: Gotham et al. name as a *short score* a set of raw materials before orchestration [25]. Somehow, in jazz/pop styles, *lead sheets* with melodies and chords may also be seen as a condensed version of a music piece – or could correspond to a sketch of the final song. MIR and AI methods aiming at (co-)creating such music often generate such lead sheets at first, then proceed with accompaniment generation [28], even if the split is debatable and an end-to-end generation is sometimes preferred [3].

2.2   Defining and Modeling a Layer Score

Inspired by such hypothetical intermediary "sketch scores" and by analytical considerations, we define here the *layer score* as a score with a variable number of *layers*, each one with a given *role*, following the description of roles we gave in [38] for classical-romantic orchestral music, but without any indication on the instrument.

On the opening of the Beethoven's 9th Symphony, we analyze three layers with different roles, ℓ^{front}, ℓ^{rhythm}, and $\ell^{harmony}$ (Fig. 2b). Those layers are distributed on several instruments in the full orchestral score. Here, one part in the layer score roughly corresponds to one layer in the orchestral score, but the actual music in the orchestral part could be more different from the content of the layer score, possibly depending on the capabilities of the instruments chosen to render a layer. For example, rhythmic motion could be rendered with a different density of notes if reproduced through a timpani roll, or through tremolo strings *sul ponticello*.

At the opposite, if we consider now a *piano reduction* of an orchestral score, the same layers will be blended together into the two staves of the piano reduction – a single pianist should be able to play it. In the layer ℓ^{rhythm} of the opening of the Beethoven's 9th Symphony, the lower strings *repeat* chords made of A and E, whereas the piano reduction by Liszt *alternates* between the same pitches (Fig. 2c). Again, the actual music in the piano reduction could be different, in the rhythm organization for instance. In practice, the octaves often differ in piano reduction, the pianist hands not having the same ambitus as the orchestra. In some cases, the actual pitches may even differ between instrumentations, for example including patterns or scales in some of them.

The layer score therefore generally has fewer staves than a full orchestral score and more staves than a piano reduction, but could be (anachronistically) viewed as the *common ancestor* between the two – or, more generally, an ancestor of any other instrumentation of the music. The layer score includes the structural, melodic, harmonic content, and some of the textural content, but not the actual music rendering. This concept of layer score allows thus to decouple composition from instrumentation and orchestration (although this decoupling may be artificial) and to devote our attention mainly to the latter ones.

Back to the question of composer sketches, a layer score is not intended to reproduce an existing compositional practice. We do not claim either that writing such layer scores would be a desirable practice. Writing music for the orchestra (or for any other instrument) can be a non-linear process, with iterations between high-level ideas and actual music content. For example, the constraints given by the ambitus of each instrument influence the composition itself, forcing the composer to rethink some of the choices they have already made, and even to rewrite major sections of the composition.

Anyway, layer scores give new perspectives on topics related to orchestration. For example, piano reduction and orchestration from piano are symmetrical tasks. Orchestration from piano usually requires "de-pianotizing" the piano music, including voice separation [42] or texture analysis [13] – as piano reduction requires "de-orchestrating" the orchestral music. Modeling such a layer score thus increases the possibilities for a human intervention.

2.3 Results on Angeles

We illustrate our methodology through the first two measures of *El Jardin Etero*, the last movement of *Angeles* (Fig. 3). In the layer score, we have separated into two different staves the melodic ℓ^{mel} and rhythmic ℓ^{rhy1} layers that emerge from the piano score. We also decided to stress the two last notes in a separate layer ℓ^{rhy2} and to add another rhythmic layer ℓ^{rhy3}. Such decisions in the analytical process have their part of subjectivity and contribute to the co-creativity, allowing to recreate new textures. The professional orchestrator states:

> PO: *My role was to make human musical choices during all stages of the AI-assisted orchestration process. As an orchestrator, identifying the texture is my first job. Analyzing the piano score allows me to deconstruct the music into different roles: the main melody, harmony, rhythm, resonance...I prepare the addition of new parts. The piano is limited by its technique. As I orchestrate, I will add what is "absent but suggested", resonance, missing registers, textures to be recreated. The first stage involved the creation of the "layer score" for this movement.*

Creative choices, such as adding a layer, can already be made at this stage, going beyond a pure analysis of the original score. For example, a music pattern

(a) **Piano score**   (d) **Full orchestral score**

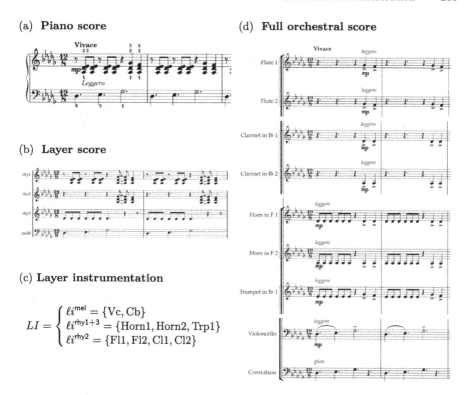

(b) **Layer score**

(c) **Layer instrumentation**

$$LI = \begin{cases} \ell i^{\mathrm{mel}} = \{\mathrm{Vc, Cb}\} \\ \ell i^{\mathrm{rhy1+3}} = \{\mathrm{Horn1, Horn2, Trp1}\} \\ \ell i^{\mathrm{rhy2}} = \{\mathrm{Fl1, Fl2, Cl1, Cl2}\} \end{cases}$$

Fig. 3. First two measures of movement 6 *"El Jardin Etereo"* from *Angeles* by Gissel Velarde, op. 7. (a) Original Piano score, provided by the composer. Clearly, the main melody is at the bass, and the right hand plays a rhythmic layer. (b) Layer Score elaborated as an intermediate step of the orchestration process. A melody in the low register ℓ^{mel}, and a rhythmic layer ℓ^{rhy1} have been directly identified by splitting the right and left hand of the piano score. Other layers (ℓ^{rhy2}, ℓ^{rhy3}) have been added to stress the importance of some notes and to have more possibilities of rhythms, departing from a typical pianistic texture. (c) Layer instrumentation, assigning instruments to each one of these three layers. The human orchestrator decided to have one rhythmic layer $\ell^{\mathrm{rhy1+3}}$, with added notes in the downbeats. (d) Orchestration of the piece by Mael Oudin. The instrumental parts have been written following the selected orchestration plan. Scores and rendered audio for selected extracts of the piece are available at http://www.algomus.fr/data.

may have at the same time a melodic and an harmonic role. In that case, the layer score should define a mixed melodic/harmonic layer (see Sect. 5).

3   Modeling, Generating, Selecting an Orchestration Plan

3.1   Orchestration Plans

Once a score is split into layers, we have to map them to the available instruments, and instrument groups, in the ensemble, in order to describe the envisioned instrumentation for the piece. Given a set of layers such as $\mathcal{L} = \{\ell^{\mathsf{mel}}, \ell^{\mathsf{rhy1}+3}, \ell^{\mathsf{rhy2}}\}$, a *layer instrumentation* is a list LI of instruments in each layer, each instrument occurring in at most one layer (see Fig. 3c).

Each orchestration change, being it progressive, or a contrast, uses a new layer instrumentation. We thus define an *orchestration plan* as a set of layer instrumentations $LI^1, LI^2, \ldots, LI^n$, one for each *orchestration segment* $s \in \{1, 2, \ldots, n\}$ of the score. In the current model, it is the musician who segments the score at relevant points where they want orchestration changes. The syntax of the orchestration plan describes, for each segment, the layers using instruments in a predefined order and details the layers and roles (Fig. 4c).

3.2 Personalized Orchestration Plan with Markov Models

Once they have set the orchestration segmentation, the artist could themselves write the layer instrumentation for every segment, and create an orchestration plan. Here, instead, we decided to have a simple knowledge-based algorithm to experiment with AI/human interaction. This model for semi-automated layer instrumentation follows two goals. The layers should include instruments that blend together, drawn from presets of *possible instrumentations*. Moreover, the "loudness" of the instrumentation at each segment should be close to the musician's desired outcome, for which they provide a *loudness profile* as input. These concepts are detailed in the following paragraphs.

Loudness Profile and Acoustic Weights. To underline the form, the musician inputs a *loudness profile* as a list of targeted loudness values $(\lambda_1, \lambda_2, \ldots, \lambda_n)$ for the n segments. The effective loudness depends on the dynamics, but also on the number and the qualities of each instrument[1]. A simple model is to consider that each instrument i has an *acoustic weight* w_i. We decided here to have higher coefficients for brasses and instrument of lower range, using the following values:

(Fl:1, Ob:1, Cl:1, Fg:1.5, Hrn:1.5, Trp:2, Vln1:1, Vln2:1, Vla:1, Vc:1, Cb:1.5)

The loudness could be estimated as the sum of the weights of the instruments involved. However, selecting only the instruments according to such values would not realize a proper orchestration, as it would ignore blending qualities and orchestrator preferences.

Possible Instrumentations. For each layer $\ell^\alpha \in \mathcal{L}$, the musician will thus define a set of *possible layer instrumentations* $p\ell i^\alpha = \{p\ell i_1^\alpha, p\ell i_2^\alpha, \ldots\}$, each one being a weighted list of instruments. For instance, a rhythmic layer ℓ^{rhy1} could be associated to two distinct instrumentations, either on woodwinds, or brasses:

$$p\ell i^{\mathsf{rhy1}} = \begin{cases} p\ell i_{wood}^{\mathsf{rhy1}} = (\mathrm{Fl}:.3, \mathrm{Ob}:.1, \mathrm{ClBb}:.2, \mathrm{Fg}:.15) & L(p\ell i_{wood}^{\mathsf{rhy1}}) = .825 \\ p\ell i_{brass}^{\mathsf{rhy1}} = (\mathrm{HrnF}:.7, \mathrm{TrpBb}:.2) & L(p\ell i_{brass}^{\mathsf{rhy1}}) = 1.45 \end{cases}$$

---

[1] We call "instrument" an instrument group. Groups may include several people (Vl1).

(a) **Loudness profiles**

```
LOUDNESS = [ 0.20, 0.40, 0.20, 0.50, 0.30, 0.40, 0.20, 0.50,
             0.30, 0.50, 0.30, 0.60, 0.40, 0.50, 0.60, 0.70 ]
```

(b) **Generated orchestration plans, with several relative loudness**

```
InstList: <Fl.Ob.ClBb.Fg|HrnF.TrpBb|Vln1.Vln2.Vla.Vc.Cb>
## Gen        104a (0.5)         104b (1.0)         104c (4.0)
[p01]    <..2.|13|...m.>    <2.2.|13|...m.>    <2.2.|13|...mm>    {0.20}
[p02]    <....|..|123mm>    <....|..|123mm>    <...3|m3|123mm>    {0.40}
[p03]    <....|13|..2m.>    <....|13|.22m.>    <....|13|.22m.>    {0.20}
```

(c) **Final orchestration plan**

```
InstList: <Fl.Ob.ClBb.Fg|HrnF.TrpBb|Vln1.Vln2.Vla.Vc.Cb>
[p01]  <2.2.|13|...mm>  1:rhy1:brass 2:rhy2:wood 3:rhy3:brass m:mel:mel2 (0)
[p02]  <....|..|123mm>  1:rhy1:string 2:rhy2:string 3:rhy3:string m:mel:mel2 (4)
[p03]  <....|13|.22m.>  1:rhy1:brass 2:rhy2:string 3:rhy3:brass m:mel:mel1 (0)
```

Fig. 4. Creating the orchestration plan of Angeles, mvt 6. (a) The score is split by the musician into 16 *instrumentation segments*, each with a target loudness. (b) The model generates, for each segment, three *layer instrumentations* taking into account the expected segment loudness and another relative loudness coefficient (0.5, 1.0, 4.0) (c) In the selected orchestration plan, for the segment [p01] (first two measures), there are four layers instrumentations ℓ_{brass}^{rhy1}, ℓ_{wood}^{rhy2}, ℓ_{brass}^{rhy3}, and ℓ_{mel2}^{mel}. The layers are mapped to the instruments appearing in the order declared in InstList: For example, the "<2.2.|" bloc in the woodwinds refers to the layer instrumentation ℓ_{wood}^{rhy2}, with here flutes (Fl) and clarinets (ClBb).

Each component (i, p_i) tells that the instrument i should have a probability p_i of being used in this $p\ell i$: The actual instruments that will be used will be a subset of that $p\ell i$. Selecting a $p\ell i$ ensures that these instruments blend together for this particular layer. The sum $\sum p_i$ of the probabilities of a $p\ell i$ is the expected number of instruments in that $p\ell i$. We rather use the *expected loudness* of the $p\ell i$, that is $L(p\ell i) = \sum w_i p_i$, weighting each probability by the acoustic weight of each instrument.

Selecting the $p\ell i$ then the Instrumentation for Each Segment. Given a layer $\ell^\alpha \in \mathcal{L}$ and a segment $s \in \{1, 2, \ldots, n\}$, the $p\ell i^{\alpha,s}$ is selected in $p\ell i^\alpha$ according to a Markov model (Fig. 5) that depends on the previous $p\ell i^{\alpha,s-1}$. In order to match the prescribed loudness λ_s, the model also tries to minimize $\delta^\alpha = |\lambda_s - L(p\ell i^{\alpha,s})|$ by applying a further $e^{|\delta^\alpha \tau|}$ coefficient with $\tau = 2.0$.

For a given segment s, once all $p\ell i^{\alpha,s}$ are selected for all layers ℓ^α, instruments are assigned following the individual probabilities p_i. At the end of this step, it may happen that either a layer has no instrument assigned, or that an instrument is assigned to more than one layer. Such cases are resolved by further random assignations, based again on the p_i in the $p\ell i$.

The personalization of the AI "to the style of the orchestrator" and the possibility to steer the AI are thus done both on the presets/$p\ell i$ selections and on

$$pli_{H1}^{mel,s-1} \longrightarrow pli_{H1}^{mel,s} : .8 \qquad pli_{H2}^{mel,s-1} \longrightarrow pli_{H1}^{mel,s} : .6 \qquad pli_{B}^{mel,s-1} \longrightarrow pli_{H1}^{mel,s} : .3$$
$$\longrightarrow pli_{H2}^{mel,s} : .6 \qquad \longrightarrow pli_{H2}^{mel,s} : .8 \qquad \longrightarrow pli_{H2}^{mel,s} : .3$$
$$\longrightarrow pli_{B}^{mel,s} : .1 \qquad \longrightarrow pli_{B}^{mel,s} : .1 \qquad \longrightarrow pli_{B}^{mel,s} : .8$$

Fig. 5. Extract of the transition table of the Markov model modeling the evolution of $pli^{mel} = \{pli_{H1}^{mel}, pli_{H2}^{mel}, pli_{B}^{mel}, \ldots\}$ for movement 6. The transition table was created through iterations between the MIR researchers and the orchestrator. These coefficients are further adjusted by a loudness factor, then normalized.

the loudness profile input. Moreover, the implemented method generated segments with three levels of relative loudness for each segment (Fig. 4b), enabling the orchestrator to further select instrumentations at each segment, but still keeping the coherency of the pli.

3.3 Results on Angeles

PO: *To develop the model that generates orchestration plans, I collaborated with the Algomus team to propose instrument combinations that I enjoy using, and to fine-tune the model.*

According to our orchestrator, there are several goals that a good orchestration should pursue: respecting and enhancing the piano composition, a good balance between layers (especially, the melody should not be muzzled by accompaniment), variation (contrasting moments in the piece should carry different orchestrations), and efficient dynamics (through loudness values and coefficients).

Respecting the piano composition means that the score brings constraints in register and dynamics that need to be reflected in the orchestration plan. For example, movement 6 was a Vivace, with the melody on the bass and an accompaniment more rhythmic than harmonic. The orchestration then had to address the character of the piece and abide by the register of each layer.

PO: *The bass melody could only be performed by three instruments in the orchestra: the cellos, the contrabasses, and the bassoons. But not every choice would give a satisfying balance to the other layers played by the rest of the orchestra. The bassoons or contrabasses alone, for instance, would not be prominent enough so cellos were necessary here. Any generated orchestration plan that would not nominate cellos for that layer would be in practice almost unusable.*

Balance in an orchestration is also reached through the separation of the orchestra into different groups (namely, the strings, the woodwinds, the brass, and the percussion). Harmonic blending is best achieved when all the notes of a chord are performed by instruments in the same group. This was a constraint to our model if we wanted to avoid too much disparity in the instrument combinations proposed in the orchestration plans. Movement 6 had a continuous 3-voice rhythmic layer and we wanted these three voices to be performed by instruments from the same group. This wasn't always the case in the orchestration plans generated so this was also a criteria in the selection of the best outputs. The ensemble of these parameters creates many constraints on orchestration possibilities, so one of the first human tasks when analyzing the output generated by the model is to remove what seems impossible, for reasons as varied as register limitations, number of voices to be played, and poor blending or contrast. Perspectives include (semi-)automatizing some of these tasks. At the same time, the challenge was to foresee the potential of each generated combination when formalized into a musical score at the next stage of the process. Some combinations, such as opening the rhythmic layer with brass only, were unexpected by the orchestrators but rather "proposed" by the model (Fig. 4c).

PO: *I worked on the outer sections of "El Jardin Eterno" using several dozens of orchestration plans generated by the model. It was my responsibility to sort through them and select the most convincing ones according to my taste (while also being open to surprises).*

4   Writing and Performing the Orchestral Score

Once the orchestration plan is decided, the orchestrator has a large space of possibilities related to the range, dynamics, and playing techniques of each instrument that can still be creatively explored. The orchestration plan only suggests the instruments to be used in every portion of the piece, but many decisions still need to be taken to get a playable score, in particular to have idiomatic patterns for each instrument of the orchestra.

In the final orchestral score, on the same first two measures (Fig. 3d), the choice has been made to fill the rest of the first beat of the rhythmic texture to provide a more efficient and easier line to the brass instruments at this fast tempo. The rhythmic layer is then rendered differently from the original pianistic texture, but it preserves the intention. Likewise, the choice of writing pizzicati for the contrabass part, to lighten the orchestral texture and express the *mezzo-piano* dynamic, was taken at that stage of the process.

Our orchestrator shared some reflections on the artistic side of this final step:

PO: *The distribution of instruments is suggested by the model, but there is still considerable freedom in the choice of notes and registers. It is also up to me to choose and indicate nuances, phrasing, playing modes and expressive indications. Some of the model's choices wouldn't have been what I would have done, like starting directly with the trumpets at the beginning of movement 6. But it's stimulating!*

Movement 2 *"Inexorable"* and the outer sections of movement 6 *"El Jardin Eterno"* have been orchestrated with this procedure whereas the middle section of movement 6 has been orchestrated with a "traditional" method. The other movements have been commissioned to orchestrators outside of our team. The whole suite has been performed by the Orquesta Kronos conducted by Andrés Guzmán-Valdez at Nuna Theatre in La Paz, Bolivia, on 19th July 2023.

The experience of working with different methods showed that the AI-assisted method starts to offer a gain in productivity once the orchestration plan is reliable. It is also a tool for creative thinking:

> AO: *Like any creative work, an orchestrator may face the anxiety of the blank page. Especially as an amateur orchestrator, I enjoyed having such suggestions. Even when they were inappropriate, they stimulated creativity through reinforcement, contrast, or opposition.*

5 Discussion and Perspectives

In [16] the use of *low-tech AI* is advocated, to ease the communication between the composer and the researchers, and to obtain tailor-made models with scarce data. We adopted a similar approach here, to focus on modeling the process of orchestration by identifying possible steps, and adapting it to the co-creation environment. The computational model used to generate orchestration plans has been conceived and designed with a continuous back and forth between the artist and the research team. It is meant to be the simplest possible, so that it can be more easily modified to experiment with different inputs and controls. Concepts like loudness profiles and coefficients have been added to the model to respond to the ideas and the necessities of the orchestrator.

We believe that we have succeeded in individuating three well-separated stages of orchestration (Fig. 1), which could be performed by three different (human or algorithmic) actors. Each step included a self-refining feedback loop. For example, the human task at the final stage of the process (3. writing orchestral score) can be described in two phases, the second of which is usual for "traditional orchestration": (3.1) interpret and adapt the orchestration plan, (3.2) write notes for the instruments. In the first stage, the orchestrator will read the output made by the machine and mentally link it to the score to find the best strategy to transform these outputs into music notes (this phase is fundamental to selecting the best outputs from the machine). The second phase consists of writing the notes idiomatically for the instruments, but also the dynamics and the phrasing and expressive instructions.

The current model has limitations. The process is not always linear: it was sometimes in phase (3.2) that choices made in phase (3.1) retrospectively appeared to be pitfalls, and the entire process had to be made again (when, for example, a particular arrangement was not compatible with the plan selected for the subsequent section). Other points could also be improved, as for example the formation of mixed layers. More generally, a challenge is to better model *large-scale orchestral thinking*. Orchestral contrast is typically achieved when the same instrument (or combination of instruments) is not used in the same way in two successive contrasting parts. This may lead orchestrators to "reserve" an instrument on purpose for a specific moment in the piece. Somehow, the $p\ell i$ presets combined to the acoustic weights and the targeted loudness profiles help such a large-scale homogeneity and steerability, but these models could be refined.

Looking back to the categories proposed by Kantosalo and Jordanous [35], we have experimented with a process rooted in the interaction between the computational models and the artist, in which the role of the model is in between *AI as a colleague* and *AI as a tool*. Dividing the orchestration process into steps has facilitated the introduction of computational models. In this way, the role of the AI is to act on a well defined and specific task, making the model an essential *tool* in the overall process. The algorithm is acting on the product of human actions (the layer score), and is enabling further human processing with its output (writing the final orchestral score). At the same time, the model is able to "suggest" unforeseen ideas to the humans, acting more as a *co-creative colleague*, who can inspire and enhance the inspiration of the human artist.

Any co-creative project confronts us with questions related to the authenticity and the ownership of the such art [46]. When dealing with AI-generated art, ethical, legal, and moral concerns emerge, questioning the status of the product itself as having artistic qualities [45]. In this research and creation project, the development of the algorithm and the artistic creation of the orchestrated score were intertwined processes. For this reason, the score have been signed with *Orchestrated by Mael Oudin and the Algomus team*, recognizing authorship to all members of the project, musicians and computer scientists.

In summary, we proposed a framework for AI assisted orchestration, in which the orchestrators craft their art in collaboration with AI algorithms. We divided the process into three steps, modeling *layer scores* and *orchestration plans* as intermediate objects. The code for generating orchestration plans is available under an open-source license at algomus.fr/code. Through this preliminary project, this approach has proven to be effective in formalizing the art of orchestration, enabling the involvement of both machine and human actors, each contributing at different moments. The possibilities for the employment of AI in the process are not limited to the ones selected for the scope of this project. Perspectives include modeling other tasks in the process with Deep Learning AI, both for texture analysis tasks related to the creation of the layer score, and for constrained notes generation, in the creation of the final score. Co-creative interactions would be allowed through model parameters, and through creative modifications of the outputs at several stages: when writing the layer score, the orchestration plan, and the final rendering of the notes. All these steps can be accomplished partly by the machine and partly by the human being, with a fruitful continuous exchange of information.

Acknowledgments. We deeply thank Gissel Velarde, Andrés Guzmán-Valdez and the Orquesta Kronos, and the Nuna Theatre in La Paz, Bolivia. We thank the valuable input and comments provided by Dinh-Viet Toan Le, the Algomus team, and the anonymous reviewers. We also extend our gratitude to the ACTOR project for facilitating collaboration among the authors of this paper and inspiring enriching discussions.

References

1. Adkins, S., Sarmento, P., Barthet, M.: LooperGP: a loopable sequence model for live coding performance using GuitarPro tablature. In: Johnson, C., Rodríguez-Fernández, N., Rebelo, S.M. (eds.) EvoMUSART 2023. LNCS, vol. 13988, pp. 3–19. Springer, Cham (2023). https://doi.org/10.1007/978-3-031-29956-8_1
2. Adler, S.: The Study of Orchestration, 1st edn. Norton (1982)
3. Agostinelli, A., et al.: MusicLM: generating music from text (2023)
4. Agres, K., Forth, J., Wiggins, G.A.: Evaluation of musical creativity and musical metacreation systems. Comput. Entertainment (CIE) 14(3), 1–33 (2016)
5. Assayag, G.: Creative symbolic interaction. In: Sound and Music Computing Conference (SMC 2014) (2014)
6. Ben-Tal, O., Harris, M.T., Sturm, B.L.: How music AI is useful: engagements with composers, performers and audiences. Leonardo 54(5), 510–516 (2021)
7. Benward, B., Saker, M.: Music in Theory and Practice, vol. 1, 8th edn. McGraw-Hill Professional (2008)
8. Berlioz, H.: Grand Traité d'instrumentation et d'orchestration Modernes, 1st edn. Novello (1844)
9. Bosch, J.J., Marxer, R., Gómez, E.: Evaluation and combination of pitch estimation methods for melody extraction in symphonic classical music. J. New Music Res. 45(2), 101–117 (2016)
10. Briot, J.P., Hadjeres, G., Pachet, F.D.: Deep Learning Techniques for Music Generation. Springer, Cham (2019). https://doi.org/10.1007/978-3-319-70163-9
11. Briot, J.P., Pachet, F.: Deep learning for music generation: challenges and directions. Neural Comput. Appl. 32(4), 981–993 (2020). https://doi.org/10.1007/s00521-018-3813-6
12. Cella, C.E.: Orchidea: a comprehensive framework for target-based computer-assisted dynamic orchestration. J. New Music Res. 51, 40–68 (2022). https://doi.org/10.1080/09298215.2022.2150650
13. Couturier, L., Bigo, L., Levé, F.: Annotating symbolic texture in piano music: a formal syntax. In: Sound and Music Computing Conference (SMC 2022) (2022)
14. Crestel, L., Esling, P.: Live Orchestral Piano, a system for real-time orchestral music generation. In: Sound and Music Computing Conference (SMC 2017), p. 434 (2017). https://hal.archives-ouvertes.fr/hal-01577463
15. Crestel, L., Esling, P., Heng, L., McAdams, S.: A database linking piano and orchestral MIDI scores with application to automatic projective orchestration. In: International Society for Music Information Retrieval Conference (ISMIR 2017) (2017)
16. Déguernel, K., Giraud, M., Groult, R., Gulluni, S.: Personalizing AI for co-creative music composition from melody to structure. In: Sound and Music Computing (SMC 2022), Saint-Étienne, France, pp. 314–321 (2022). https://doi.org/10.5281/zenodo.6573287
17. Dong, H.W., Donahue, C., Berg-Kirkpatrick, T., McAuley, J.: Towards automatic instrumentation by learning to separate parts in symbolic multitrack music. In: International Society for Music Information Retrieval Conference (ISMIR 2021) (2021). https://doi.org/10.5281/zenodo.5624447
18. Dunsby, J.: Considerations of texture. Music. Lett. 70(1), 46–57 (1989)
19. Esling, P., Carpentier, G., Agon, C.: Dynamic musical orchestration using genetic algorithms and a spectro-temporal description of musical instruments. In: Di Chio, C., et al. (eds.) EvoApplications 2010. LNCS, vol. 6025, pp. 371–380. Springer, Heidelberg (2010). https://doi.org/10.1007/978-3-642-12242-2_38

20. Esling, P., Devis, N.: Creativity in the era of artificial intelligence. arXiv:2008.05959 (2020)
21. Fernández, J.D., Vico, F.: AI methods in algorithmic composition: a comprehensive survey. J. Artif. Intell. Res. **48**(1), 513–582 (2013)
22. Fernández, J.M., Köppel, T., Lorieux, G., Vert, A., Spiesser, P.: GeKiPe, a gesture-based interface for audiovisual performance. In: New Interfaces for Musical Expression Conference (NIME 2017), pp. 450–455 (2017)
23. Forsyth, C.: Orchestration, 1st edn. Courier Corporation (1914, 1935)
24. Ghisi, D.: Music across music: towards a corpus-based, interactive computer-aided composition. Ph.D. thesis, Paris 6 (2017)
25. Gotham, M.R.H., Song, K., Böhlefeld, N., Elgammal, A.: Beethoven X: Es könnte sein! (It could be!). In: Conference on AI Music Creativity (AIMC 2022) (2022). https://doi.org/10.5281/zenodo.7088335
26. Handelman, E., Sigler, A., Donna, D.: Automatic orchestration for automatic composition. In: Artificial Intelligence and Interactive Digital Entertainment Conference (AIIDE 2012) (2012)
27. Herremans, D., Chuan, C.H., Chew, E.: A functional taxonomy of music generation systems. ACM Comput. Surv. **50**(5), 69:1–69:30 (2017). https://doi.org/10.1145/3108242
28. Huang, C.Z.A., Koops, H.V., Newton-Rex, E., Dinculescu, M., Cai, C.J.: AI song contest: human-AI co-creation in songwriting. In: International Society for Music Information Retrieval Conference (ISMIR 2020) (2020)
29. Huang, J.L., Chiu, S.C., Shan, M.K.: Towards an automatic music arrangement framework using score reduction. ACM Trans. Multimedia Comput. Commun. Appl. **8**(1), 8:1–8:23 (2012). https://doi.org/10.1145/2071396.2071404
30. Huron, D.: Characterizing musical textures. In: International Computer Music Conference (ICMC 1989), pp. 131–134 (1989)
31. Ji, S., Luo, J., Yang, X.: A comprehensive survey on deep music generation: multi-level representations, algorithms, evaluations, and future directions. arXiv:2011.06801 (2020)
32. Johnson, D.P., Tyson, A., Winter, R.: The Beethoven Sketchbooks: History, Reconstruction, Inventory. University of California Press, Berkeley (1985)
33. Jordanous, A.: A standardised procedure for evaluating creative systems: computational creativity evaluation based on what it is to be creative. Cogn. Comput. **4**(3), 246–279 (2012). https://doi.org/10.1007/s12559-012-9156-1
34. Jordanous, A.: Has computational creativity successfully made it "beyond the fence" in musical theatre? Connection Sci. **29**, 350–386 (2017). https://doi.org/10.1080/09540091.2017.1345857
35. Kantosalo, A., Jordanous, A.: Role-based perceptions of computer participants in human-computer co-creativity. In: AISB Symposium of Computational Creativity (CC@AISB 2020) (2020)
36. Kendall, R.A., Carterette, E.C.: Identification and blend of timbres as a basis for orchestration. Contemp. Music. Rev. **9**(1–2), 51–67 (1993)
37. Koechlin, C.: Traité de l'orchestration. Max Eschig (1941 (completed), 1954-1959 (posthumous ed))
38. Le, D.V.T., Giraud, M., Levé, F., Maccarini, F.: A corpus describing orchestral texture in first movements of classical and early-romantic symphonies. In: Digital Libraries for Musicology (DLfM 2022), pp. 22–35 (2022)
39. Liu, J., et al.: Symphony generation with permutation invariant language model. In: International Society for Music Information Retrieval Conference (ISMIR 2022) (2022). https://doi.org/10.48550/arXiv.2205.05448

40. Louie, R., Coenen, A., Huang, C.Z., Terry, M., Cai, C.J.: Novice-AI music co-creation via AI-steering tools for deep generative models. In: Conference on Human Factors in Computing Systems (CHI 2020), pp. 1–13 (2020)

41. Lubart, T.: How can computers be partners in the creative process: classification and commentary on the special issue. Int. J. Hum Comput Stud. **63**(4–5), 365–369 (2005)

42. Makris, D., Karydis, I., Cambouropoulos, E.: VISA3: refining the voice intergration/segregation algorithm. In: Sound and Music Computing Conference (SMC 2016) (2016)

43. Mateja, D., Heinzl, A.: Towards machine learning as an enabler of computational creativity. IEEE Trans. Artif. Intell. **2**(6), 460–475 (2021). https://doi.org/10.1109/TAI.2021.3100456

44. McAdams, S., Goodchild, M., Soden, K.: A taxonomy of orchestral grouping effects derived from principles of auditory perception. Music Theory Online **28**(3), 55 (2022)

45. McCormack, J., Cruz Gambardella, C., Rajcic, N., Krol, S.J., Llano, M.T., Yang, M.: Is writing prompts really making art? In: Johnson, C., Rodríguez-Fernández, N., Rebelo, S.M. (eds.) EvoMUSART 2023. LNCS, vol. 13988, pp. 196–211. Springer, Cham (2023). https://doi.org/10.1007/978-3-031-29956-8_13

46. McCormack, J., Gifford, T., Hutchings, P.: Autonomy, authenticity, authorship and intention in computer generated art. In: Ekárt, A., Liapis, A., Castro Pena, M.L. (eds.) EvoMUSART 2019. LNCS, vol. 11453, pp. 35–50. Springer, Cham (2019). https://doi.org/10.1007/978-3-030-16667-0_3

47. McKay, G.F.: Creative Orchestration. Allyn and Bacon (1963)

48. Miller, A.I.: The Artist in the Machine: The World of AI-Powered Creativity. The MIT Press, Cambridge (2019). https://doi.org/10.7551/mitpress/11585.001.0001

49. Miron, M., Carabias-Orti, J.J., Bosch, J.J., Gómez, E., Janer, J.: Score-informed source separation for multichannel orchestral recordings. J. Electr. Comput. Eng. **2016**, 8363507 (2016)

50. Miron, M., Carabias-Orti, J.J., Janer, J.: Audio-to-score alignment at the note level for orchestral recordings. In: International Society for Music Information Retrieval Conference (ISMIR 2014), pp. 125–130 (2014). https://doi.org/10.5281/zenodo.1416150

51. Nattiez, J.J.: Fondements d'une Sémiologie de La Musique. Dufrenne (1975)

52. Nordgren, Q.R.: A measure of textural patterns and strengths. J. Music Theory **4**(1), 19–31 (1960). https://doi.org/10.2307/843045

53. Pachet, F.: A joyful ode to automatic orchestration. ACM Trans. Intell. Syst. Technol. **8**(2), 18:1–18:13 (2016). https://doi.org/10.1145/2897738

54. Parmentier, A., Déguernel, K., Frei, C.: A modular tool for automatic Soundpainting query recognition and music generation in Max/MSP. In: Sound and Music Computing Conference (SMC 2021) (2021)

55. Piston, W.: Orchestration. Norton (1955)

56. Reybrouck, M.M.: Musical creativity between symbolic modelling and perceptual constraints: the role of adaptive behaviour and epistemic autonomy. In: Musical Creativity, pp. 58–76. Psychology Press (2006)

57. Rimsky-Korsakov, N.: Principles of Orchestration: With Musical Examples Drawn from His Own Works, vol. 1. édition russe de musique (1873 (begun), 1912 (posthumous ed))

58. Rosselló, L.B., Bersini, H.: Music generation with multiple ant colonies interacting on multilayer graphs. In: Johnson, C., Rodríguez-Fernández, N., Rebelo, S.M.

(eds.) EvoMUSART 2023. LNCS, vol. 13988, pp. 34–49. Springer, Cham (2023). https://doi.org/10.1007/978-3-031-29956-8_3

59. Sallis, F.: Music Sketches. Cambridge University Press, Cambridge (2015)
60. Sandell, G.J.: Roles for spectral centroid and other factors in determining "blended" instrument pairings in orchestration. Music. Percept. **13**(2), 209–246 (1995)
61. Sarmento, P., Kumar, A., Chen, Y.H., Carr, C.J., Zukowski, Z., Barthet, M.: GTR-CTRL: instrument and genre conditioning for guitar-focused music generation with transformers. In: Johnson, C., Rodríguez-Fernández, N., Rebelo, S.M. (eds.) Evo-MUSART 2023. LNCS, vol. 13988, pp. 260–275. Springer, Cham (2023). https://doi.org/10.1007/978-3-031-29956-8_17
62. Schnittke, A.: Timbral relationships and their functional use. In: Orchestration: An Anthology of Writings, pp. 162–175 (2006)
63. de Sousa, D.M.: Textural design: a compositional theory for the organization of musical texture. Ph.D. thesis, Universidade Federal do Rio de Janeiro (2019)
64. Takamori, H., Sato, H., Nakatsuka, T., Morishima, S.: Automatic arranging musical score for piano using important musical elements. In: Sound and Music Computing Conference (SMC 2017), pp. 35–41 (2017)
65. Vechtomova, O., Sahu, G.: LyricJam sonic: a generative system for real-time composition and musical improvisation. In: Johnson, C., Rodríguez-Fernández, N., Rebelo, S.M. (eds.) EvoMUSART 2023. LNCS, vol. 13988, pp. 292–307. Springer, Cham (2023). https://doi.org/10.1007/978-3-031-29956-8_19
66. Zembylas, T., Niederauer, M.: Composing Processes and Artistic Agency: Tacit Knowledge in Composing. Routledge (2017)

Evolving User Interfaces: A Neuroevolution Approach for Natural Human-Machine Interaction

João Macedo[1]([⊠]) [ID], Habtom Kahsay Gidey[2] [ID], Karina Brotto Rebuli[3] [ID], and Penousal Machado[1] [ID]

[1] University of Coimbra, CISUC/LASI, DEI, Coimbra, Portugal
{jmacedo,machado}@dei.uc.pt
[2] Technische Universität München, Munich, Germany
habtom.gidey@tum.de
[3] University of Torino, Turin, Italy
karina.brottorebuli@unito.it

Abstract. Intelligent user interfaces for human-machine interaction should be intuitive, invisible, and embodied in the user's natural physical environment. Despite recent advances, most computing systems still lack interfaces that can perceive complex visual and auditory stimuli.

In this study, we propose a neuroevolution approach, employing artificial neural networks optimized by evolutionary algorithms to evolve and accurately translate user inputs into system commands. This methodology adapts and refines system responses by evolving to accommodate diverse user inputs into precise system commands.

Our findings confirm the effectiveness of this approach, particularly in scenarios involving high-dimensional input spaces. The evolving interfaces developed herein show potential for improved user experiences and could pave the way for intelligent systems with natural user interfaces that understand and respond to user needs effectively.

The successful application of neuroevolution methods underscores their utility in creating natural, intuitive, and personalized interfaces. The study illustrates the potential of such approaches to revolutionize human-computer interaction by allowing organic and unobtrusive interfaces that adapt to individual user behaviors and preferences.

Keywords: neuroevolution · intelligent interfaces · interface evolution

1 Introduction

In the 80's, few people had access to video games in their homes. Instead, they would frequently go to establishments, known as arcades, where they could play popular games such as Pacman, Donkey Kong, or Space Invaders. To play, the users had to pay, by inserting a coin into the machine. When no one was playing, these machines would go into a demonstration mode, which showed a bot playing the game. As a result, it was common to see children who were out of coins, interacting with the machine, pretending to play, i.e., pretending to play the machine while it was in demo mode. This sight inspired the main idea behind this paper: what if machines could adapt their interface to the user's intent, rather than the user having to learn the correct way to interact with the machine? This would potentially reduce the learning curve of interacting with new systems and enable those systems to adapt to people with various disabilities. These issues have already been noted in the Human-Computer Interaction (HCI) literature, as studies showed that traditional interaction systems often require users to conform to rigid input methods, which can be counter-intuitive and inaccessible for some users [1].

Despite the remarkable advances in adaptive user interfaces (AUI) [2] and immersive experiences, such interface designs have significant limitations in incorporating representations of the physical natural world. These approaches attempt to construct worlds that range from simple 2D interfaces to sophisticated models of reality. Yet, they struggle to integrate seamlessly and intuitively with the natural world humans innately understand. The user's sense of presence, tactile feedback, along with the full spectrum of human senses, are often poorly emulated, resulting in a disconnect between the user and the interface. This disconnect manifests in several ways, from the cognitive load required to learn and operate within these abstracted systems to the physical discomfort experienced in virtual reality (VR) interfaces [2]. Furthermore, by design, these approaches prioritize the digital replication of reality over a natural integration with it, leading to a form of escapism rather than an augmentation of real-world experience. Consequently, users must adapt to the constraints and peculiarities of these interfaces instead of the interfaces adapting and evolving to accommodate natural, intuitive human behaviors and the environments they inhabit. On the other hand, natural user interfaces (NUI) aim to enable the user to interact with the computer in a more natural manner. These interfaces enhance accessibility, reduce the cognitive load often associated with traditional interfaces, and facilitate natural interaction. They significantly assist individuals with disabilities, enabling more inclusive design and interaction paradigms that support various input modalities [1]. Such interfaces can also enable bidirectional adaptability, allowing real-time evolution of interactions that align with user preferences and interface capabilities within systems [3]. Nevertheless, unlike human-to-human interactions, computers still lack intricate multimodal perceptual capabilities encompassing complex visual perceptions, such as gestures, facial expressions, and body poses, complementing speech and voice-assisted conversations [3]. Moreover, intelligent human-computer interaction (HCI) aspires to create interfaces that not only process all these multimodal perceptions but also meaningfully understand the user's intent and adapt personalized interactions for every user [3]. As a result, such interfaces can become natural and invisible, leading to context-sensitive, intuitive user experiences [4].

The aforementioned adaptability may be achieved through a machine learning model, that maps the user interactions to the commands expected by the system. Artificial Neural Networks (ANNS) are currently one of the most popular machine learning models and have proved to be suitable for problems from various domains. In recent years, they have been combined with Evolutionary Algorithms (EAs), creating a field of research known as Neuroevolution [5]. In this work, we propose to apply neuroevolution techniques to create adaptive user interfaces for a popular retro video game. This approach could significantly impact assistive technology, gaming, and, generally, intelligent software systems design. As a result, we investigate and further develop invisible interfaces to contribute to advances in NUIs.

The study has two objectives. The first aims to empirically validate the effectiveness of neuroevolution methods in user interface evolution. The second objective is to demonstrate the practical benefits of such approaches in creating more natural and intelligent user interfaces.

The subsequent sections are structured as follows: Sect. 2 provides the foundational background. Next, Sect. 3 describes the approach employed, detailing the neuroevolution approach and experimental setup. Section 4 presents the results and validates the approach. Sections 5 and 6 discuss our findings and conclude the paper, summarizing our work's key points and contributions.

2 Background

The experience of HCI has evolved through several stages, from the first command line interfaces to question-and-answer graphical user interface wizards and the recurring trends of conversational interfaces and interface agents [6–8]. The fundamental premise guiding this evolution is to enable a more natural and intuitive interaction between humans and computers. To build such an intuitive HCI, the user interface (UI) plays a fundamental role. Specifically, the main changes in the UI paradigms for achieving this objective are: (I) improving the ability of a UI to adapt to the user's needs [9], (II) developing UI devices and computational approaches capable of capturing user communication in immersive or natural ways [3], and (III) in some cases, assisting the user with interface agents [10,11]. The first two paradigms are of interest to this study. The former refers to Intelligent UIs, and the latter to natural UIs. The present work proposes to combine these two strategies into an Intelligent Natural UI, evolved with neuroevolution. Hereafter, we explore the foundations of these domains, describing the principles and methods that advance the paradigms of HCI.

2.1 Intelligent User Interfaces (IUI)

An intelligent system has two essential features: goal-directedness and degree of autonomy in performing a specific task [12,13]. Similarly, intelligent user interfaces (IUI) autonomously tailor the UI in response to individual user behavior, adapting to the user modes of interaction and actions [3,14–16]. They encompass several approaches to make HCIs more intuitive and adaptable to the individual user context and needs [17]. These approaches include: (i) adaptive user interfaces (AUIs) and (ii) interface agents.

In AUIs, the objective of creating intelligent interfaces is achieved by changing the dynamics of static elements, content, and interaction layouts in user interfaces [9,18]. Changes are introduced based on user models, the context or domain of the user environments, and the capabilities of the computing elements and their application environments for action and interaction [9,19]. Consequently, AUIs change and tailor the particulars of user interface displays, their specific components, and the layouts of individual interface elements in real-time to facilitate personalized user experiences [17].

A notable use case of IUIs, in such cases, can be a platform for online courses and e-learning that learns a student's learning style, progress, and areas of difficulty and then modify the content presentation accordingly. An advanced adaptive learning platform might present more visual information to a user who learns best through visual representations. At the same time, another might offer interactive problem-solving exercises to a user who benefits from hands-on learning. For example, Mahdi et al. [20] proposed an IUI for an online mobile academy. They used a support vector machine to adapt the system UI to guarantee the cross-cultural usability of the platform. In fact, application of IUIs can be found in many, *e.g.*, gaming [21], web image visualizations for people with visual impairments [22], uncertainty visualization for decision making [23], among others.

Interface agents, on the other hand, aim to extend interfaces by adding a separate layer of intelligence, i.e., using software agents that operate on user interfaces [10]. These agents interact with users, often through a conversational interface, learning their needs, behaviors, and context [7,9]. Interface agents serve as intermediaries between users and the platforms they operate on, aiming to facilitate tasks, provide information, or enhance the user experience in a more personalized and context-aware assistance [7,11]. Although most interface agents are simple user agents and assistants that perform simple task automation or user behavior cloning, they can also be autonomous, user-like bots capable of perceptual skills and cognitive behaviors that closely emulate human users on digital platforms [8].

Interface agents are beneficial in expert systems, like Computer-Aided Design (CAD) tools, where adaptivity in UIs may not be feasible; instead, users may require assistance navigating UIs with such agents' support. Familiar examples include calendar assistants that learn user preferences for tailored information and task management, such as an agent observing a user's scheduling habits to suggest appointments or reminders proactively [24]. The development of interface agents often involves various methods and approaches, including natural language processing and machine learning algorithms like reinforcement learning for behavior cloning and decision-making [9,25]. Such agents can also leverage knowledge from APIs for improved perception and action. Current efforts in autonomous navigation and interaction with interface agents employ multimodal approaches that combine vision-language-action models to emulate user-like behavior [26,27].

However, significant limitations remain in how IUIs embody the natural interactions humans use intuitively. In this context, approaches such as neuroevolution learning present significant potential in realizing intelligent and personalized HCI [28].

2.2   Natural User Interfaces (NUI)

Software and cyber-physical systems are progressing towards Natural User Interfaces (NUIs), intuitive interactions using natural human actions such as gestures, voice, and touch. Alongside this evolution towards NUIs, the hardware components of such interfaces are also advancing for greater flexibility. This is exemplified by using physical and virtual input and output devices, leading to the emergence of natural, or so-called invisible, UIs. This advancement is evident in two distinct interface design approaches: embodied interfaces that seamlessly blend with the physical world and immersive interfaces that transport users into fully-realized virtual environments.

Embodied interfaces focus and capitalize on capturing and interpreting natural user inputs like voice, gestures, and facial expressions, enabling intuitive interactions with real-world environments [29–31]. They represent a possibility for integrating computational processes into physical reality, rendering HCIs an intuitive extension of natural behaviors and environments [4,30,31].

In the case of immersive interfaces employed in augmented and virtual reality, the focus is on creating interfaces that immerse users into the computing elements and virtual realities in such environments. Immersive interfaces create a sense of presence and immersion, often utilizing AR and VR technologies to integrate digital experiences with the physical world in mixed reality or to transport users to entirely virtual environments for them to explore [2,32]. These interfaces achieve the sense of immersion using head-mounted displays, auditory devices, various sensors, and haptics to simulate the sense of perception and interaction in a virtual environment [32].

NUIs offer the user more context-sensitive and intuitive experiences [4], as it does not require the manipulation of specialized devices but instead capture the user movements and expressions as inputs for the computer system. It is worth noting that the distinction between virtual and natural (or invisible) interfaces is not always clearly defined in the existing literature. However, we believe that the distinction is noteworthy. In the former case, the HCI is done through projected virtual interfaces, whereas in the latter, the machine captures the user's actions. Indeed, the naturalness of NUI refers to its ability to allow users to interact with the real world in an intuitive way by capturing and giving meaning to users' voices, gestures, facial expressions, and body movements [29]. In recent work, Erra et al. [33] proposed a system combining NUI and VR for exploring graph visualizations. The VR was composed by a 3D graph explorer. For the NUI component, they used a device that tracks the user's body and hand movements in the physical space. They evaluated the system's perceived usefulness, ease, and playfulness by applying it to the users' questionnaires based on the Technology Acceptance Model (TAM) [34]. Their results suggest that these upcoming technologies are more challenging than the traditional ones but enable users to be more involved during graph interaction and visualization tasks, given the enjoyable experience [33].

NUIs are particularly beneficial in enhancing accessibility and reducing cognitive load, as seen in applications for smart home systems, educational tools, robotics, smart factories, and support for individuals with disabilities [1,35–37]. To mention some interesting examples in the existing literature, Liao et al. [29] presented a systematic review on NUI for smart home systems, Cárdenas-Sainz et al. [35] studied the use of NUIs for educational systems, Hsiao et al. [36] analyzed the NUI for improving the seniors' HCI experience, and Hemery et al. [37] conceptualized a gestured-based NUI for a music learning system.

2.3 Neuroevolution

Neuroevolution is a field of research where Evolutionary Algorithms (EAs) are applied to evolve Artificial Neural Networks (ANNs) [38]. ANNs are Machine Learning models inspired by natural structures in brains, which have been successfully applied to a wide range of problems [39]. They consist of layers of interconnected units that intend to mimic biological neurons. Given the input, an ANN produces its output by passing data between its neurons, multiplying it by the connection weights, and subjecting it to non-linear activation functions. ANNs are typically trained by backpropagation, a method that evaluates the network's performance in a dataset, computes the error gradients, and adjusts the network's parameters accordingly [40].

A different strand of work, which is used in our proposal, uses EAs to evolve neural networks [41]. EAs are a family of stochastic search heuristics loosely based on the principles of evolution through natural selection and Mendel's genetics [42]. Genetic Algorithms (GAs) [43] are a family of population-based EAs that have demonstrated notable effectiveness in addressing challenges across a diverse spectrum of application domains. One of such application is the evolution of ANNs, yielding the field of Neuroevolution [5]. There are three main ways to evolve ANNs: (1) evolving only the parameters (connection weights and biases), keeping the architecture fixed [38]; (2) evolving the ANN architecture but using traditional methods to train its parameters [44,45]; and (3) evolving the parameters and the network topology simultaneously (*e.g.*, NEAT [28]). As, to the best of our knowledge, this is the first application in the literature of Neuroevolution for evolving UIs, the approach (1) was adopted, *i.e.*, only the connection weights and biases were evolved.

3 Approach and Setup

3.1 Approach

The proposed approach aims to enable a user to engage with a computer system using a NUI without a predefined knowledge of the interaction process. To do so, we propose to use an Artificial Neural Network (ANN) to capture the user's inputs and map them to the commands expected by the machine. As a proof-of-concept of this proposal, in this work, we use a computer game to demonstrate the capability of enabling a user to play the game through an Intelligent NUI without previous knowledge of the correct game inputs.

Computer games are usually demanding in terms of motor, sensor, and mental skills needed for interaction control and often require mastering inflexible and complicated input devices and techniques [21] Therefore, they represent a good choice for studying the proposed system. In this work, we used the Space Invaders (SI) game [46] (Fig. 1). This is a popular arcade game released in the late 70's, but it has recently regained popularity in OpenAI's Gymnasium [47] as a reinforcement learning benchmark. In this game, the user controls a space ship, and the goal is to destroy all alien invaders before they reach the bottom of the screen whilst avoiding their shots. The ship can only move sideways and shoot, and thus, there are only four commands: (i) move left, (ii) move right, (iii) shoot, and (iv) standstill.

In this work, we subsample the keyboard keys to emulate an arcade machine with 5 input buttons. The A, S, D, W, and SPACE keys are mapped so that A and D respectively move the ship left and right and W shoots. Neither S nor SPACE have any function assigned and are only included to enable a wider range of user interactions.

The details of the training data, the ANN, and the GA used are described below.

Training Data. In order to evolve the ANNs, the fitness of each candidate solution must be measured, which consists of computing the Mean Squared Error between the predictions of the ANN and the target output, i.e., the command to

Fig. 1. The green spaceship (on the bottom) is controlled by the player. The light green objects (5 × 6 group of objects on the top half of the screen) are the alien invaders that shoot and move horizontally. The three red objects between the player's ship and the aliens represent rocks that can be used as cover from the invaders' shots. The goal of the game is to eliminate as many invaders as possible by shooting them. (Color figure online)

be fed into the game. This implies that we must build a dataset, containing samples of the user's and expected interactions. To do so, we go back to the analogy presented in the introduction, where the computer is in a demonstration mode, showing a bot playing the game, while the user interacts with the computer, pretending to play the game. During this process, the user interactions are logged along with the command that the bot uses to make each action.

Artificial Neural Network. As described in the background section, the ANNs can be evolved in various ways. In this work, we opt to evolve only the weights and biases of the ANN, keeping its topology fixed. This is the simplest form of neuroevolution and was chosen due to having shown to perform satisfactorily in preliminary experimentation. The number of neurons at the input layer was equal to the size of the input data received from the user. It varied according to the UI, as explained below, in the Experiments Section (Sect. 3.2). The output layer had the number of neurons equal to the number of buttons in our emulated arcade machine, i.e., 5, as described in Subsection Approach (Sect. 3.1). Following the recommendations in [48], the topology of the ANN has a single hidden layer whose size is set to the mean of the number of input and output neurons. Moreover, all neurons use the Sigmoid activation function (Eq. 1).

$$S(x) = \frac{1}{1 + e^{(-x)}} \qquad (1)$$

This topology was set based on empirical experimentation, as during the first trials, it was possible to verify that this simple topology performed adequately. For this reason, more complex approaches, such as NEAT, were left for future work.

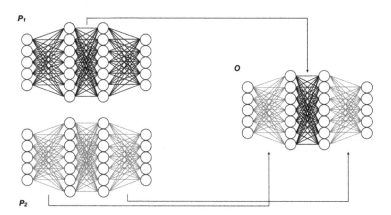

Fig. 2. Crossover operator used in this work. It exchanges the whole set of connection weights and biases of a layer, thus preserving the modularity of the Artificial Neural Network.

Genetic Algorithm. A genetic algorithms was devised to evolve a population of ANNs over a pre-defined number of generations. The initial population was generated as proposed by Glorot and Bengio [49]. On each generation, individuals were selected through tournament selection to generate offspring through crossover or mutation with a fixed probability. The next generation population was built through elitist selection. The proposed algorithm represents each individual as two real-valued matrices, one containing the connection weights between the neurons and the other containing the biases of each layer. The choice of representing this information in matrices rather than in a single vector is due to the crossover operator devised. This operator is inspired in the uniform crossover [42]. However, while the uniform crossover randomly selects a parent to donate each gene, the proposed crossover randomly selects a parent to donate an entire layer of connection weights or biases. As a result, this crossover aids in preserving the modularity of the ANNs, leading to less disruptive modifications. Figure 2 graphically presents one example of a crossover operation, where two parents (P_1 and P_2) produce an offspring (O) with P_1 donating the hidden layer and P_2 donating the input and output layers. The offspring may also be subjected to mutation, in which a zero-mean Gaussian noise with standard deviation σ is probabilistically applied to each connection weight and bias. Finally, each individual was evaluated using a subset of the training data, with the user and the system inputs.

The performance of each individual is assessed with the Mean Squared Error (MSE, Eq. 2) using training or test data.

$$MSE = \frac{1}{N} \sum_{i=1}^{N} \left(Y_i - \hat{Y}_i\right)^2 \tag{2}$$

where N is the number of dataset instances, Y_i is the target of the ith data instance, and $\hat{Y}_i$ is the ANN output for the ith data instance. The remaining parameters of the proposed algorithm are presented in Table 1.

Table 1. Genetic Algorithm parameters.

Parameter	Value
Population size	200
Generations	500
Tournament size	2
Elite size	1
Crossover rate	0.7
Mutation rate	0.01
Mutation σ	0.03

Fig. 3. Mediapipe face landmarks.

3.2 Experiments

Two different experiments were conducted, both of which started with the user visualizing a bot playing the game and attempting to match its actions. In the first, called *keyboard experiment*, the user interacted with the game through a keyboard with no knowledge of the effect of each button. This experiment was conducted to verify if it would be possible to evolve an ANN that maps between the user's own keyboard commands to the correct game inputs. In the second, called *camera experiment*, instead of using the keyboard, the user interacted with the game through head movements and facial expression, attempting to match the bot's actions. It should be noted that the user was free to interact in the most natural way, as there was no predefined set of movements or expressions to execute the game commands. In order to avoid using more complex models, such as convolutional neural networks, the MediaPipe [50] library was used to extract 468 face landmarks, whose x and y coordinates were fed into the ANN, yielding a total of 936 inputs. The face landmarks consist on keypoints of the human face that can be used for detecting facial expressions. Figure 3 presents an example of the landmark mesh overlaid on top of a human face.

This experiment was conducted to assess the feasibility of evolving an ANN that maps between the user's head movements and facial expressions to the correct game inputs, thus producing a natural user interface.

Due to the difference in the size of the inputs, the networks had different sizes in the two experiments. In the keyboard experiment, the ANN had 5 input neurons, 5 hidden neurons, and 5 output neurons. In turn, in the camera experiment, the ANN had 936 input neurons, 470 hidden neurons, and 5 output neurons. For each experiment, 5856 data samples were generated and split 70% for training and 30% for testing. 30 independent trials were made for each approach, with the data in the train and test sets being re-sampled for each trial.

4   Results

This section presents the experimental results for both the keyboard and the camera experiment. Thirty independent runs of the evolutionary algorithm were made for each experiment, each using a different seed and a different batch of train/test data. We start by assessing the performance of the proposed neuroevolution approach in both experiments and later we compare it to backpropagation.

4.1   ANN Evolution

Figure 4 presents the mean and standard deviation of the fitness of the best individuals from all runs for each generations. As can be seen, in both experiments, the system approximates the data quite well, achieving good performance in both train and test. Due to the overlap between the train and test curves, it can also be concluded that the models do not suffer from overfitting. Moreover, while in the keyboard experiment (left) the evolution seems to have stagnated, in the camera experiment (right) the fitness keeps improving until the last generation, so it is likely that even better results are achievable. It is also interesting to note that, despite the camera experiment being harder than the keyboard experiment, its networks achieve better fitness, which is likely due to their increased size (1411 neurons vs 15 neurons).

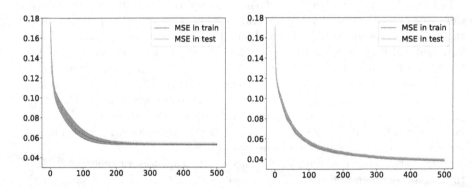

Fig. 4. Train/test performance over the generations for the keyboard *(left)* and camera *(right)* remapping experiments.

4.2   Evolutionary vs. Backpropagation

To further assess the usefulness of neuroevolution for this problem, we compared it to adapting the parameters of the ANN through backpropagation. The parameters of this approach are set to provide fair comparisons to the neuroevolution approach and are presented in Table 2. Figure 5 presents the boxplots of the

Table 2. Parameters of the backpropagation approach.

Parameter	Value
Optimiser	Adam
Learning rate	0.001
Epochs	1000 (same to EA)
Activation function	Sigmoid

Table 3. p-values of the Kolmogorov-Smirnov test for the normality of the data.

	e_{train}	b_{train}	e_{test}	b_{test}
Keyboard	4.6705e−6	4.4358e−4	7.2513e−5	2.0124e−5
Camera	4.1077e−5	1.8057e−5	1.5507e−4	4.3751e−5

distribution of the MSE of the best individuals attained from 30 runs of the evolutionary (e) and backpropagation (b) approaches in the train and test datasets. As can be seen, the evolutionary approach seems to produce better results than the backpropagation in the keyboard experiment. In the camera experiments, the differences in performance of the two approaches are clearer, favouring the use of neuroevolution. In order to draw statistically-validated conclusions, we proceed to conduct statistical hypothesis tests, setting the significance level $\alpha = 5e - 2$. In order to choose an adequate test, the first step is to assess the normality of the data, using the Kolmogorov-Smirnov test (Table 3). As can be seen, the p-values of this test applied to all sets of data are below the chosen significance value of 0.05 and thus the data cannot be considered to follow normal distributions. As a result, a non-parametric test must be applied and thus we resort to the Wilcoxon test. The test results are presented in Table 4 and show that, when using the keyboard, there are no statistically significant differences between neuroevolution and backpropagation, both in training and in test. In turn, when using the camera, the Wilcoxon test shows that there are statistically significant differences between the performance of the two approaches in both train and test. Analysing the boxplots in Fig. 5, it can be concluded that those differences are due to the proposed evolutionary approach producing significantly better performance than backpropagation, in the camera experiments. These results are likely due to backpropagation being easily trapped in local optima, as discussed in [38].

Importantly, it is also possible to see in these results that the ANNs generalize adequately to unseen data, as the performance in test is not very different from the performance in train.

Table 4. p-values of the Wilcoxon test comparing the results of evolutionary and backpropagation approaches in the keyboard and camera experiments.

	Train	Test
Keyboard	**8.4065e−2**	**9.6102e−2**
Camera	6.2866e−5	1.3406e−3

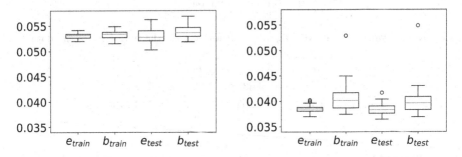

Fig. 5. Train/test performance of the best models produced through evolution (e) and backpropagation (b) for the keyboard (left) and camera (right) remapping experiments.

4.3 Playability

Independently of the numerical results, the essential questions concern the effects on the game's playability. Is it possible to play the game through the evolved interfaces? How does it affect gameplay? Is it fun to play?

This paper does not include formal user testing, but we can offer insights based on our experiences and those of our colleagues. It is indeed possible to play the game with the new interfaces, including controlling the ship, shooting, and selecting strategic positions. However, precise control is a challenge. This limitation affects the gameplay; players can't rely on accurate shots or small movements. We believe that highly skilled players might not prefer this, but in our observation, many people find the lack of fine control to add an interesting twist to the game, a new challenge to overcome.

Interestingly, a significant number of people found the game incredibly fun, which was unexpected. The new interface transforms the original game, introducing new challenges and a refreshed experience for the user.

A small video illustrating the user experience can be seen here: https://cdv. dei.uc.pt/wp-content/uploads/2024/01/evostar2024.mp4.

5 Discussion

The results of our experiments, presented in Sect. 4, indicate that neuroevolution is a valid approach for optimizing the parameters of ANNs applied to the creation of Intelligent NUIs. This was true, especially in scenarios where the input data dimensionality is high, as in the case of the camera input experiment. GA demonstrated a robust ability to generate a model that accurately maps the user's unconventional inputs to the correct game controls, which could have broader implications for designing intelligent NUIs or UIs in general.

5.1 Comparison with Backpropagation

While backpropagation is a well-established method for training ANNs, the experiments performed in this work demonstrated that performance was comparable to the evolutionary approach when using the low-dimensional keyboard as input but significantly less effective when dealing with high-dimensional data from the camera. The evolutionary approach's outstanding performance with the camera input might be attributed to its ability to explore a more diverse set of solutions [41,44], which is essential when navigating complex and high-dimensional input spaces. The traditional gradient descent process can struggle in such spaces, often getting trapped in local minima or failing to navigate flat regions effectively [44,51].

Furthermore, the evolutionary approach exhibits a form of inherent regularization, as it continuously evaluates a population of candidate solutions, avoiding overfitting, which might occur in backpropagation [52–54]. This is an advantage over backpropagation, which relies on error gradient optimization of network weights and can be trapped in a poor local optimum [38], especially when faced with complex error surfaces or suboptimal initial weight selection.

5.2 Applications on Intelligent Natural User Interface Design

The experimental setup also provided insights into the application of neuroevolution in Intelligent NUI design, which aims to enable seamless, invisible, and intuitive interaction. In the context of the IUI represented by the keyboard experiment, the ANN successfully learned the mapping between the user's inputs and the desired output commands. This showed the feasibility of this approach in making interfaces more accessible without the need for explicit instruction.

The camera experiment, on the other hand, represented an Intelligent NUI in which the user interacts with the game in a non-traditional and more natural manner through facial expressions and head movements. The success of the combined neuroevolution approach in this context showcases its potential for creating more immersive and natural user experiences. As discussed previously in Sect. 2, this could be especially beneficial for individuals with disabilities, as it can allow for alternative methods of interaction that do not rely on traditional and predefined input devices.

5.3 Implications for Future Research

The findings in this study suggest several implications for developing IUIs in various fields. First, the approach can be generalized to other domains where there is a need for mapping unconventional inputs to system commands, such as in assistive technologies, robotics, and various HCI scenarios. Second, the successful experimental results with the neuroevolution underscore the value of exploring alternative machine learning methods. This is particularly important when the data is not well-suited for established methods. With its global search properties, the neuroevolution approach provides a compelling alternative for scenarios where the input space is vast and complex. Third, neuroevolution approaches in NUIs can vastly improve technology accessibility, providing more inclusive, intelligent, natural, and adaptable systems that cater to a broader range of human behavior and preferences. The ability to map non-standard inputs to system commands paves the way for developing interfaces that adapt to the user rather than forcing the user to adjust to the interfaces.

Overall, this research demonstrates the effectiveness of neuroevolution in training ANNs for complex input spaces, paving the way for intelligent NUIs. This also has significant implications for designing user-centered, accessible, and intuitive systems for natural HCI.

5.4   Limitations and Future Work

Despite the results, our study's experimental setup and evaluation architecture have limitations that future research could further explore. The study's experiment evaluation was set in a controlled environment with a specific set of inputs and a single use case. Future research could explore the method's applicability across domains and with varied input modalities. Besides, the ANN architectures used were relatively simple. Evaluating complex architectures and topologies, such as deep Learning models, might further enhance the performance and capability of the system. The evolutionary strategy could also be augmented with local search techniques to refine the solutions it discovers, combining the global search capability of evolution with the local optimization efficiency of gradient-based methods. Finally, as the intelligent NUI is user-centered, future research should also perform controlled experiments with users evaluating the usefulness and enjoyability of the system, for example, with questionnaires like the Technology Acceptance Model [55].

6   Conclusion

Intelligent systems pose challenges that demand advanced interface designs and interactive capabilities. Addressing such challenges also requires considering various unconventional approaches to tackle the diverse problems in the domain. Thus, this study presented empirical results for the viability of neuroevolution approaches for training artificial neural networks, particularly in handling high-dimensional input spaces. The experimental results demonstrated that our approach could outperform traditional backpropagation in scenarios with complex and unconventional input types, such as those provided by camera inputs for interactive interfaces.

These findings have substantial implications for intelligent NUIs, suggesting a shift towards more natural invisible interfaces capable of interacting with users with advanced levels of adaptability to the context and capabilities of the user. By leveraging evolutionary algorithms, systems can evolve to fit the users' unique interaction patterns, offering a more accessible, inclusive experience that is particularly beneficial to individuals with disabilities.

Exploring neuroevolution extends prospects in interface designs for intelligent systems that are not only responsive to user behavior but also capable of evolving with the users over time. As an outlook, future studies could expand on this foundation by exploring diverse application domains, integrating more neural network architectures, and refining evolutionary techniques for greater efficiency and applicability.

Acknowledgement. We would like to thank the SPECIES Society for granting us the opportunity to take part in the SPECIES Summer School, which ignited this research. This work was partially supported by the Portuguese Recovery and Resilience Plan (PRR) through project C645008882-00000055, Center for Responsible AI, by the FCT, I.P./MCTES through national funds (PIDDAC), within the scope of CISUC R&D Unit - UIDB/00326/2020 or project code UIDP/00326/2020 and by Project "NEXUS Pacto de Inovação - Transição Verde e Digital para Transportes, Logística e Mobilidade". ref. No. 7113, supported by the Recovery and Resilience Plan (PRR) and by the European Funds Next Generation EU, following Notice No. 02/C05-i01/2022.PC645112083-00000059 (project 53), Component 5 - Capitalization and Business Innovation - Mobilizing Agendas for Business Innovation.

References

1. Stephanidis, C.: User Interfaces for All: Concepts, Methods, and Tools. CRC Press, Boca Raton (2000)
2. Akçayır, M., Akçayır, G.: Advantages and challenges associated with augmented reality for education: a systematic review of the literature. Educ. Res. Rev. **20**, 1–11 (2017)
3. Lew, M., Bakker, E.M., Sebe, N., Huang, T.S.: Human-computer intelligent interaction: a survey. In: Lew, M., Sebe, N., Huang, T.S., Bakker, E.M. (eds.) HCI 2007. LNCS, vol. 4796, pp. 1–5. Springer, Heidelberg (2007). https://doi.org/10.1007/978-3-540-75773-3_1
4. Gross, M., Do, E.: Toward design principles for invisible interfaces. In: Workshop on Invisible and Transparent Interfaces in Conjunction with AVI Conference on Advanced Visual Interfaces, pp. 623–644. Citeseer (2004)
5. Zhang, M., Banzhaf, W., Machado, P.: Handbook of Evolutionary Machine Learning. Springer, Singapore (2024). https://doi.org/10.1007/978-981-99-3814-8
6. Dahlbäck, N., Jönsson, A., Ahrenberg, L.: Wizard of Oz studies: why and how. In: Proceedings of the 1st International Conference on Intelligent User Interfaces, pp. 193–200 (1993)
7. Lebeuf, C., Storey, M.-A., Zagalsky, A.: Software bots. IEEE Softw. **35**(1), 18–23 (2017)
8. Gidey, H.K., Hillmann, P., Karcher, A., Knoll, A.: Towards cognitive bots: architectural research challenges. In: Hammer, P., Alirezaie, M., Strannegård, C. (eds.)

AGI 2023. LNCS, vol. 13921, pp. 105–114. Springer, Cham (2023). https://doi.org/10.1007/978-3-031-33469-6_11

9. Jameson, A.: Adaptive interfaces and agents. In: The Human-Computer Interaction Handbook, pp. 459–484. CRC Press (2007)

10. Maes, P., Kozierok, R.: Learning interface agents. In: AAAI, vol. 93, pp. 459–465 (1993)

11. Lashkari, Y., Metral, M., Maes, P.: Collaborative interface agents. In: AAAI, vol. 94, pp. 444–449. Citeseer (1994)

12. Pfeifer, R., Scheier, C.: Understanding Intelligence. MIT Press, Cambridge (2001)

13. Doctor, T., Olaf, W., Solomonova, E., Bill, D., Levin, M.: Biology, Buddhism, and AI: care as the driver of intelligence. Entropy **24**(5), 710 (2022)

14. Langley, P.: Machine learning for adaptive user interfaces. In: Brewka, G., Habel, C., Nebel, B. (eds.) KI 1997. LNCS, vol. 1303, pp. 53–62. Springer, Heidelberg (1997). https://doi.org/10.1007/3540634932_3

15. Amershi, S., et al.: Guidelines for human-AI interaction. In: Proceedings of the 2019 CHI Conference on Human Factors in Computing Systems, pp. 1–13 (2019)

16. Maybury, M.: Intelligent user interfaces: an introduction. In: Proceedings of the 4th International Conference on Intelligent User Interfaces, pp. 3–4 (1998)

17. Jalil, N.: Introduction to intelligent user interfaces (IUIS). In: Software Usability. IntechOpen (2021)

18. Mitchell, J., Shneiderman, B.: Dynamic versus static menus: an exploratory comparison. ACM SigCHI Bull. **20**(4), 33–37 (1989)

19. Gullà, F., Cavalieri, L., Ceccacci, S., Germani, M., Bevilacqua, R.: Method to design adaptable and adaptive user interfaces. In: Stephanidis, C. (ed.) HCI 2015, Part I. CCIS, vol. 528, pp. 19–24. Springer, Cham (2015). https://doi.org/10.1007/978-3-319-21380-4_4

20. Miraz, M.H., Ali, M., Excell, P.S., Khan, S.: Ai-based culture independent pervasive m-learning prototype using UI plasticity design. Comput. Mater. Continua **68**, 1021–1039 (2021)

21. Grammenos, D., Savidis, A., Stephanidis, C.: Designing universally accessible games. Comput. Entertain. (CIE) **7**(1), 1–29 (2009)

22. Alam, M.Z.I., Islam, S., Hoque, E.: SeeChart: enabling accessible visualizations through interactive natural language interface for people with visual impairments. In: Proceedings of the 28th International Conference on Intelligent User Interfaces, IUI 2023, New York, NY, USA, pp. 46–64. Association for Computing Machinery (2023)

23. Prabhudesai, S., Yang, L., Asthana, S., Huan, X., Vera Liao, Q., Banovic, N.: Understanding uncertainty: how lay decision-makers perceive and interpret uncertainty in human-AI decision making. In: Proceedings of the 28th International Conference on Intelligent User Interfaces, IUI 2023, New York, NY, USA, pp. 379–396. Association for Computing Machinery (2023)

24. Kozierok, R., Maes, P.: A learning interface agent for scheduling meetings. In: Proceedings of the 1st International Conference on Intelligent User Interfaces, pp. 81–88 (1993)

25. Gidey, H.K., Hillmann, P., Karcher, A., Knoll, A.: User-like bots for cognitive automation: a survey. arXiv preprint arXiv:2311.12154 (2023)

26. Zitkovich, B., et al.: Rt-2: vision-language-action models transfer web knowledge to robotic control. In: 7th Annual Conference on Robot Learning (2023)

27. Gur, I, et al.: A real-world webagent with planning, long context understanding, and program synthesis. arXiv preprint arXiv:2307.12856 (2023)

28. Stanley, K.O., Miikkulainen, R.: Evolving neural networks through augmenting topologies. Evol. Comput. **10**(2), 99–127 (2002)
29. Liao, L., Liang, Y., Li, H., Ye, Y., Guangdong, W.: A systematic review of global research on natural user interface for smart home system. Int. J. Ind. Ergon. **95**, 103445 (2023)
30. Fishkin, K.P., Moran, T.P., Harrison, B.L.: Embodied user interfaces: towards invisible user interfaces. In: Chatty, S., Dewan, P. (eds.) Engineering for Human-Computer Interaction. IIFIP, vol. 22, pp. 1–18. Springer, Boston, MA (1999). https://doi.org/10.1007/978-0-387-35349-4_1
31. Dourish, P.: Where the Action is: The Foundations of Embodied Interaction. MIT Press, Cambridge (2001)
32. Dede, C.: Immersive interfaces for engagement and learning. Science **323**(5910), 66–69 (2009)
33. Erra, U., Malandrino, D., Pepe, L.: A methodological evaluation of natural user interfaces for immersive 3D graph explorations. J. Vis. Lang. Comput. **44**, 13–27 (2018)
34. Davis, F.D.: Perceived usefulness, perceived ease of use, and user acceptance of information technology. MIS Q. **13**, 319–340 (1989)
35. Cárdenas-Sainz, B.A., Barrón-Estrada, M.L., Zatarain-Cabada, R., Ríos-Félix, J.M.: Integration and acceptance of natural user interfaces for interactive learning environments. Int. J. Child-Comput. Interact. **31**, 100381 (2022)
36. Hsiao, S.-W., Lee, C.-H., Yang, M.-H., Chen, R.-Q.: User interface based on natural interaction design for seniors. Comput. Hum. Behav. **75**, 147–159 (2017)
37. Hemery, E., Manitsaris, S., Moutarde, F., Volioti, C., Manitsaris, A.: Towards the design of a natural user interface for performing and learning musical gestures. Procedia Manuf. **3**, 6329–6336 (2015). 6th International Conference on Applied Human Factors and Ergonomics (AHFE 2015) and the Affiliated Conferences, AHFE 2015
38. Yao, X.: Evolving artificial neural networks. Proc. IEEE **87**(9), 1423–1447 (1999)
39. Russell, S.J., Norvig, P.: Artificial Intelligence a Modern Approach. London (2010)
40. Humelhart, D., Hinton, G., Williams, R.: Learning representations by back-propagating errors. Nature **323**, 533–536 (1986)
41. Stanley, K.O., Clune, J., Lehman, J., Miikkulainen, R.: Designing neural networks through neuroevolution. Nat. Mach. Intell. **1**(1), 24–35 (2019)
42. Eiben, A.E., Smith, J.E.: Introduction to Evolutionary Computing. Springer, Heidelberg (2015). https://doi.org/10.1007/978-3-662-44874-8
43. Goldberg, D.E.: Genetic Algorithms in Search, Optimization, and Machine Learning. Addison Wesley series in artificial intelligence. Addison-Wesley (1989)
44. Floreano, D., Dürr, P., Mattiussi, C.: Neuroevolution: from architectures to learning. Evol. Intell. **1**, 47–62 (2008)
45. Assunçao, F., Lourenço, N., Machado, P., Ribeiro, B.: DENSER: deep evolutionary network structured representation. Genet. Program Evolvable Mach. **20**, 5–35 (2019)
46. Arcade. Space invaders. [Video Game] (1978)
47. Towers, M., et al.: Gymnasium (2023)
48. Heaton, J.: Introduction to Neural Networks with Java. Heaton Research Inc. (2008)
49. Glorot, X., Bengio, Y.: Understanding the difficulty of training deep feedforward neural networks. In: Proceedings of the thirteenth international conference on artificial intelligence and statistics, pp. 249–256. JMLR Workshop and Conference Proceedings (2010)

50. Lugaresi, C., et al.: MediaPipe: a framework for building perception pipelines. arXiv preprint arXiv:1906.08172 (2019)
51. Galván, E., Mooney, P.: Neuroevolution in deep neural networks: current trends and future challenges. IEEE Trans. Artif. Intell. **2**(6), 476–493 (2021)
52. Such, F.P., Madhavan, V., Conti, E., Lehman, J., Stanley, K.O., Clune, J.: Deep neuroevolution: genetic algorithms are a competitive alternative for training deep neural networks for reinforcement learning. arXiv preprint arXiv:1712.06567 (2017)
53. Liang, J., Gonzalez, S., Shahrzad, H., Miikkulainen, R.: Regularized evolutionary population-based training. In: Proceedings of the Genetic and Evolutionary Computation Conference, pp. 323–331 (2021)
54. Ding, L., Spector, L.: Evolving neural selection with adaptive regularization. In: Proceedings of the Genetic and Evolutionary Computation Conference Companion, pp. 1717–1725 (2021)
55. Davis, F.D.: Perceived usefulness, perceived ease of use, and user acceptance of information technology. MIS Q. **13**(3), 319–340 (1989)

Evolving Visually-Diverse Graphic Design Posters

João Macedo$^{(\boxtimes)}$, Daniel Lopes$^{(\boxtimes)}$, João Correia , Penousal Machado ,
and Ernesto Costa

University of Coimbra, CISUC/LASI, DEI, Coimbra, Portugal
{jmacedo,dfl,jncor,machado,ernesto}@dei.uc.pt

Abstract. Finding unconventional visual solutions that stand out and draw attention is frequently one of the goals of Graphic Design (GD). However, to save time, graphic designers often adhere to design trends and templates, resulting in creations that frequently lack distinctive qualities. To speed up the creative process, among other techniques, researchers have been using evolutionary algorithms to generate and present novel GD solutions to end-users, so they can save time by working over the generated ideas. Nevertheless, state-of-the-art approaches often converge to a final group of results that look too similar to each other. In this paper, we test the application of niching techniques to generate a set of visually varied GD solutions and present it to end-users. GD posters were evolved as a proof of concept. The results suggest that implementing the tested niching technique in evolutionary systems can be a viable approach to present users with a wider range of ideas, at least for poster design.

Keywords: Niching · Evolutionary Algorithms · Graphic Design · Poster

1 Introduction

The Graphic Design (GD) process is often driven by the pursuit of creative and unconventional visual solutions that captivate audiences. However, the demands of the modern creative industry often leave graphic designers with limited time to explore innovative visual solutions. As a result, designers frequently resort to using design trends and templates, compromising the distinctive qualities of their work.

To speed up the GD creative process, multiple computational techniques have been explored by researchers and practitioners, such as generative design, machine learning, or evolutionary algorithms (EA). Generative design can be useful to speed up the creation of design variations for specific design projects. However, it is often necessary to develop specific algorithms for each different project, requiring an initial time overhead and programming background. Machine learning models can be useful to speed up a number of creative tasks,

---

J. Macedo and D. Lopes—These authors contributed equally.

© The Author(s), under exclusive license to Springer Nature Switzerland AG 2024
C. Johnson et al. (Eds.): EvoMUSART 2024, LNCS 14633, pp. 265–278, 2024.
https://doi.org/10.1007/978-3-031-56992-0_17

such as generating high-resolution images [19], which can be included in design artefacts. Nevertheless, such models often generate pastiche [24], i.e., imitations of existing styles, making machine learning not well-suited for generating innovative visual solutions. On the other hand, evolutionary algorithms [3] explore solutions in a semi-stochastic way, allowing unforeseen results to show up. Thus, for the generation of innovative visual solutions, evolutionary techniques might be better suited.

As an example, such techniques can be used to provide designers with a pool of unforeseen ideas and allow them to build upon these foundations to save time in the design workflow. However, evolutionary algorithms driven only by fitness often converge to solutions that are too similar to each other, thereby limiting the diversity of the ideas presented to the designers, i.e., if all the generated images look alike, only one or a few might be useful to inspire the designer.

In response to this problem, this paper explores the application of niching techniques in the realm of GD. More specifically, as a proof of concept, this study focuses on the evolution of simple GD posters. In future work, experiments must be conducted to evolve more complex posters. The evolutionary engine implemented is based on a state-of-the-art system [12] also designed to evolve GD posters.

The results suggest that, by leveraging the presented niching techniques, it is possible to contradict the generation of homogeneous populations of posters, providing end-users with a wider spectrum of design suggestions.

In the remainder of this paper, existing evolutionary systems for generating GD artefacts are reviewed along with existing niching techniques. Thereafter, the tested approach is presented, going through a comprehensive description of both the evolutionary engine and the tested niching method. Lastly, the experimental setup and results are presented, and conclusions are drawn.

2   Related Work

The present paper aims to contribute by proposing the application of a niching technique to the task of evolving sets of visually varied GD artefacts, so graphic designers can save time in the creative process, by getting inspired by the generated results and building on top of them. As a proof of concept, we focus on the generation of simple GD posters. In that sense, this section reviews existing evolutionary systems for generating GD artefacts, especially posters. Furthermore, existing diversity preservation techniques are reviewed, including explicit fitness sharing [6], novelty search [10] and map-elites [18].

2.1   Evolutionary Poster Design

Evolutionary algorithms can be roughly divided into two different approaches: interactive and automatic. Interactive systems require the users to manually drive the generation by picking the individuals (posters) they prefer. On the other hand, automatic systems must perform the evolutionary process without human intervention by relying on fitness functions to evaluate individuals.

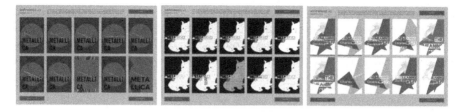

Fig. 1. Three examples of final populations of ten individuals each, generated using *Graphagos* along 11, 34 and 40 generations, from left to right, respectively. Retrieved from www.graphagos.com/portfolio.

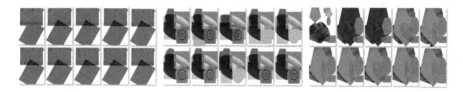

Fig. 2. Three examples of final populations of ten individuals each, generated using *EvoDesigner* along 10, 30 and 40 generations, from left to right, respectively.

Due to the difficulty in objectifying and automating aesthetics, in GD, interactive approaches are often endorsed [4,8,9,16,21]. Nevertheless, work on the automatic evolution of GD artefacts, such as posters [12,20] or typography [14], has also been done.

Considering either interactive or automatic techniques, for most of the review systems, it is often difficult to assess the diversity of the generated populations since the published results often do not focus on that matter. Nevertheless, one can verify the existing lack of diversity within final populations, for example, by looking at systems like *Graphagos* [16], which offers an interactive online demo as well as a portfolio of example results.

For instance, Fig. 1 showcases three final populations taken from *Graphagos'* portfolio (concerning different numbers of generations). Judging from the examples, one might consider that, between different runs, the phenotypes of the individuals can vary enough for them to be deemed visually distinct. However, within each final population, the ten phenotypes showcased can be deemed visually similar to each other.

Also, an identical issue can be noticed by testing *EvoDesigner* [12], a more recent automatic system, also meant to evolve GD posters. For instance, refer to Fig. 2 which showcases three examples of final populations evolved using *EvoDesigner* along different numbers of generations.

Similarly to the aforementioned systems, diversity has been lacking in numerous evolutionary contexts, especially when the process is driven only by fitness metrics. Apart from the desire to obtain diverse solutions, the use of fitness-guided search has been shown to cause the population to converge to a region of the search space, which may not contain the global optima [5]. This is due to a phenomenon known as deception, which states that successive improvements in fitness may lead the search away from the global optima. For that reason, since early [6,7], there has been an interest in the research for effective diversity-preserving methods.

2.2 Diversity Preservation Algorithms

Diversity preservation (DP) algorithms differ from traditional EAs in the way they guide the search: while traditional EAs guide their search solely through fitness, DP algorithms attempt to find a set of good performing, yet diverse, solutions for a given problem [1]. Currently, various DP algorithms have been proposed, but two main classes emerge—Niching and Quality-Diversity. While both classes aim to provide a set of diverse good quality solutions, they differ in the way they do so.

Niching. Niching [2,17] algorithms simply aim to provide a set of good diverse solutions by encouraging the emergence of different species in the population. Crowding [2] is one of the earliest niching methods and was initially proposed for a steady state algorithm, where the offspring were created only through muta-tion of the parents. Every time an offspring is created, C_f parents are randomly selected and the most similar to the offspring is replaced. Explicit Fitness Shar-ing [6,7,22] is another early niching method, and was also initially proposed as a mechanism to promote diversity and prevent premature convergence in evo-lutionary algorithms. This is done by adjusting the fitness values of individuals based on their proximity to other individuals in the search space, with the goal of penalising solutions in densely populated regions. If two solutions are too similar to one another their fitness scores are penalised according to the sharing function described in Eq. 1:

$$Sh(d) = \begin{cases} 1 - (\frac{d}{\sigma})^{\alpha}, & d \leq \sigma \\ 0, & d > \sigma \end{cases} \tag{1}$$

where d is the distance between two individuals, σ is the distance threshold, under which two solutions are considered to be neighbours and *alpha* is a coeffi-cient that is typically set to 1. As a result, the adjusted fitness of the ith individ-ual in the population ($F_a(i)$) is set considering all individuals in the population, according to Eq. 2.

$$F_a(i) = \frac{F(i)}{\sum_{j=1}^{N} Sh(d_{i,j})} \tag{2}$$

where $F(i)$ is the fitness of the ith individual, N is the number of individuals in the population and $d_{i,j}$ is the distances between the ith and jth individuals. Note that from this definition, the distance may be measured over various spaces, such as the genotypic, phenotypic or even semantic (or behavioural) spaces.

This way, explicit fitness sharing discourages the dominance of specific solutions and fosters a more even distribution of individuals across diverse regions of the solution space, i.e. encourages the evolution of a wider range of possibilities. Clearing [17] is another diversity maintenance mechanism employed in evolutionary algorithms that groups individuals into species based on their similarity. However, unlike explicit fitness sharing, each species is assigned a dominant individual (i.e., the fittest) whose fitness remains unchanged. All other individuals in the population have their fitness reset, effectively reducing their chances of surviving into the next generations. The idea is to create space in crowded regions, allowing for the exploration of new, unoccupied areas in the solution space. By periodically clearing out densely populated regions, the algorithm encourages a more uniform distribution of individuals.

In both explicit fitness sharing and clearing, it is necessary to define a distance threshold (σ) under which the individuals are considered to belong to the same niche. This requirement poses one of the main drawbacks of these approaches. Similarly, choosing an adequate value for the crowding factor (C_f) of crowding also poses some difficulties.

Novelty search [10] differs from the previous approaches by completely disregarding the fitness of the individuals throughout evolution. Instead, focuses on promoting solutions that exhibit unique and previously unexplored characteristics within the current population. This is done by making small modifications to the standard evolutionary algorithms, simply replacing the fitness score with a novelty score. The novelty of each individual is simply computed through its distance to the other individuals in the population as well as to its predecessors, which are stored in an archive. Note that not all individuals make it to the archive, but rather those deemed to have sufficiently novel behaviours (higher than a predefined threshold) at the time of their creation. Novelty search with local competition [11] is an improvement over the original method that attempts to combine fitness with novelty to guide the search. This is done by using a multi-objective evolutionary algorithm that simultaneously optimises the novelty of each individual as well as how well it performs relative to the other members of its species.

Quality-Diversity. Quality-diversity methods go one step further than Niching, attempting to illuminate the semantic (or behavioural) space to produce the best possible individual for each location [18]. It does so by first defining a set of characteristics (commonly known as behaviour descriptors) that describe the behaviour (or semantics) of the individuals. The search space is then discretised based on those characteristics and, for each cell of the space, Map-Elites attempts to find the most fit solution. This approach has been typically applied to robotics, to evolve repertoires of diverse robotic behaviours to perform a given task [15].

In summary, diversity preservation methods offer various advantages in evolutionary algorithms. Explicit fitness sharing, crowding and clearing aim to evolve diverse sets of fit solutions by treating fitness as a resource that is shared among the individuals within each species. In turn, novelty search completely disregards the quality of the individuals and focuses on promoting the exploration of solutions that are innovative compared to other individuals. Lastly, Map-Elites aims to illuminate the search space, creating a map based on predefined characteristics and providing the best individual for each location of that map.

Although any of the aforementioned techniques could be worth testing for poster generation, in this paper, we selected Explicit Fitness Sharing due to its simplicity and computational efficiency, compared to some other diversity approaches. Also, using this technique, it can be relatively simple to retrieve a given number of qualified but diverse individuals, presenting to the user a given number of different ideas at the end of the process. In further studies, the several other approaches must be tested against each other.

3   Approach

To assess whether Explicit Fitness Sharing could fit the poster generation task, we started by implementing a simplified version of a state-of-the-art automatic evolutionary engine to generate GD posters [12]. Thereafter, we continued by evolving posters guided only by fitness and then proceeded to evolve posters with explicit fitness sharing. Similarly to other methods from the literature (see Figs. 1 and 2), the goal of the EA in both approaches is to produce a set of visually diverse individuals in a single run, i.e., once the evolution terminates, the user must be presented with a given number of different posters, For the following tests, we generated 5 different ideas.

3.1   Evolutionary Engine

As mentioned before, to generate posters, an evolutionary engine based on state-of-the-art work [12] was implemented. The original engine was implemented in *JavaScript* and *ExtendScript* and the phenotypes were rendered directly by *Adobe InDesign*. However, to ease the implementation, our engine was developed in *Python*, and the phenotypes were rendered in *JavaScript* using a *Node.js* implementation of the HTML canvas. The similarities between the two engines rely primarily on the representation of the individuals, selection, and variation.

The system must be initialised by defining the following page settings: (i) grid information, i.e., margin size, number of columns, number of rows, and gutter size; and (ii) background colour. Moreover, the user must set up the number, order, and type of items to include in the evolved pages. Two types of items can be passed: geometric shapes and text boxes. Shapes require no extra parameters, but for text boxes, some text content must be provided. These initialisation settings are fixed during evolution.

After initialisation, the engine evaluates the individuals and checks for termination criteria, i.e., whether a given number of generations has been completed. If no termination criteria were matched, a new population is created, starting with sampling a pool of mates from the current individuals using a tournament method of size 3. These mates then reproduce through crossover and mutation,

```
page {
    grid: {
        margins: {
            top: percentage_of_page_width,
            bottom: percentage_of_page_width,
            left: percentage_of_page_width,
            right: percentage_of_page_width
        },
        cols: number_of_cols,
        rows: number_of_rows,
        gutter: percentage_of_page_width
    },
    background: [hue 0-100, saturation 0-100, brightness 0-100],
    items: [
        {
            type: "text"|"shape"|"image",
            x: col_number,
            y: row_number,
            width: number_of_cols,
            height: number_of_rows,
            fill: [hue 0-100, saturation 0-100, brightness 0-100],
            opacity: 0-100,
            ... (other properties)
        },
        ... (other items)
    ]
}
```

Fig. 3. Schematic representation of the input passed to the renderer.

creating a set of offspring that are also evaluated. Finally, a new population is built with individuals from the current population and the offspring, through elitist selection with a size of 5 (equal to the number of final posters to present to the user). This process is repeated in a loop until the termination criteria are met. At such point, the final elite individuals are presented to the user.

Representation. As in the original paper, in our system, a genotype consists of a JSON object containing information about each page item (see Fig. 3).

Besides the initialisation information mentioned before, page items are defined by positioning, geometry, and style properties. More specifically, position (i.e. column and row), size (i.e. number of columns and rows), fill colour and opacity. For shapes, the type of shape (ellipse or rectangle) must also be defined. For text boxes, the following additional properties are available: text size, text colour and text opacity.

Phenotypes consist of PNG images. To render these, the initialisation information and the item properties of a poster are compiled into a single JSON object (see Fig. 3) and passed via an HTTP request to a *Node.js* server in which the poster is rendered using a *Node.js* implementation of the HTML canvas, as mentioned before. The items are rendered on the page in the order predefined during initialisation, i.e. in case of overlapping, the later items will be rendered on top of the foremost. Lastly, the rendered image is sent back to the *Python* engine for the evolutionary process to proceed.

Variation. Variation-wise, both crossover and mutation operations are applied. As in the original paper, crossover operations consist of passing whole items to the offspring without changing any item properties. Initially, the standard uniform crossover was selected to create a pair of offspring for each pair of parents. However, empirical testing revealed that the standard crossover operator was overly destructive. As such, the standard operator was modified so that the first offspring has a 90% probability of inheriting each gene from the first parent and a 10% probability of inheriting it from the second parent. Conversely, for the second individual, these probabilities are reversed. This modification effectively creates directed mutations of each parent towards the other, resulting in a less disruptive operator that aids the evolutionary process.

Mutation operations consist of changing each item property mentioned in the sub-section *Representation*. To do that, uniform mutation is applied. More specifically, each property can be changed within a 1% likelihood. Such a high mutation probability was tested to understand whether it was sufficient to introduce enough diversity in the final showcase individuals. Different item properties can refer to one of three different types of values. Numeric values will change randomly within the specified domain. Strings and booleans will change for another value present in an array of possible values.

Fitness Assignment. Fitness assignment is primarily determined using a state-of-the-art metric to assess visual pleasantness—NIMA (Neural Image Assessment) [23]. The NIMA model has been trained on the AVA dataset, comprising 255 thousand images gathered from photography competitions. These images were rated on a scale from 1 to 10 by an average of 200 people. On average, the evaluations for these images fall closely to a value of 5.5. Therefore, artefacts with evaluations approximate to this value can be deemed visually pleasant. Since this model was trained on photos, it might not be fully suited for evaluating GD artefacts, as posters differ from photographs in the way they convey information. Nevertheless, to the best of our knowledge, there are no similar models trained on posters. Future work must comprise training and testing a homologous model specifically trained to evaluate posters.

Encouraging Diversity. As previously described, the main goal of this study is to encourage the evolution of more diverse posters. Therefore, two versions of the evolutionary engine are devised: one that is guided solely by fitness ($Engine_f$) and another that is guided by Explicit Fitness Sharing ($Engine_{fs}$). The fitness-guided version operates as previously described. In turn, in the fitness sharing version, the fitness of each individual is adjusted with Eq. 2, effectively reducing it depending on the number and closeness of individuals in its species. As a result, a distance metric must be selected, along with the values for α and σ. Explicit Fitness Sharing allows for the distance between individuals to be measured over various spaces, namely the genotypic, phenotypic or semantic.

In the domain of GD, measuring the distance between posters in the phenotypic space is more meaningful, as it effectively avoids the competing conventions problem (i.e., the phenomenon where there is more than one genotypic that maps to the same phenotype). As a result, the Root Mean Squared Error (RMSE) was selected, as it is computationally efficient.

The next issue is to select the value for σ, which typically requires knowledge of the fitness landscape. To overcome that issue, in this work, we dynamically set the value of σ. We do so by, on each generation, computing the RMSE between the phenotypes of all individuals in the population. Moreover, considering we want to provide five posters, σ is set to the quantile 0.2. Finally, following what is commonly done in the literature, the value of α is set to one, creating a linear penalisation function.

4   Experimental Setup and Results

The conducted experiments comprised: (i) evolving posters using NIMA to assess fitness without employing fitness sharing; and (ii) evolving posters using both NIMA and Explicit Fitness Sharing. 20 independent runs were done for each setup, using the EA and initialisation parameters presented respectively in Tables 1 and 2.

Table 1. Parameters of the Evolutionary Algorithms

Parameter	Value
Population size	25
Generations	1000
Tournament size	3
Elite size	5
Crossover rate	0.7
Mutation rate	0.01

Table 2. Initialisation Parameters

Parameter	Value
Number of columns	4
Number of rows	6
Number of Shapes	10
Number of text boxes	1
Text content	MANY DIFFERENT IDEAS!
Margins	0.05
Gutter	0

Figure 4 showcases the five best individuals of three different runs of $Engine_f$. We cherry-picked three runs we believe are representative of the typical results. As illustrated in run c), on a few runs some diversity can already be noticed, probably due to the high mutation likelihood (1%). However, oftentimes, most

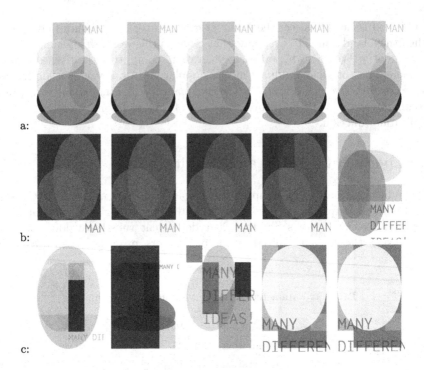

Fig. 4. Examples of the five best individuals in the final population of three different runs—a), b) and c) —, evolved using NIMA alone.

of the final individuals are too similar to each other, as demonstrated by runs a) and b). Therefore, although the high mutation likelihood can help produce more diverse results, this approach is not adequate to guarantee the diversity of the population, especially if one aims to generate a given number of different individuals.

Figure 5 presents the five best individuals of three different runs of our $Engine_{fs}$. For these second experiments, although we cherry-picked the presented three runs, all the runs produced considerably visually different final individuals (see Fig. 5a, 5b and 5c). For that reason, the use of Explicit Fitness sharing can be considered to be beneficial for GD systems where the goal is not to find the global optima (i.e., the absolute best poster), but rather to provide the user with a diverse set of good quality starting points for his work.

To better sustain the previous claim, the fitness and diversity of the final individuals of each run must be analysed. Figure 6 shows the boxplots of the fitness of the best individuals evolved with $Engine_f$ (left) and $Engine_{fs}$ (right). The first conclusion that can be drawn is that the fitness of the posters is not far off the average aesthetics value in the NIMA dataset, with both approaches producing median values over 5, implying that these posters can be considered to be aesthetically pleasing. The second conclusion is that, even though $Engine_{fs}$ is not focusing solely on finding the overall best solution, there is some overlap

Fig. 5. Examples of the five best individuals in the final population of three different runs—a), b) and c) —, evolved using NIMA for fitness assignment and Explicit Fitness Sharing for fostering diversity.

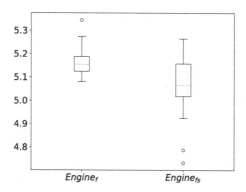

Fig. 6. Fitness of the best individuals evolved with $Engine_f$ (left) and $Engine_{fs}$ (right).

between its fitness values and those of $Engine_f$. Contrary to what was expected, in this work, Explicit Fitness Sharing does not produce better quality solutions than simply optimising for fitness, which may be due to the small population size preventing the EA from finding all the optima in each species. Nevertheless, since the goal of this system is to provide a diverse set of good solutions, the slightly inferior aesthetic quality of the posters evolved by $Engine_{fs}$ is not troublesome.

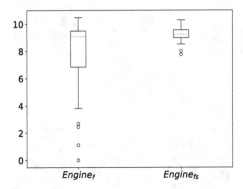

Fig. 7. Phenotypic distance (RMSE) of the best individuals evolved with $Engine_f$ (left) and $Engine_{fs}$ (right).

We now move to analyse the diversity of the best posters produced by each approach in each run. We do so by computing the RMSE between each pair of the 5 posters and computing the respective boxplots, presented in Fig. 7. As can be seen, there is little overlap between the two approaches, i.e. the inclusion of explicit fitness sharing leads to much more diverse posters, matching what could be expected from the analysis of the phenotypes (see Figs. 4 and 5). In fact, the plots show that $Engine_{fs}$ not only attains a higher median diversity than $Engine_f$, but also that its diversity values are much less dispersed. The lowest RMSE values exhibited by $Engine_{fs}$ are of approximately 8 and are also outliers. In turn, $Engine_f$ has its first quartile at approximately 7 and lower whisker set at 4. The fitness-guided approach also has several outliers lower than 4, including some at 0, indicating the output of repeated posters.

5   Conclusion

Finding creative and unconventional visual solutions is of the utmost importance for making GD stand out to the target public's attention. However, the increasing demand for faster and cheaper results often leads professionals to adopt design trends and templates, compromising the distinctive qualities of their designs.

Computer techniques such as evolutionary algorithms have been used to aid the exploration of distinctive visual solutions which designers may use to get inspiration from. However, most of the existing systems (at least the ones publicly available) often converge to populations of very similar individuals, making most of the generated individuals not informative to the creative process. Thus, to generate different ideas, the users must run these systems several times, spending unnecessary time on the process.

In response to that, in this paper, we implemented and tested the inclusion of niching techniques in the generation GD posters using an evolutionary system based on state-of-the-art work [12,13]. The experiments comprised evolving a set of posters that were evaluated for their aesthetic quality through a Machine Learning model from the state of the art. Two variants of the evolutionary system were developed, differing only in the inclusion of explicit fitness sharing.

Firstly, a visual analysis of the posters evolved by both systems was conducted, revealing that the use of explicit fitness sharing effectively produces a diverse set of posters in all runs, while the evolution solely guided by fitness typically led to a converged set of repeated posters. Secondly, the fitness and diversity of the outputted posters was analysed, showing that: (i) both systems evolved posters that could be considered aesthetically pleasing according to the metric used; (ii) the use of explicit fitness sharing led to slightly worse fitness values, which may be due to the parametrisation of the EA; and (iii) the use of explicit fitness sharing improves the diversity of the resulting posters, never occurring any repetitions. Thus, the results suggest that the tested niching technique can help contradict the generation of homogeneous populations of poster designs, providing end-users with a greater spectrum of ideas.

Future work must comprise (i) exploring the use of other distance metrics from the literature to obtain more meaningful distances; (ii) exploring the applicability of other diversity preservation methods, such as novelty search, for evolving diverse sets of posters from a single run; and (iii) exploring different fitness metrics to control design quality, such as ensuring the legibility of the text-contents, enabling the evolution of more complex posters.

Acknowledgements. This work has been partially supported by Project *"NEXUS Pacto de Inovação – Transição Verde e Digital para Transportes, Logística e Mobilidade"*. ref. No. 7113, supported by the Recovery and Resilience Plan (PRR) and by the European Funds Next Generation EU, following Notice No. 02/C05-i01/2022.PC645112083-00000059 (project 53), Component 5 - Capitalization and Business Innovation - Mobilizing Agendas for Business Innovation and by the FCT - Foundation for Science and Technology, I.P./MCTES through national funds (PIDDAC), within the scope of CISUC R&D Unit - UIDB/00326/2020 or project code UIDP/00326/2020 and grant SFRH/BD/143553/2019.

References

1. Chatzilygeroudis, K., Cully, A., Vassiliades, V., Mouret, J.-B.: Quality-diversity optimization: a novel branch of stochastic optimization. In: Pardalos, P.M., Rasskazova, V., Vrahatis, M.N. (eds.) Black Box Optimization, Machine Learning, and No-Free Lunch Theorems. SOIA, vol. 170, pp. 109–135. Springer, Cham (2021). https://doi.org/10.1007/978-3-030-66515-9_4
2. De Jong, K.A.: An Analysis of the Behavior of a Class of Genetic Adaptive Systems. University of Michigan (1975)
3. Eiben, A.E., Smith, J.E.: Introduction to Evolutionary Computing. Springer, Heidelberg (2015). https://doi.org/10.1007/978-3-662-44874-8
4. Geigel, J., Loui, A.: Using genetic algorithms for album page layouts. IEEE Multimed. **10**(4), 16–27 (2003). https://doi.org/10.1109/MMUL.2003.1237547
5. Goldberg, D.E.: Simple genetic algorithms and the minimal, deceptive problem. Genet. Algorithms Simul. Annealing 74–88 (1987)
6. Goldberg, D.E., Richardson, J., et al.: Genetic algorithms with sharing for multimodal function optimization. In: Genetic Algorithms and Their Applications: Proceedings of the Second International Conference on Genetic Algorithms, vol. 4149. Lawrence Erlbaum, Hillsdale (1987)

7. Holland, J.H.: Adaptation in Natural and Artificial Systems. An Introductory Analysis with Applications to Biology, Control and Artificial Intelligence. University of Michigan Press, Ann Arbor (1975)
8. Kitamura, S., Kanoh, H.: Developing support system for making posters with interactive evolutionary computation. In: 2011 Fourth International Symposium on Computational Intelligence and Design, vol. 1, pp. 48–51 (2011). https://doi.org/10.1109/ISCID.2011.21
9. Klein, D.: Evolving layout: next generation layout tool (2016). evolvinglayout.com
10. Lehman, J., Stanley, K.O.: Abandoning objectives: evolution through the search for novelty alone. Evol. Comput. 19(2), 189–223 (2011). https://doi.org/10.1162/EVCO_a_00025
11. Lehman, J., Stanley, K.O.: Evolving a diversity of virtual creatures through novelty search and local competition. In: Proceedings of the 13th Annual Conference on Genetic and Evolutionary Computation, pp. 211–218 (2011)
12. Lopes, D., Correia, J., Machado, P.: EvoDesigner: aiding the exploration of innovative graphic design solutions. In: Johnson, C., Rodríguez-Fernández, N., Rebelo, S.M. (eds.) EvoMUSART 2023. LNCS, vol. 13988, pp. 383–398. Springer, Cham (2023). https://doi.org/10.1007/978-3-031-29956-8_25
13. Lopes, D., Correia, J., Machado, P.: Towards the automatic customisation of editable graphics. In: Proceedings of the 14th International Conference on Computational Creativity, ICCC 2023, Waterloo, 19–23 June 2023. Association for Computational Creativity (ACC) (2023)
14. Martins, T., Correia, J., Costa, E., Machado, P.: EvoType: from shapes to glyphs. In: Proceedings of the Genetic and Evolutionary Computation Conference 2016, pp. 261–268. ACM (2016)
15. Nilsson, O., Cully, A.: Policy gradient assisted map-elites. In: Proceedings of the Genetic and Evolutionary Computation Conference, pp. 866–875 (2021)
16. Onduygu, D.C.: Graphagos: evolutionary algorithm as a model for the creative process and as a tool to create graphic design products. Ph.D. thesis, Sabanci University (2010). https://research.sabanciuniv.edu/24145/, https://www.graphagos.com/portfolio/
17. Pétrowski, A.: A clearing procedure as a niching method for genetic algorithms. In: Proceedings of IEEE International Conference on Evolutionary Computation, pp. 798–803. IEEE (1996)
18. Pugh, J.K., Soros, L.B., Stanley, K.O.: Quality diversity: a new frontier for evolutionary computation. Front. Robot. AI 3, 40 (2016)
19. Ramesh, A., Dhariwal, P., Nichol, A., Chu, C., Chen, M.: Hierarchical text-conditional image generation with clip latents (2022). https://doi.org/10.48550/ARXIV.2204.06125, https://arxiv.org/abs/2204.06125
20. Rebelo, S., Martins, P., Bicker, J., Machado, P.: Using computer vision techniques for moving poster design. In: 6.ª Conferência Internacional Ergotrip Design (2017)
21. Rebelo, S., Fonseca, C.M., Bicker, J., Machado, P.: Experiments in the development of typographical posters. In: 6th Conference on Computation, Communication, Aesthetics and X (2018)
22. Stanley, K.O., Miikkulainen, R.: Evolving neural networks through augmenting topologies. Evol. Comput. 10(2), 99–127 (2002)
23. Talebi, H., Milanfar, P.: NIMA: neural image assessment. IEEE Trans. Image Process. 27(8), 3998–4011 (2018). https://doi.org/10.1109/TIP.2018.2831899
24. Toivonen, H., Gross, O.: Data mining and machine learning in computational creativity. Wiley Int. Rev. Data Min. Knowl. Disc. 5(6), 265–275 (2015). https://doi.org/10.1002/widm.1170

No Longer Trending on Artstation: Prompt Analysis of Generative AI Art

Jon McCormack$^{(\boxtimes)}$ ⓘ, Maria Teresa Llano ⓘ, Stephen James Krol ⓘ, and Nina Rajcic ⓘ

SensiLab, Monash University, Caulfield East, VIC 3145, Australia
{Jon.McCormack,Teresa.Llano,Stephen.Krol,Nina.Rajcic}@monash.edu

Abstract. Image generation using generative AI is rapidly becoming a major new source of visual media, with billions of AI generated images created using diffusion models such as Stable Diffusion and Midjourney over the last few years. In this paper we collect and analyse over 3 million prompts and the images they generate. Using natural language processing, topic analysis and visualisation methods we aim to understand collectively how people are using text prompts, the impact of these systems on artists, and more broadly on the visual cultures they promote. Our study shows that prompting focuses largely on surface aesthetics, reinforcing cultural norms, popular conventional representations and imagery. We also find that many users focus on popular topics (such as making colouring books, fantasy art, or Christmas cards), suggesting that the dominant use for the systems analysed is recreational rather than artistic.

Keywords: Generative AI · Prompting · Visual Arts & Culture

1 Introduction

In just a few years, generative AI has become a dominant new paradigm for the creation of high-quality digital media, including images, video and text. One recent estimate put the number of AI generated images created to date at over 15 billion – exceeding the number of photographs taken in the first 150 years of photography [37]. Text-to-image (TTI) systems, such as Stable Diffusion, Midjourney or DALL-E allow the creation of high-quality images or illustrations just by supplying a short text description (prompt) of the desired image. These systems have become so popular, that "prompting" is now being considered as a part of the skillset of successful commercial art production.

However, as is the case with most recent generative AI advances, these systems raise many cultural, ethical and conceptual issues. Exploring and understanding these issues is increasingly important, given the rapid update and normalisation of generative AI imagery into mainstream software tools[1]. Important

---

[1] Adobe's generative AI system "firefly" is now integrated into Photoshop, with users already generating over 2 billion images [4].

C. Johnson et al. (Eds.): EvoMUSART 2024, LNCS 14633, pp. 279–295, 2024.
https://doi.org/10.1007/978-3-031-56992-0_18

concerns raised by generative AI include bias and the reinforcement of cultural stereotypes, concerns regarding censorship, issues of data laundering [2] and training on copyrighted data, "style theft" and the automation of specific artistic styles, questions of authenticity, "hallucinations" and "deep-fake" imagery, loss of traditional human skills, cultural and aesthetic homogenisation [11,23,35]. At the centre of many of these issues are the prompts themselves, the language they cultivate, and the *type* of images they produce. By understanding how – collectively – people are expressing their ideas through prompts, we can improve our understanding of how generative AI is impacting visual art and culture.

In this paper we examine several large prompting datasets, using Natural Language Processing (NLP) and data visualisation to draw a broad understanding of how users of TTI systems utilise language in generative AI systems. Our overall aim is to improve understanding of generative AI's influence on how we think when making a prompt and in particular, our use of language in relation to art-making and creativity. Hence, we focus on how artistic concepts are represented in prompt language; tracing their evolution as TTI systems continue to become more prolific and technically sophisticated. To this end, we compare prompting datasets from mid-2022[2] to November 2023, by which point, TTI systems had become widely adopted. We also explore how TTI systems have and continue to impact human artists and illustrators.

2   Background and Related Work

2.1   Text-to-Image Generation

The introduction of diffusion models marked a significant leap in image generation quality over previous methods such as Generative Adversarial Networks (GANs) [13]. The fundamental innovation of recent prompt-based image generators is the fusion of two independently powerful models – CLIP (Contrastive Language-Image Pre-training) [29] for text-based understanding, and Diffusion Models for image synthesis – into a cohesive system capable of generating images from textual prompts.

CLIP was trained on dataset of 400 million images and their associated textual descriptions. Unlike predecessor models that focused on direct object recognition, CLIP also set out to understanding the context, style, and other abstract concepts conveyed through language. The training set was scraped from online sources: social media posts, image-caption pairs, alt text descriptions, among others. For TTI models, CLIP serves as a mechanism to interpreting prompts. When a user inputs a textual description, CLIP analyses this text to construct the visual elements and themes that should be present in the output image.

---

[2] In mid-2022, many of the current systems were in pre-release, or just gaining a sizeable body of users.

2.2 Prompt and Image Analysis

With the introduction of TTI systems, in addition to being a creative tool, generative AI has become a social phenomenon [30]. Writing prompts is primarily an iterative, trial and error process; an exploratory practice [40]. More importantly, specific terms that coax the model into generating high-quality results have become a significant aspect of successful prompting.

Existing literature in prompt analysis has focused on aiding prompt engineering and making improvements to the underlying models [41]; for instance, by identifying popular or "most useful" keywords and model parameters that would allow users to generate higher quality images [21,28,42]. Xie et al. [40] further propose using higher rating prompts – in which their analysis pointed at longer prompts, and prompts including artists' names – as a way to further train the models, feeding back to the systems what users want. These approaches to prompt analysis pay less attention to large-scale analysis from a cultural perspective, and often favour feedback approaches that tend to homogenise content and reinforce popular clichés, amplifying concerns about biases, stereotypes, authorship and authenticity. Additionally, they reinforce the artistic styles of just a handful of specific artists (see our analysis below).

A recent study undertook a more exploratory approach to prompt analysis [30] using the DiffusionDB database [39]. Applying topic analysis, the author identified a taxonomy of prompt specifiers and developed a model for identifying the categories from this taxonomy from TTI prompts. Although the main goal of this study was the development of an interactive tool to assist with prompting, the results from this work open up possibilities for better understanding of what topics are being expressed in prompts. We build upon this work, performing topic analysis on a more recent dataset, and examining how current trends in prompting may inform thinking about art and visual culture.

Visual analysis of AI-generated images has also become a topic of significant interest, with current studies focused on identifying biases (e.g. under-representing certain race groups [3,26]), cultural gaps (e.g. over-representing specific nations [26]), and the reinforcement of stereotypes (e.g. "a photo of a lawyer" consistently showing a white male [5]). An analysis of 3,000 AI images depicting national identities also highlighted these tendencies towards bias and stereotypes of TTI systems (e.g. New Delhi's streets were mostly portrayed as polluted and littered) [35]. This perpetuates cultural norms that are prevalent in training datasets while under-representing less stereotypical and non-Western aspects of culture, art and society. Although some researchers have proposed ways to mitigate these effects, such as adding specific phrases (e.g. "irrespective of gender" [3]) or through the use of multilingual prompts (e.g. "a photo of a king" – appending a Russian character to the prompt [38]), these mitigation strategies are often ineffective (e.g. despite explicitly mentioning words such as "white", "wealthy" or "mansion", the authors in [5] report that Stable Diffusion continues to associate poverty with people of colour). In this paper we follow a different perspective from the aforementioned studies and perform the first large-scale analysis to understand the impact of prompting on art practice and art by analysing how prompting has changed over a 1 year timeframe.

3   Datasets

To perform our analysis we used three different datasets:

1. **DiffusionDB**: a text-to-image prompt dataset containing 14 million images generated by Stable Diffusion, 1.8 million unique prompts, and hyperparameters [39]. Collected over a 2 week period in August, 2022, the dataset is publicly available at: https://poloclub.github.io/diffusiondb;
2. **Midjourney 2022 Discord dataset**: 248k prompts and their associated generated images obtained by scraping ten public Discord channels over 28 days in June 2022 [34];
3. **Midjourney 2023 dataset**: 2.84M prompts and associated generated images obtained by scraping public Discord channels over 16 days in October-November 2023. This dataset was created by the authors of this paper and is publicly available [24].

Due to the stochastic nature of the image generation process, it is common for prompting interfaces to produce a number of image variants from a single user supplied prompt. In Midjourney for example, the default is to generate four images, from which the user has options to regenerate, upscale, or produce four new variants from one of the generated images. For our prompt analysis we only consider the initial generation, removing identical prompts that appear in upscaling or variant generation from our analysis. For the DiffusionDB dataset, tiled images were automatically split into separate images by the dataset's authors (one of the reasons why the number of images is much greater than the number of unique prompts).

Both the DiffusionDB and Midjourney 2022 Discord dataset (hereafter *MJ-2022*) were generated in the early stages of development and public release of these systems (around mid 2022). DiffusionDB was obtained by scraping public Discord channels when Stable Diffusion was in public beta testing. Midjourney has always used Discord as its interface for prompt generation.

Following an initial analysis of the DiffusionDB and MJ2022 datasets, we opted to create a new, more recent dataset to compare how prompting might have changed over the course of one year[3].

All the datasets include the full prompt text, user id, generated image URL and timestamp, the DiffusionDB dataset also includes other metadata, including NSFW scores for images and prompts, configuration parameters specific to stable diffusion and image dimensions. These additional fields were not used in our analysis.

---

[3] Both Midjourney and Stable Diffusion have undergone several major developments since 2022. At the time of writing the current versions are Midjourney 5.2 and Stable Diffusion XL.

3.1   Initial Processing

Before analysing the datasets we performed some basic cleanups to help ensure the reliability of the analysis. This included removing any records with missing data or empty prompts, incomplete requests, and non-image generating prompts. Table 1 gives a comparison of the three datasets used in our analysis. Prompts without validated users were removed from user statistics.

Table 1. Comparison of the three datasets used in this paper

Dataset	DiffusionDB	MJ2022	MJ2023
Data collection period	6-20 Oct 2022	1-28 Jul 2022	25 Oct-9 Nov 2023
Raw records	2.00M	250k	2.84M
Prompts	1,528,513	145,080	936,589
Mean Prompt Length	162	150	101
Users	10,173	1,681	34,429
Median prompts/user	54	12	8
Max prompts/user	5,630	2,493	3,666

3.2   Dataset Statistics and Initial Analysis

The most obvious change in the overall dataset comparisons shown in Table 1 is the growth in the number of users – Midjourney data from 2022 has less than 1.7k unique users, whereas just over a year later there are over 34k users, a twenty-fold increase. While there are more people using these TTI systems, they are not using the system as often or as much: the distribution of users vs. prompts shows a small number of "power users" (people spending a large majority of their time on the system), and a large number of casual users, with 75% of users entering less than 15 prompts over the collection period in 2023.

Our analysis was performed using language tools based on English. However, we analysed the datasets to determine the language composition of prompts, using the fastText model [16] to automatically classify each prompt by language. The results are shown in Table 2. As can be seen, English is by far the dominant language used in prompting within these systems, with over 99% of prompts in each dataset identified as English[4]. However, we note a significant increase in non-English prompts in languages such as French and Spanish between 2022 and 2023.

Prompts were divided into components (*specifiers*), separated by commas or periods, ignoring letter case and removing a small set of stopwords, system commands and punctuation (similar to [30]). We also ensured that terms utilising

---

[4] We classified non-text prompts – such as Emojis – as English. As the number of these type of prompts are very small, it did not impact significantly on the overall results.

Table 2. Top five languages for prompts and the total number of languages detected in each dataset

	DiffusionDB		Midjourney2022		Midjourney2023	
	Lang	Freq (%)	Lang	Freq (%)	Lang	Freq (%)
1	English	99.914	English	99.893	English	99.13
2	French	0.027	French	0.023	French	0.380
3	German	0.012	German	0.016	Spanish	0.172
4	Spanish	0.007	Spanish	0.013	Italian	0.091
5	Italian	0.007	Japanese	0.012	German	0.072
Total	50		19		14	

periods (e.g. f1.2) are captured. Table 3 shows the most popular individual specifiers for each dataset and the total number of unique specifiers. Frequency values are given as a percentage of the total number of specifiers in each dataset's corpus. For the Midjourney datasets, we differentiate between *initial* and *variation* prompts (typically initiated using the /imagine command) and other requests such as up-scaling or panning/zooming. For the prompt analysis we only consider the initial and variation prompts.

Table 3. The 10 most frequent specifiers and total number of unique specifiers for each dataset

	DiffusionDB		MJ2022		MJ2023	
	Term	Freq (%)	Term	Freq (%)	Term	Freq (%)
1	highly detailed	11.19	cinematic	7.21	white background	2.76
2	artstation	10.00	octane render	5.98	8k	2.72
3	sharp focus	8.80	8k	5.43	cinematic	2.11
4	trending on artstation	8.39	artstation	4.66	photorealistic	1.99
5	concept art	8.34	ar 16:9	4.40	realistic	1.82
6	octane render	6.56	4k	3.95	4k	1.75
7	intricate	6.54	detailed	3.30	black and white	1.35
8	digital painting	6.41	trending on artstation	2.51	hyper realistic	1.24
9	illustration	6.06	realistic	2.29	raw	1.21
10	8 k	6.02	highly detailed	2.17	cinematic lighting	1.12
Total	1,723,539		156,045		512,634	

Phrases to cajole the image quality or specific style dominate, reflecting common folklore in prompting that phrases such as "cinematic", "highly detailed" and "8k" result in better images than those without such qualifiers. Interestingly, the terms "artstation"[5] and "trending on artstation" feature prominently in the 2022

---

[5] Artstation is a portfolio showcase site, popular for games and commercial entertainment artists.

datasets, but are no longer in the top 10 by 2023. The likely reason for this is that early versions of Stable Diffusion suggested using these (and other terms in the top 10) in their introductory prompt guides. In early versions of Stable Diffusion in particular, these terms were necessary to get images of good quality, but are no longer required in the most recent releases.

Table 4. Popular prompt specifiers related to style or stylistic direction

DiffusionDB	MJ2022	MJ2023
style of beksinski	machinarium	cartoon
anime	pixar	anime
fantasy	manga	pixar
pixar	in stylized style	watercolor
cyberpunk	anime	comic (book)
vogue cover	anime waifu style character	vector
smooth style beeple	photographic	banana fish anime

We analysed the datasets for references to styles. A selection of the most popular styles for each dataset are shown in Table 4. As can be seen from the table, illustration styles ("anime") or names of popular studios ("Pixar") are the most common. We observe a shift from the general ("anime", "fantasy") to the specific ("anime waifu style character", "banana fish anime"), suggesting that users are after more specific stylistic references in the images produced. While not shown in the table, cultural memes of the time also feature strongly, with references to "Barbie" featuring in the MJ2023 dataset, for example. The final observation is the diversity of styles increases with the number of users of the system, with more generic styles dominating the MJ2023 dataset ("cartoon", "watercolor", "comic book").

4 Analysis

4.1 Use of Artist Names

"Stability has a music generator that only uses royalty free music in their dataset. Their words: "Because diffusion models are prone to memorisation and overfitting, releasing a model trained on copyrighted data could potentially result in legal issues." Why is the work of visual artists being treated differently?"[1] (Lois Van Baarle – #3 artist in DiffusionDB).

It has become a common practice to use artist's names in prompts as a way of generating images in the style of that artist – what has been called "style theft" [23]. We used the SpaCy[6] NLP library to perform identification of people's

---

[6] https://spacy.io.

names in the datasets. We also looked at specifiers that contain the term "style", as often desired styles reference artworks or organisations rather than creators (e.g. "Banana Fish anime style" or "Pixar style" – Table 4).

Tables 5, 6 and 7 show demographic information for the top ten artists mentioned in prompts, for DifussionDB, MJ2022 and MJ2023 datasets respectively. In all datasets, male artists dominate requests for specific artists' styles. Similarly, artists from the United States form the largest group, and artists from Western cultures dominating in general.

In the earlier datasets (DiffusionDB and MJ2022), fantasy, comic, and game art styles dominates across the top artists. This trend reflects the demographics, aims and use-cases of early adopters of the system. In the MJ2023 dataset, we see that the user base has grown significantly. Accordingly, the style of top artists has moved more into the realm of mainstream art, seen by the increase in the range of art styles including film, abstract painting, photography, sculpture, and architecture (as evidenced by Tables 5, 6 and 7) – with realism emerging as the prevalent artistic style. We also see a shift towards a greater balance between contemporary and modern artists in the later dataset (as evidenced by the active years of artists in the latter dataset). We believe this difference follows the more widespread adoption of the tool over time.

Table 5. Top 10 artists named in prompts from the DiffusionDB dataset, the **Freq** column shows the frequency, in percent, that the name appears in all prompts that mention a name. Names with a * symbol denote artists that are deceased.

Artist Name	Freq (%)	Gender	Based in	Age	Medium
Greg Rutkowski	9.58	M	POL	35	Illustrator & Concept Artist
James Jean	1.43	M	USA	44	Illustrator & Painter
Lois Van Baarle	1.12	F	NLD	38	Digital Artist & Illustrator
Peter Mohrbacher	1.10	M	USA	40	Concept artist & Illustrator
Tom Bagshaw	0.86	M	GBR	46	Illustrator
Craig Mullins	0.80	M	USA	59	Concept artist & Digital Painter
Ross Tran	0.75	M	USA	31	Concept Designer & Illustrator
Ruan Jia	0.75	M	CHN	40	Concept & Digital artist
Dan Mumford	0.74	M	UK	32	Illustrator & Concept Artist
Norman Rockwell*	0.68	M	USA	84	Painter & Illustrator

As images 'in the style of' specific artists become prolific, some artists are pursuing legal action or voicing their concerns and disapproval about the non-consensual use of their work in training generative AI systems[7]. A major issue – currently disputed – is that of copyright [10]. An argument put forward is that

---

[7] In late 2022 artists started a public mass protest against AI-generated artwork on ArtStation https://arstechnica.com/information-technology/2022/12/artstation-artists-stage-mass-protest-against-ai-generated-artwork/.

Table 6. Table of the top 10 artists mentioned in the MidJourney 2022 dataset (names with a * symbol denote artists that are deceased).

Artist Name	Freq (%)	Gender	Based in	Age	Medium
Craig Mullens	4.10	M	USA	59	Digital Painter & Illustrator
Peter Mohrbacher	2.73	M	USA	40	Digital Painter & Illustrator
Tsutomu Nihei	1.61	M	JPN	52	Manga Artist
Robert Mapplethorpe*	1.42	M	USA	43	Photographer
James Jean	1.22	M	USA	44	Painter & Illustrator
Albert Bierstadt*	0.94	M	GER/USA	72	Painter
Katsuhiro Otomo	0.91	M	JPN	69	Manga Artist & Illustrator
Marc Simonetti	0.91	M	FRA	46	Conceptual Artist & Illustrator
Ross Tran	0.89	M	USA	31	Concept Artist & Illustrator
Mike Mignola	0.80	M	USA	63	Comic Artist & Writer

Table 7. Table of the top 10 artists mentioned in the MidJourney 2023 dataset (names with a * symbol denote artists that are deceased).

Artist Name	Freq (%)	Gender	Based in	Age	Medium
Wes Anderson	0.51	M	USA	54	Film director & Producer
Frank Frazetta*	0.50	M	USA	82	Illustrator
Gertrude Abercrombie*	0.28	F	USA	68	Painter
William Baziotes*	0.27	M	USA	50	Painter
David Smith*	0.27	M	US	59	Sculptor & Painter
Yayoi Kusama	0.25	F	JPN	94	Sculptor, Art Installation & Painter
Tim Burton	0.24	M	USA	65	Film Maker & Animator
Alphonse Mucha*	0.24	M	CZE	78	Painter, Illustrator & Graphic Artist
Max Dunbar	0.23	M	CAN	–	Illustrator, Comic & Concept Artist
Anne Holtrop	0.23	M	NLD	46	Architect

industries such as music have long-established rights in place to prevent on-line copying and protect copyright. In response, companies such as Stability AI claim that generative AI is not about copying artworks but creating new ones [15,19], as is the common practice of human artists who seek inspiration by viewing or copying other artists' work. Whether or not this dispute rules in favour of visual artists, it may involve more than just removing their original work from the training data, it may also require to remove all images that have been influenced by this artist's style as TTI systems work from the collective use of data [12].

Artists whose styles are being copied by AI also face criticism due to the perceived ease by which AI can replicate their style, with critics arguing that this "exposes weaknesses" of human-made art [32] and emphasising the possibility that AI art may lead individuals to be "more rigorous enjoyers/audiences of art" [6]. Although art critique is a standard practice in the art world, this fails to acknowledge what is beyond the images themselves. As Wes Anderson (#1 artist in MJ2023) puts it, "An artist or illustrator has a particular hand that they've developed and they find their set of ideas – they find their voice... And I don't know how good AI is at creating a voice" [9]. Some argue that an increased engagement of the artist with the audience would be needed in order to mitigate these issues [22].

Contrary to the above views, some of the artists whose work has been heavily featured in training datasets have identified potential roles for TTI systems in their creative practice. Peter Mohrbacher (#2 artist in MJ2022) mentioned using Midjourney as a kind of "personalised" image generation model [6], while artist Ruan Jia (#8 artist in DifussionDB) suggested the use of TTI systems as a communication and brainstorming tool within artistic teams [18]. When thinking about the artistic possibilities of TTI systems, emphasising the dialogue between human artists and machines rather than just the final artefact may be more insightful [27]; in this sense, creating a unique style would require more than just entering a prompt.

Looking beyond the individual, one may ask how stylistic referencing might impact art and culture more generally. The more people are exposed to certain bodies of work, the more these works influence their preferences: "acquired cultural competences play a role in art preference; the more familiar the art and artist are to the consumer and the more s/he knows about them, the higher satisfaction the art gives." [36]. The ease with which one can generate content using these systems, and the ease of faithfully replicating a visual style, may play a significant role in influencing the prominence of certain artists or artistic styles.

4.2   Topic Analysis

Building on a previous topic analysis of the DiffusionDB dataset [30], we conducted a topic analysis on prompt specifiers in the MJ2023 dataset. Prompt specifiers with 100 or more uses were included in the analysis, resulting in 1700 individual prompt specifiers. The topic modelling adhered to the approach outlined in [14] and involved first utilising the MPnet (Masked and Permuted Pre-training for Language Understanding) model [33] to encode prompt specifiers into vector representations. MPnet, a language model specialising in language understanding, is capable of embedding phrases into a latent space that proves effective for diverse language tasks, including clustering and sentence similarity. As in [30], we utilised the *all-mpnet-base-v2* pre-trained weights and the sentence-transformer library [31] for encoding. The UMAP [25] dimensionality algorithm was then applied to the vector embeddings to reduce them to a 5-dimensional space in order to mitigate the negative effect of high dimensionality on clustering performance [7]. We utilised the hierarchical density-based spatial clustering (HDBSCAN) algorithm [8] to identify topic clusters within the embedding space resulting in an initial collection of 40 topics. We present these topics in Fig. 1 where we utilised UMAP to reduce the original embedding to 2D. It is worth noting that there are some limitations to this method, with some outliers being assigned to incorrect clusters. However, as in [30], we utilised a class-specific term frequency-inverse document frequency (c-TF-IDF) to identify the most important specifiers in each group and utilised them to accurately label the cluster. An interactive version of Fig. 1 can be found at[8].

Topic Clusters from MJ2023 Dataset

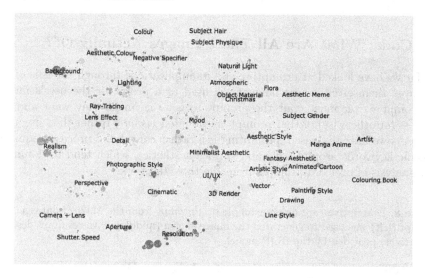

Fig. 1. Visualisation of the MPnet embeddings of 1700 prompt specifiers and the 40 topics identified using the HDBSCAN clustering algorithm.

The five most referenced topics were *Resolution, Realism, Mood, Detail* and *Lighting*, which account for roughly a third of prompt specifiers. This is expected, given the fundamental nature of these topics in the context of images. Many topics found in this analysis were also present in [30], such as the fantasy aesthetic and photography themes.

A prominent theme identified in our analysis was photography, encompassing topics such as *Camera + Lens, Shutter Speed, Aperture, Photographic Style, Perspective.* This, combined with the widespread use of *Realism* specifiers, indicates that users are utilising MidJourney to generate images that are photo realistic. Moreover, technical camera functions, such as those related to lens aperture and camera shutter speed, are being applied as if they were operating on a physical camera. This may be an attempt to control visual features such as depth of field and motion blur in both photorealistic and non-realistic images.

Art was another identified theme and covered topics such as *Painting Style, Drawing, Artistic Style* ect. This suggests that the system is still being used by some as a tool to explore aesthetics or to generate images that mimic a particular artistic style or medium.

Some unexpected topics discovered in our analysis included *Christmas* and *Colouring Book.* We hypothesise that *Christmas* emerged as a topic due to the

---

[8] https://sensilab.github.io/NoLongerTrendingOnArtstation/.

data being collected in proximity to Christmas, perhaps inspiring users to generate Christmas themed images. *Colouring Book* illustrates how users are prompting the system to generate assets that have value beyond their static form and can be used in more recreational settings, such as colouring for children and adults.

5 Coda: What Are All These Images Actually Of?

So far we have looked at prompting and its implications through the lens of the prompts themselves, where language is used as a proxy for the user's intent. A prompt writer must craft the prompt to get the image they want and the success (or otherwise) of that prompt is evaluated *visually* rather than linguistically. As we have discussed, prompting is a rather convoluted process, requiring specific keywords around style and surface aesthetics, which tend to dominate our analysis of over 3M prompts from the three datasets.

Table 8. Descriptive captioning example of a prompt from the MJ2023 dataset. The prompt (left) was used to generated the image shown (middle), with resultant description (right) provided by the BLIP model.

Prompt	Generated Image	Description
Imagine a dream-like scene where reality blurs and the boundaries between woman and peacock dissolve. Sketch a woman's body full of delicate vulnerability, her features soft and poetic. Let the peacock's head emerge, seamlessly integrating with its essence, symbolizing the deep connection with the world of colors of the peacock's tail. Use the empasto technique to add a tactile quality, allowing the viewer to visually feel the texture of the artwork. Set against a deep, velvety canvas of dark blue on a black background, this ethereal combination creates a sense of enchantment, encouraging viewers to explore the depths of their imagination		*painting of a woman with peacock feathers on her head*

While language analysis tells us something about the prompt-author's intent, it does not tell us if that intent was successfully realised in the resultant image. To better understand this aspect, we turned our attention to "upscale" requests in the dataset. The reasoning is that you will only bother to upscale an image if it is of interest, i.e. one or more of the image variants from the prompt you specified may be approaching what you intended (or like). This is a typical workflow for Midjourney (recall that our prompt analysis was performed on initial or variation requests, not on upscale requests).

We used the upscale requests from the MJ2023 dataset. After removing duplicates and failed requests, a total of 536,014 prompts and associated AI generated images were found. To discover what images actually depict, we ran each image through a state-of-the-art BLIP (Bootstrapping Language-Image Pre-training for Unified Vision-Language Understanding and Generation) image captioning model [20], generating descriptive captions for each image. The captioning model gives an overall description of the image, much like the way a human would, as the model recognises not only the objects in a scene but the relationships and basic surface aesthetic properties. We limited the description to a maximum of 20 words to avoid overly wordy descriptions with unnecessary detail. An example is shown in Table 8.

We then analysed each description to find the most popular depictions in the generated images. After removing adjectives and basic aesthetic or stylistic words (e.g. references to colours) the top subjects depicted were: *woman* (22.26% of images), *man* (16.2%), *dress* (6.92)%, *hair* (5.51)% *room* (5.44)% *flower* (5.33)%.

We next performed topic modelling on the description data using BERTopic [31]. Similar to the prompt Topic Analysis, image descriptions were converted to embeddings then dimensionally reduced using UMAP before clustering with HDBSCAN. BERTopic uses cTF-IDF to get topic weightings before an additional fine-tuning to optimise the number of topic categories. Figure 2 shows the eight most important topics and relative frequency of keywords.

Fig. 2. The 8 most popular topics for upscaled images in MJ2023

Our analysis shows that the most dominant image subject for MidJourney users are illustrations of Women and Anime Girls.

6   Summary and Conclusion

6.1   Limitations of the Study

Our datasets represent results only for two systems: Stable Diffusion and Midjourney. Other popular systems, such as DALL-E, Leonardo, Crayion, Firefly are all propriety, making access to prompts difficult. As Stable Diffusion is an open-source model, it is no longer run via Discord and users are free to download their own versions. One possibility to address this issue would be to scrape data from the many "prompt showcase" websites that communities of people can post AI-generated images and their associated prompts (e.g. sites such as Prompt Hero https://prompthero.com), we leave this as future work.

6.2 Findings

As shown in Table 1, Midjourney's user base has grown significantly between 2022 and 2023. The median prompt length is reducing, probably due to changes in the software that require less specialist prompting (hinting at the tool's accessibility). While there is a significant increase in users, the majority of those users do not prompt very much, suggesting that for many people, the use of TTI systems is largely recreational (something confirmed informally by a poll in [17]). This observation is further reinforced from our topic analysis of both prompts and images, with topics such as "colouring books" and "Christmas" being popular (MJ2023 data was collected in November, leading to the Christmas period).

Our analysis shows that prompting emphasises and privileges popular styles and surface aesthetic appearances ("cinematic lighting", "photorealistic", "ultra detailed"). It forces a fixation on surface aesthetics, potentially at the expense of other important factors in art making, including narrative, realism, authenticity and individuality.

While we identified a wide variety of artistic styles, TTI systems tend towards the popular, reinforcing stylistic norms and aesthetic "sameness". Our image analysis showed what seems like common knowledge when viewing websites such as the Midjourney Showcase or various "Prompt Art" sites: that most of the images are closeups or medium shots of young women. Genres of fantasy art, game art and comic or anime illustration dominate the specifiers used in prompting. As prior research has demonstrated, TTI systems continue to promote and reinforce racial, gender and other cultural biases. Despite the vast volume of new synthetic images generative AI has brought us, its seems that most people using these systems are after "more of the same". In the process of working with these datasets and being exposed to many thousands of images, our view is that few if any AI images are memorable in the way that human art is, and its difficult to see how they can be aesthetically unique in the way that human art can.

The agency exerted by TTI systems onto the human user, while difficult to quantify, is an important topic for future investigation. When viewing TTI systems as a creative medium, their inherent properties will inevitably shape and contribute to the future of image production. The precise way in which this occurs is multiple and nuanced. In this paper, we set out to trace this effect through a qualitative interpretation of a statistical analysis of the language employed in TTI systems and the images they produce. This leads us to conclude that generative AI imagery, at least in its current form, is probably not a serious threat to human art. After all, surface mimicry of a popular style does not constitute an artistic innovation or practice. A physical painting will always embody the act and intention of a human artist, something that no synthetic AI image can ever do.

Acknowledgements. This research was supported by an Australian Research Council Grant, DP220101223.

References

1. Baarle, L.V.: Loish blog: no to AI generated images (2022). https://blog.loish.net/post/703723938473181184/theres-a-protest-going-on-against-ai-art-over-on
2. Baio, A.: AI data laundering: how academic and nonprofit researchers shield tech companies from accountability (2022). https://waxy.org/2022/09/ai-data-laundering-how-academic-and-nonprofit-researchers-shield-tech-companies-from-accountability/
3. Bansal, H., Yin, D., Monajatipoor, M., Chang, K.W.: How well can text-to-image generative models understand ethical natural language interventions? In: Goldberg, Y., Kozareva, Z., Zhang, Y. (eds.) Proceedings of the 2022 Conference on Empirical Methods in Natural Language Processing, pp. 1358–1370. Association for Computational Linguistics, Abu Dhabi, United Arab Emirates (2022). https://doi.org/10.18653/v1/2022.emnlp-main.88, https://aclanthology.org/2022.emnlp-main.88
4. Beck, C.: Adobe releases new firefly generative AI models and web app; integrates firefly into creative cloud and adobe express (2023). https://news.adobe.com/news/news-details/2023/Adobe-Releases-New-Firefly-Generative-AI-Models-and-Web-App-Integrates-Firefly-Into-Creative-Cloud-and-Adobe-Express/default.aspx
5. Bianchi, F., et al.: Easily accessible text-to-image generation amplifies demographic stereotypes at large scale. In: Proceedings of the 2023 ACM Conference on Fairness, Accountability, and Transparency, pp. 1493–1504. FAccT 2023, Association for Computing Machinery (2023). https://doi.org/10.1145/3593013.3594095
6. Biles, P.: What is art without the human mind? (2022). https://mindmatters.ai/2022/12/what-is-art-without-the-human-mind/
7. Van den Bussche, J., Vianu, V., van Leeuwen, J.: On the surprising behavior of distance metrics in high dimensional space. In: Van den Bussche, J., Vianu, V. (eds.) ICDT 2001. LNCS, vol. 1973, pp. 420–434. Springer, Heidelberg (2000). https://doi.org/10.1007/3-540-44503-x_27
8. Campello, R.J.G.B., Moulavi, D., Sander, J.: Density-based clustering based on hierarchical density estimates. In: Pei, J., Tseng, V.S., Cao, L., Motoda, H., Xu, G. (eds.) PAKDD 2013. LNCS (LNAI), vol. 7819, pp. 160–172. Springer, Heidelberg (2013). https://doi.org/10.1007/978-3-642-37456-2_14
9. Cavna, M.: Social media keeps spoofing Wes Anderson. Here's why he's not watching (2023). https://www.washingtonpost.com/arts-entertainment/2023/06/24/wes-anderson-tiktok-ai-asteroid-city/
10. Chesterman, S.: Good models borrow, great models steal: intellectual property rights and generative AI. Great Models Steal: Intellectual Property Rights and Generative AI (2023)
11. Crawford, K.: Atlas of AI: Power, Politics, and the Planetary Costs of Artificial Intelligence. Yale University Press, London (2021)
12. Evans, B.: Generative AI and intellectual property (2023). https://www.ben-evans.com/benedictevans/2023/8/27/generative-ai-ad-intellectual-property
13. Goodfellow, I., et al.: Generative adversarial networks. Commun. ACM **63**(11), 139–144 (2020)
14. Grootendorst, M.: Bertopic: neural topic modeling with a class-based TF-IDF procedure. arXiv preprint arXiv:2203.05794 (2022)
15. Guadamuz, A.: A scanner darkly: Copyright liability and exceptions in artificial intelligence inputs and outputs. In: A Scanner Darkly: Copyright Liability and Exceptions in Artificial Intelligence Inputs and Outputs: Guadamuz, Andrés. [Sl]: SSRN (2023)

16. Joulin, A., Grave, E., Bojanowski, P., Mikolov, T.: Bag of tricks for efficient text classification. arXiv preprint arXiv:1607.01759 (2016)

17. Kelly, K.: Picture limitless creativity at your fingertips (2022). https://www.wired.com/story/picture-limitless-creativity-ai-image-generators/

18. King, G.: We chatted with Ruan Jia about the whole process of challenging AI painting: AI was overestimated before (2023). https://inf.news/en/tech/a46b12c491243e9bd257ddb36ce53457.html

19. Lemley, M.A.: How generative AI turns copyright upside down (2023)

20. Li, J., Li, D., Xiong, C., Hoi, S.: Blip: bootstrapping language-image pre-training for unified vision-language understanding and generation (2022)

21. Liu, V., Chilton, L.B.: Design guidelines for prompt engineering text-to-image generative models. In: Proceedings of the 2022 CHI Conference on Human Factors in Computing Systems. CHI 2022, Association for Computing Machinery (2022). https://doi.org/10.1145/3491102.3501825

22. Lyu, Y., Wang, X., Lin, R., Wu, J.: Communication in human-AI co-creation: perceptual analysis of paintings generated by text-to-image system. Appl. Sci. **12**(22) (2022). https://www.mdpi.com/2076-3417/12/22/11312

23. McCormack, J., Cruz Gambardella, C., Rajcic, N., Krol, S.J., Llano, M.T., Yang, M.: Is writing prompts really making art? In: Johnson, C., Rodríguez-Fernández, N., Rebelo, S.M. (eds.) EvoMUSART 2023. LNCS, pp. 196–211. Springer, Cham (2023). https://doi.org/10.1007/978-3-031-29956-8_13

24. McCormack, J., Rodriguez, M.T.L., Krol, S., Rajcic, N.: Midjourney 2023 Dataset (2024). https://doi.org/10.26180/25038404.v1, https://bridges.monash.edu/articles/dataset/Midjourney_2023_Dataset/25038404

25. McInnes, L., Healy, J., Melville, J.: Umap: uniform manifold approximation and projection for dimension reduction (2020)

26. Naik, R., Nushi, B.: Social biases through the text-to-image generation lens. In: Proceedings of the 2023 AAAI/ACM Conference on AI, Ethics, and Society, pp. 786–808. AIES 2023, Association for Computing Machinery (2023). https://doi.org/10.1145/3600211.3604711

27. Oppenlaender, J.: Prompt engineering for text-based generative art. arXiv preprint arXiv:2204.13988 (2022)

28. Pavlichenko, N., Ustalov, D.: Best prompts for text-to-image models and how to find them. In: Proceedings of the 46th International ACM SIGIR Conference on Research and Development in Information Retrieval, pp. 2067–2071. SIGIR 2023, Association for Computing Machinery (2023). https://doi.org/10.1145/3539618.3592000

29. Radford, A., et al.: Learning transferable visual models from natural language supervision (2021)

30. Sanchez, T.: Examining the text-to-image community of practice: why and how do people prompt generative AIs? In: Proceedings of the 15th Conference on Creativity and Cognition, pp. 43–61 (2023)

31. sbert.net: Pretrained models sentence-transformers documentation. https://www.sbert.net/docs/pretrained_models.html?highlight=pretrained. Accessed 13 Nov 2023

32. Smee, S.: Kehinde wiley is selling kitsch (2023). https://www.washingtonpost.com/entertainment/art/2023/07/21/kehinde-wiley-obama-san-francisco-exhibit/

33. Song, K., Tan, X., Qin, T., Lu, J., Liu, T.: Mpnet: masked and permuted pre-training for language understanding. CoRR **abs/2004.09297** (2020). https://arxiv.org/abs/2004.09297

34. Turc, I., Nemade, G.: Midjourney user prompts & generated images (250k) (2022). https://doi.org/10.34740/KAGGLE/DS/2349267
35. Turk, V.: How AI reduces the world to stereotypes (2023). https://restofworld. org/2023/ai-image-stereotypes/
36. Uusitalo, L., Simola, J., Kuisma, J.: Perception of abstract and representative visual art. In: Proceedings of AIMAC, 10th Conference of the International Association of Arts and Cultural Management, pp. 1–12. Dallas, TX (2009)
37. Valyaeva, A.: AI has already created as many images as photographers have taken in 150 years. statistics for 2023. Everypixel J. (2023). https://journal.everypixel. com/ai-image-statistics
38. Ventura, M., Ben-David, E., Korhonen, A., Reichart, R.: Navigating cultural chasms: exploring and unlocking the cultural POV of text-to-image models (2023)
39. Wang, Z.J., Montoya, E., Munechika, D., Yang, H., Hoover, B., Chau, D.H.: DiffusionDB: a large-scale prompt gallery dataset for text-to-image generative models. In: Proceedings of the 61st Annual Meeting of the Association for Computational Linguistics (Volume 1: Long Papers), pp. 893–911. Association for Computational Linguistics, Toronto, Canada (2023). https://doi.org/10.18653/v1/2023.acl-long. 51, https://aclanthology.org/2023.acl-long.51
40. Xie, Y., Pan, Z., Ma, J., Jie, L., Mei, Q.: A prompt log analysis of text-to-image generation systems. In: Proceedings of the ACM Web Conference 2023, pp. 3892–3902 (2023)
41. Xie, Y., Pan, Z., Ma, J., Jie, L., Mei, Q.: A prompt log analysis of text-to-image generation systems. In: Proceedings of the ACM Web Conference 2023, pp. 3892–3902. WWW 2023, Association for Computing Machinery, New York, NY, USA (2023). https://doi.org/10.1145/3543507.3587430
42. Zamfirescu-Pereira, J., Wong, R.Y., Hartmann, B., Yang, Q.: Why johnny can't prompt: how non-AI experts try (and fail) to design LLM prompts. In: Proceedings of the 2023 CHI Conference on Human Factors in Computing Systems. CHI 2023, Association for Computing Machinery (2023). https://doi.org/10.1145/3544548. 3581388

AI-Driven Meditation: Personalization for Inner Peace

Peter Nguyen[1]($\boxtimes$), Javier Fdez[1], and Olaf Witkowski[1,2]

[1] Cross Labs, Cross Compass, Kyoto, Japan
petern0408@gmail.com
[2] University of Tokyo, Tokyo, Japan
https://www.crosslabs.org

Abstract. Meditation is a mindful practice known for its difficulties, requiring focused attention despite distractions. Many people have traditionally relied on meditation apps with calming audio for support. This paper introduces an innovative AI-driven system aimed at improving the meditation experience, personalized to each user's needs. This system consists of three core parts. First, it uses a language model to create meditation scripts that match user preferences. Second, it converts these scripts into audio, accompanied by selected background music to create a serene atmosphere. Lastly, a Compositional Pattern-Producing Network (CPPN) generates visually appealing videos featuring intricate patterns influenced by sentiment analysis and input audio. One of the system's strengths is its adaptability since users can indicate their preferences among generation options and inform the system about their feelings and thoughts. An experiment, involving 14 participants, demonstrated comparable content and audio quality as traditional methods. Participants perceived the system as more personalized, expressing a preference for tailored meditation practices and indicating potential for increased user engagement. In summary, this AI-powered meditation system represents a significant advancement in the field, providing personalized, immersive experiences that integrate text, audio, and visuals. Its ability to adapt to users' preferences holds promise for enhancing meditation outcomes and fostering inner peace.

Keywords: Meditation · AI · Text generation · TTS · Video generation

1 Introduction

Meditation is one of the modalities used in Ayurveda (Science of Life), the comprehensive, natural health care system that originated in the ancient Vedic times of India [21]. It usually refers to a formal practice that can calm the mind and enhance awareness of ourselves, our minds, and our environment [2].

Meditation has witnessed a remarkable surge in popularity in recent years. Its virtues extend far beyond stress relief and relaxation, encompassing heightened

C. Johnson et al. (Eds.): EvoMUSART 2024, LNCS 14633, pp. 296–310, 2024.
https://doi.org/10.1007/978-3-031-56992-0_19

focus, mindfulness, and even tangible improvements in overall health. Despite its manifold benefits, adopting meditation into daily routines remains challenging for many. While the percentage of adults in the United States who experimented with meditation rose significantly from 4.1% in 2012 to 14.2% in 2017 [5], the number of regular practitioners remains comparatively modest.

The diverse range of practices that fall under the term *meditation* presents a unique challenge in classification and understanding [1, 8]. It is often used to describe not only the mental techniques employed by meditators but also the altered states of consciousness that result from these practices [4, 17].

In their endeavor, Matko and Sedmeier [15] revealed two fundamental categories along which meditation techniques can be classified: *activation* and *amount of body orientation*. The former spans practices like cultivating compassion, contemplating spiritual questions, and chanting mantras on one end. The other end includes activities such as scanning the body and manipulating the breath. Practices emphasizing bodily experience received higher values on this dimension, contrasting with those with more abstract or conceptual focuses. The body orientation dimension ranges from silent sitting, focusing on paradoxes, and observing thoughts or emotions on one end. The other end includes active practices like meditation with movement and sensory observation. Lower values on this dimension indicate passive and contemplative meditation, while higher values denote physically engaged practices. These dimensions offer a structured framework for understanding the diversity of meditation techniques based on their levels of bodily involvement and activity.

Within this framework, seven distinct clusters of meditation techniques have been identified (cf., Fig. 1). The first cluster, known as *mindful observation* emphasizes the practice of mindfully observing oneself in a state of stillness, which includes lying meditation, sitting in silence, and observing thoughts and emotions. The second cluster, labeled *body-centered meditation*, focuses on heightened awareness of the body, breath, and sensory perceptions, sometimes involving concentration on specific body locations or energy centers. *Visual concentration*, the third cluster, involves techniques like visualizations and concentration on visual objects to achieve a centered state of mind. The fourth cluster, *contemplation*, encourages deep reflection on contradictions, paradoxes, or spiritual questions. The fifth cluster, termed *affect-centered meditation*, encompasses practices that cultivate positive emotions such as compassion, loving-kindness, and equanimity. The sixth cluster, *mantra meditation*, involves the repeating syllables, words, or phrases, often with sounds and mantras, to facilitate meditation. Finally, *meditation with movement*, the seventh cluster, combines mindfulness with physical activity, including manipulating the breath, walking, and sensory observation.

Traditional meditation unfolds through various forms, accommodating both solitary and collective practices. As observed by [6], a combined 48.4% of individuals seeking a personal journey turn to guided meditation apps, videos, or other internet sources, and the percentage of internet-based use is only expected to grow, as they offer a structured approach to mindfulness and relaxation. Books authored by experienced meditation guides provide additional insights for those embarking on solitary meditation. On the communal front, meditation classes,

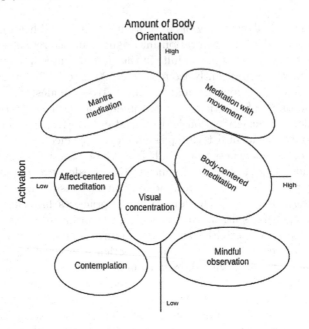

Fig. 1. Categorization of meditation types based on their activation and amount of body orientation.

workshops, and group meetups provide an inclusive environment for shared practice [10]. Online platforms facilitate virtual group meditation sessions, offering a sense of connection for those who prefer the collective experience. In the pursuit of making meditation accessible to a broader audience, the challenge lies in finding meditation methods that truly resonate with individuals. This task can be daunting, often discouraging many from engaging in the practice. To address this barrier, we acknowledge the necessity for guidance that goes beyond the complexities of various meditation techniques and traditions.

Borh and Memarzadeh's research [3] highlights the possibility of using the data collected by major corporations to create detailed personal profiles. This data, initially intended for behavioral analysis, holds the potential for understanding individuals and predicting healthcare trends. Specifically, the researchers discussed several healthcare applications where Artificial Intelligence (AI) stands ready to provide support, utilizing techniques such as natural language processing and the integration of personal records.

In tandem with these insights, recent studies have explored innovative avenues to enhance meditation experiences. For instance, [13] delves into the concept of "wandering voices", a novel approach where voice-based virtual assistants move with the user, enhancing the immersive experience of mindfulness meditation. Additionally, [23] delves into the realm of Virtual Reality (VR), showcasing the potential benefits of meditative practices in VR environments.

We venture to the intersection of computer science, technology, and the age-old practice of meditation. Our mission is to harness the power of AI to guide individuals into the meditative state they seek as we believe that with proper guidance, more people could experience meditation and accept it into their routine. Adopting meditation on a larger scale would play a major role in improving relationships and reducing conflict and violence across the world. By capitalizing on existing text and video generation models, this project aims to progress us toward this idea by using AI to guide people into a meditative state.

2 Methods

The AI system consists of three key components: (1) a language model crafting personalized meditation scripts, (2) the conversion of these scripts into audio with background music, and (3) the use of a Compositional Pattern-Producing Network (CPPN) to create visually engaging videos. Figure 2 illustrates the system's design. Please refer to the source code for further details in any of the following sections.

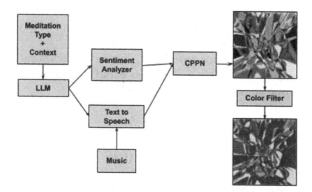

Fig. 2. System's design created by concatenating text generation, text-to-speech, and video generation models.

2.1 Text Generation

The first step is the creation of personalized meditation scripts using a Large Language Model (LLM). For the LLM, we have chosen the LLaMA-2 7B model, released by Meta in 2023 [22]. The model architecture of LLaMA 2 is an autoregressive language model that uses an optimized transformer architecture. The tuned versions use supervised fine-tuning (SFT) and reinforcement learning with human feedback (RLHF) to align with human preferences for helpfulness and safety.

This model is accompanied by an open license, enabling its utilization for both research and commercial purposes. The decision to choose the 7B version over larger variants of LLaMA was made to prioritize a lighter and faster model.

To personalize the generated script, the user provides a meditation type and context for the meditation. The context prompt is open-ended, so the user may be as broad or specific as they wish. For example, the user could indicate their thoughts and feelings of being stressed or sad. They could also indicate their general mood or goals for the meditation. The input prompt to the model indicates the provided context, meditation type, and a short description of the goal for the specified type of meditation. This allows the text generation model to personalize the script for the user.

2.2 Text-to-Speech

The creation of the speech has is carried out with the Google Text-to-Speech (gTTS) library [7]. This library provides us with a diverse set of capabilities, including multi-language support, adjustable speech parameters, and automatic language. Furthermore, gTTS' proficiency in generating human-like intonation and rhythm creates a calming experience, allowing users to effortlessly connect with the meditation content. Lastly, gTTS provides a variety of high-quality voices in the form of different accents. This provides the system with flexibility in producing audio that the user enjoys. The user may optionally provide input into the system to choose the accent they prefer.

To enhance the meditation experience, we have incorporated the Pydub library to add longer pauses after each sentence and to overlay harmonious background music to our audio-guided sessions [20].

2.3 Video Generation

Compositional Pattern Producing Networks (CPPNs) serve as the fundamental framework in our video generation system, demonstrating an impressive ability to craft intricate and visually captivating patterns. The model requires no training. It essentially takes a basic coordinate tensor and a latent vector as input, and transforms the tensor using the latent vector and a variety of smooth activation functions. Following these mathematical transformations, the output of the model is a tensor where each element represents a color intensity value for each pixel, spanning from grayscale to the complete RGB color spectrum.

The foundation of our CPPN is inspired by Ratzlaff's work [19]. While his base model's structure is capable of generating standalone images using a latent vector, as described by Ha [9], our research advances this approach by shaping the latent vector using the speech and sentiment of the text at various time points. Inspired by [14], we enhance our video generation process by incorporating audio input. To achieve this integration, we employ Mel Frequency Cepstral Coefficients (MFCCs) extracted from the audio using the Librosa library in Python [16]. MFCCs act as concise representations of key audio features, primarily capturing spectral characteristics like amplitude distribution. These MFCCs are converted into vector representations at discrete time segments, serving as latent vector inputs for our CPPN model. This fusion of audio and visual elements introduces a pioneering dimension to our video generation, enabling synchronized and harmonious audio-visual experiences.

The latent vector plays a pivotal role in shaping the model's output. Notably, feeding similar vectors into the system results in visually consistent images. Leveraging this characteristic, we employ interpolation between such vectors, creating a seamless sequence of closely related vectors that form a continuous path connecting the original ones. Since comparable latent vectors yield similar images, passing this vector sequence through the CPPN facilitates a smooth transition between the initial images associated with the original vectors.

To construct a video, we extend this process across numerous latent vectors, generating a sequence of images, each serving as an individual frame in the video. This inherent mechanism establishes the foundation for a coherent and fluid video sequence. Yet, our video's dynamism and responsiveness are further honed through the strategic selection of the latent vector inputs. These decisions profoundly influence the evolving patterns and movements within the video, empowering us to craft captivating and visually engaging video content within the CPPN framework.

Scaling Through Sentiment Analysis. To further enhance the movement of the generated videos, we incorporate sentiment analysis features to modify the input into the CPPN. We chose the Valence Aware Dictionary for Sentiment Reasoning (VADER) model as our sentiment analysis model [11]. VADER is a lexicon and rule-based Natural Language Processing (NLP) model known for its exceptional performance in analyzing sentiments from social media text. It was chosen because it not only indicates the polarity (positive or negative) but also the intensity (strength) of the emotion. Using the NLTK library, we tokenize the text into sentences and input each sentence into VADER to receive a sentiment score in the range $[-1, 1]$. We subtract this score from 1 to get a range that is centered around 1, sufficient for scaling. We then take this sentiment value to scale the latent vector input and the coordinates that are inputted into the CPPN. At the end of each sentence, we interpolate between the coordinate scaling values to create the effect of smoothly zooming in or out of the frame. This creates the illusion of diving in or jumping out to expand the palette of colors.

Applying Color Schemes. To allow for customization, we provide an optional color scheme feature. If the user selects one of the color scheme options, the RGB values of each CPPN output tensor will be altered before saving it as an image. To do so, we take each of the three RGB channels and compare them to create masks to select only part of the pixels in the image. We then scale the RGB values accordingly to fit the color scheme desired. For example, the "blue-yellow" color scheme defines a blue mask that selects the pixels where the blue value is greater than the green value. From there, we modify the pixels in the blue mask to electric blue by setting the red value to zero and setting the blue value equal to the green value. A similar process is performed on the pixels excluded from the blue mask to turn a shade of yellow. This process was used to implement six color schemes: blue-yellow, red-orange, blue-green, black-white, warm, and cool. However, there is wide potential to expand for more (Fig. 3).

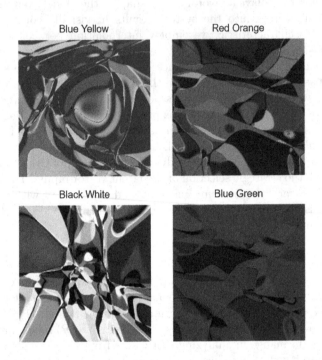

Fig. 3. Images of a variety of color scheme options that are available.

2.4 Experimental Setup

The primary objective of our experiment was twofold: first, to conduct a comparative analysis between a traditional meditation system and an AI-generated meditation system, and second, to evaluate specific features of the AI-generated system based on participant feedback. To ensure a comprehensive evaluation, our focus extended beyond the mere presentation of the AI-generated system. We aimed to directly compare the effectiveness of its innovative approach with the well-established traditional meditation methods. Additionally, we sought to assess the system's various features, examining participant responses to inform our understanding of its strengths and areas for potential improvement.

The experiment was conducted remotely, with each participant participating from the comfort of their homes. Participants were provided with detailed instructions via an online form, guiding them through the entire experimental process. The study encompassed two sessions: one involving our proposed system and the other featuring the traditional meditation system, which consisted of audio only. To ensure a consistent comparison, both sessions followed the body-centered meditation type and lasted approximately 3 min. To eliminate order-related biases, the sequence of these sessions was systematically alternated among the participants. Following the completion of the sessions, participants were administered a questionnaire that delved into various aspects. This questionnaire collected metadata such as participants' prior meditation experience and gender. Additionally, participants rated both systems on content quality, audio clarity, and overall satisfaction, providing numerical ratings ranging from 1 to 10. Furthermore, participants expressed preferences on the personalization of the systems, offering ratings between 1 and 5, where 1 indicated a strong preference for the traditional system, and 5 indicated a preference for the proposed meditation system.

Regarding the audio-only content, the traditional system utilized a freely available resource from UCLA Health [12]. In contrast, the proposed system offered participants a selection of three pre-generated videos, each designed to evoke distinct emotional contexts: (1) a state of deep sadness due to losing a loved one (sad), (2) extreme tiredness and sleepiness after a long day of school (sleepy), and (3) high-stress levels related to school (stressed). Participants were given the autonomy to choose from these contexts, allowing for a more tailored and personalized meditation experience.

3 Results and Discussion

In this section, the experiment results are initially presented. Subsequently, an analysis of the primary strengths and weaknesses of the current model is conducted, along with key findings and potential future enhancements that could be integrated.

3.1 Experimental Results

In total, fourteen participants participated in the experiment, where 57.1% were men. Additionally, half of the people came from backgrounds of both high and low levels of meditation experience. Throughout the experiment, the order of the meditation sessions was deliberately alternated among participants, ensuring a balanced approach to data collection. In particular, half of the participants started with the traditional meditation system, while the rest started with the AI-generated one. Notably, 64.3% of our participants selected the option "I'm very stressed about school (stressed).", likely caused by the fact that the majority of participants were university students (aged between 18 and 22 years old).

In our comparative analysis between the traditional audio-only system and our innovative proposed approach within the context of meditation, several key parameters were examined, including content quality, audio quality, and overall rating. The mean content quality score for the audio-only system was 6.14 ± 2.38, slightly higher than our proposed system's score of 5.86 ± 2.07. Although the difference was not statistically significant (paired t-test p=0.6452) due to large standard deviations within each group, this result is promising as it indicates that our system achieved content quality comparable to that of human-created content. Similarly, the audio quality mean for the audio-only system was 5.93 ± 3.12, while our proposed system scored 5.14 ± 2.44. Again, the absence of statistical significance (paired t-test p=0.4629) suggests that our system effectively matches the audio quality of the conventional approach. Furthermore, the overall rating for the audio-only system was 6.64 ± 1.55, only slightly higher than our proposed system's rating of 6.50 ± 1.74, with no significant difference (paired t-test p=0.8436) between them. These results highlight that our proposed system, performs on par with the traditional audio-only system in terms of content quality, audio quality, and overall user experience. This parity underscores the viability and effectiveness of our innovative approach, offering a comparable meditation experience to that of manually crafted content.

The comprehensive evaluation of our meditation system provided valuable insights into user experiences and preferences. Participants moderately appreciated our system's addition of music to the meditation experience, rating it at 3.57 ± 1.09 on a scale from 1 to 5. However, there is room for improvement in music selection, volume adjustment, and synchronization with audio. The existing feature that allows participants to choose background music could be further optimized to enhance user satisfaction in this regard.

A significant concern emerged concerning the synchronization of audio and video components, as users felt they did not align seamlessly, resulting in a score of 2.71 ± 1.27 for audio-video synchronization on a scale from 1 to 5. Future enhancements should focus on refining the coordination between audio and video elements to ensure a seamless experience. This could involve enhancing the precision of audio cues and corresponding visual elements, making the meditation experience more engaging and coherent for users.

The results also suggest the visuals did not enhance the user experience, resulting in a score of 2.57 ± 1.09 for the effectiveness of visuals on a scale from 1 to 5. However, given images of potential color schemes that could be applied, participants indicated a favoring towards applied color schemes, resulting in a score of 3.57 ± 1.40 on a scale of 1 to 5, where 5 indicates color schemes would improve the experience. This is accompanied by multiple participants' feedback that a color scheme would make the visuals "easier on the eyes" and "more relaxing." This could be a major cause of participants' low satisfaction with the visuals. Additionally, it suggests a color scheme should be applied by default in the future to prevent the visuals from being too distracting. Participants' feedback about color schemes varied greatly. Some participants indicated the blue-yellow color scheme would be the most relaxing, while another indicated a preference for the red-orange one. One suggested a purple scheme because "it's [their] favorite color," and another preferred the blue-yellow one because it "reminds [them] of a calm sunny day at the beach." Clearly, different users have different opinions about their color scheme preferences, suggesting the color scheme option is a valuable feature. Additionally, this indicates that whether the visuals add to the experience is not objectively a yes or no answer. Instead, it's specific to the user, so users should be encouraged to open or close their eyes according to what makes them most comfortable. Gathering more specific qualitative feedback from users regarding problematic synchronization aspects would further provide valuable insights for targeted improvements. Exploring innovative techniques, such as real-time adjustments or adaptive algorithms, could dynamically enhance synchronization based on user interactions and preferences.

Participants perceived the system as marginally more personalized than traditional methods, with a score of 3.14 ± 1.29 on a scale where 1 represented the audio-only system and 5 represented the proposed system. It is noteworthy that participants could only choose from three distinct contextual options (Sleepy, Stressed, and Sad), indicating the system's potential for personalization. Encouragingly, participants expressed a willingness to engage more regularly in meditation if practices were tailored to their preferences, scoring it at 3.71 ± 1.33. Additionally, participants showed openness to diverse meditation approaches, suggesting receptivity to practices beyond the body-centered method used in the study (3.71 ± 1.33). This suggests that our system's ability to produce different types of meditation has great potential to improve user's meditation habits. Further research and improvements in customization, audio-visual synchronization, and user-specific meditation options could significantly enhance user engagement and satisfaction in the proposed system.

3.2   System Components

Text Generation. Our investigation into the text generation component within our meditation system showcased LLaMA-2 7B's ability to create meditation scripts tailored to specific contexts. However, participant feedback shed light on an intriguing aspect: despite the specificity of the generated videos, users did not perceive them as significantly more personalized. This observation underscores a critical factor - the range of context options provided during the experiment. Our research highlights a clear demand for diverse scripts that align closer to user preferences, while still enabling the creation of text for various meditation types. By offering more tailored scripts, we could enhance user engagement as the meditation experience may resonate deeply with the individual preferences of each user. Although this feature is currently available, the connection between open-ended context and increased user engagement requires further evaluation.

Despite this observation, the quality of the generated content stood tall, as it aligned with the standards set by human-created scripts. This encouraging outcome underscores the potential inherent in our system, indicating that with the right adjustments, it can deliver truly personalized meditation experiences. To steer our system towards this objective, future efforts should concentrate on refining the underlying text generation model. One promising avenue involves fine-tuning the model with extensive and diverse data sourced from the vast expanse of the internet. This approach not only infuses the system with richer meditation styles and vocabulary but also ensures a more nuanced and varied output. By immersing our model in the wealth of meditation resources available online, we would empower it to generate scripts that cater to an extensive array of user preferences, fostering a sense of individuality and personal connection.

A particularly notable challenge emerged during our experiments, centering around the generation of meditation scripts for the 'Mantra' type. This specific category posed unique difficulties for our system, unveiling areas that require focused attention and potential enhancements in our model's capabilities. Mantra meditation involves the continuous repetition of a mantra (word, phrase, or set of syllables), and can be done silently or aloud [18]. Due to the user-focused nature of mantra meditation, improving the scripts alone would not suffice. One potential approach would be to use the LLM to instead generate a mantra, along with a few guiding phrases, and have the system add long pauses between phrases during the audio generation. This would provide the user with more time to focus as they repeat their mantra. By dissecting the intricacies of generating mantra meditation guides, we can unravel the complexities at play, paving the way for targeted improvements. This focused approach ensures that even the most intricate meditation styles are accurately captured and represented within our system, offering users a comprehensive and authentic meditation experience.

In conclusion, our exploration into the realm of text generation has not only highlighted the need for more tailored scripts but also underscored the unwavering potential of our system. By combining the insights gleaned from participant feedback with strategic enhancements in our model, we aim to achieve a deeper level of personalization. As we continue to refine our approach, guided by user experiences and cutting-edge advancements, our meditation system is well-positioned to redefine individualized meditation, offering users a transformative and deeply immersive journey toward well-being and mindfulness.

Text-to-Speech. Our study delved into the nuanced realm of audio integration, uncovering valuable insights into user preferences and challenges faced in creating a harmonious auditory experience for meditation. Participants consistently voiced a preference for the incorporation of music. Our study underscores the positive influence of music on user engagement: 9 of the 15 participants indicate it added to the experience, while only 2 indicate they strictly disagree. Paradoxically, our system's audio quality was not universally perceived as superior, revealing a multifaceted challenge. One avenue for enhancement lies in meticulous volume adjustments for both the speech and music components. Achieving a delicate equilibrium between these elements is crucial for harmonizing the auditory experience.

In addition, it is crucial to recognize the importance of how participants' perceptions may be subtly influenced by biases against specific accents. While our system currently offers diverse accent options, a more in-depth assessment of these biases and their impact on overall satisfaction is needed. Embracing this diversity and tailoring accent choices can promote inclusivity and user acceptance. Additionally, a nuanced approach to accent selection is vital, and employing machine learning algorithms to analyze preferences would allow dynamic adjustments, ensuring a personalized audio experience. This adaptability aligns with technology's personalization trend, enhancing our system's user-centricity.

In summary, our findings underscore the importance of balancing audio elements and addressing biases in accent preferences to create a seamless meditation experience. As we move forward, leveraging advanced technologies and embracing user diversity will be pivotal, allowing us to tailor our system to individual preferences and deliver a meditation experience that resonates deeply with users, creating a sense of connection and engagement.

Video Generation. Our research delves into a crucial aspect of participant feedback, emphasizing the importance of synchronizing video and audio components during meditation sessions. Participants' perception of misalignment between video and audio was significantly influenced by their specific meditation context. Notably, in the 'Sleepy' context, the narrator directed the participants to keep their eyes closed, whereas in the 'Stressed' context, they were given the option to close their eyes lightly. Interestingly, the 'Sad' context lacked specific eye-closure instructions, resulting in a distinct experience. These variations likely hindered participants from fully appreciating the video component's potential to enhance their meditation experience. To address this, future sessions could explore practices where eye closure is more flexible, providing the user with more opportunity to experience the visual component of the session. Additionally, the synchronization of audio and video emerged as a prominent concern, profoundly impacting the overall user experience. Precision in aligning audio cues with corresponding visual elements is crucial. Gathering specific qualitative feedback on synchronization issues can provide valuable insights. Innovative strategies like real-time adjustments and adaptive algorithms show promise in enhancing synchronization dynamically, ensuring a seamless and engaging meditation experience tailored to individual preferences.

Additionally, our study suggests the intriguing prospect of enhancing user engagement through adapting the system's color schemes to different meditation types. Vibrant and energetic colors, such as bright red or intense orange, can effectively symbolize meditation practices with a strong focus on physical activity and the body ('High Body Orientation and High Activation'). For practices that are physically oriented yet less mentally stimulating ('High Body Orientation and Low Activation'), employing calming colors like deep blue or forest green may be particularly beneficial. On the other hand, bright and vivid color choices, such as bold yellow or electric blue, can aptly signify meditation practices that are mentally stimulating with less emphasis on the body ('Low Body Orientation and High Activation'). Lastly, employing neutral and contemplative color schemes, like soft gray or pale lavender, is a viable strategy for practices that do not demand high physical or mental activity ('Low Body Orientation and Low Activation'). It's important to note that the selection of these colors is based on our conceptualization of the associations between color and meditation types, and we acknowledge the need for further research and references to validate these connections. This versatile color palette caters to a broad spectrum of user preferences, thereby enhancing the overall user experience.

In summary, our research explores multiple dimensions of the user experience within meditation sessions. Employing innovative solutions and addressing concerns regarding audio-visual synchronization and eye-closure guidelines promises an enhanced meditation practice. Additionally, the customization of color schemes aligned with meditation types offers users a tailored experience, advancing the holistic nature of our meditation system.

4   Conclusion

Improving the accessibility and personalization of meditation is crucial due to its numerous benefits across diverse situations. This study underscores the vital significance of tailored and accessible meditation experiences in different contexts. By integrating text and video generation models, our research showcases the development of audio-visual environments, illustrating the capacity for customized outputs aligned with various audio inputs. Involving 14 participants, the experiment not only confirmed the system's ability to match the quality of the traditional content and audio but also highlighted its adaptability and heightened user engagement. These findings extend the system's relevance beyond meditation, encompassing areas such as hypnosis and personalized sensory therapies.

Looking ahead, future research should explore incorporating additional sensory inputs like smell, taste, and touch, allowing the model to generate diverse outputs based on varied sensory experiences. Moreover, expanding these inputs into other models could revolutionize the medical industry, potentially leading to the design of personalized treatments and cures based on individual metrics.

Beyond its applications in healthcare, this study paves the way for innovative developments in generative art. Integrating additional elements such as audio and sentiment into the CPPN model opens exciting possibilities for artists, enabling the creation of multidimensional artworks that resonate with audiences on multiple sensory levels. By embracing audio as a creative influence, this approach transcends conventional boundaries, offering a dynamic canvas for artistic expression and enriching the intersection of technology and art. Overall, the findings underscore the extraordinary potential of personalized sensory integration, paving way for a new era in meditation practices[1].

Acknowledgments. We would like to acknowledge the support of Templeton World Charity Foundation (TWCF) grant No. 0470. The opinions expressed in this publication are those of the authors and do not necessarily reflect the views of the TWCF.

Code Availability. To encourage further research on these topics, we have made the source code of this work freely accessible to all.

---

[1] https://github.com/petern48/meditation_induction_ai.

References

1. Awasthi, B.: Issues and perspectives in meditation research: in search for a definition. Front. Psychol. **3**, 613 (2013)
2. Behan, C.: The benefits of meditation and mindfulness practices during times of crisis such as COVID-19. Irish J. Psychol. Med. **4**, 256–258 (2020)
3. Bohr, A., Memarzadeh, K.: The rise of artificial intelligence in healthcare applications. Artif. Intell. Healthcare 25–60 (2020). https://doi.org/10.1016/B978-0-12-818438-7.00002-2
4. Bond, K., et al.: Defining a complex intervention: the development of demarcation criteria for "meditation". Psychol. Relig. Spiritual. **1**(2), 129 (2009)
5. Clark, T., Barnes, P., Black, L., Stussman, B., Nahin, R.: Use of yoga, meditation, and chiropractors among U.S. adults aged 18 and over. NCHS Data Brief **325** (2018)
6. Cramer, H., et al.: Prevalence, patterns, and predictors of meditation use among us adults: a nationally representative survey. Sci. Rep. **6**(1), 36760 (2016). https://doi.org/10.1038/srep36760
7. Durette, P.N.: Google text-to-speech (2023). https://gtts.readthedocs.io/en/latest/
8. Fox, K.C., et al.: Functional neuroanatomy of meditation: a review and meta-analysis of 78 functional neuroimaging investigations. Neurosci. Biobehav. Rev. **65**, 208–228 (2016)
9. Ha, D.: Generating abstract patterns with tensorflow. blog.otoro.net (2016). https://blog.otoro.net/2016/03/25/generating-abstract-patterns-with-tensorflow/
10. Hanley, A.W., Dehili, V., Krzanowski, D., Barou, D., Lecy, N., Garland, E.L.: Effects of video-guided group vs. solitary meditation on mindfulness and social connectivity: a pilot study. Clin. Soc. Work J. **50**(3), 316–324 (2021). https://doi.org/10.1007/s10615-021-00812-0
11. Hutto, C.J., Gilbert, E.: Vader: a parsimonious rule-based model for sentiment analysis of social media text (2014). https://doi.org/10.1609/icwsm.v8i1.14550, https://www.semanticscholar.org/paper/bcdc102c04fb0e7d4652e8bcc7edd2983bb9576d
12. Institute, U.S.: Body scan meditation. https://www.uclahealth.org/programs/marc/free-guided-meditations/guided-meditations#english
13. Ku, B., Itagaki, T., Seaborn, K.: Dis/immersion in mindfulness meditation with a wandering voice assistant. In: Extended Abstracts of the 2023 CHI Conference on Human Factors in Computing Systems. ACM (2023). https://doi.org/10.1145/3544549.3585627
14. Markus, N.: Using cppns to generate abstract visualizations from audio data. Published on the web at https://nenadmarkus.com/p/visualizing-audio-with-cppns/ (2018)
15. Matko, K., Sedlmeier, P.: What is meditation? proposing an empirically derived classification system. Front. Psychol. **10**, 2276 (2019). https://doi.org/10.3389/fpsyg.2019.02276
16. McFee, B., et al.: librosa/librosa: 0.10.1 (2023)
17. Nash, J.D., Newberg, A.: Toward a unifying taxonomy and definition for meditation. Front. Psychol. **4**, 806 (2013)

18. Álvarez Pérez, Y., et al.: Effectiveness of mantra-based meditation on mental health: a systematic review and meta-analysis (2022). https://www.ncbi.nlm. nih.gov/pmc/articles/PMC8949812/#:~:text=Mantra%2Dbased%20meditation %20(MBM),or%20without%20religious%2Fspiritual%20content.
19. Ratzlaff, N.: Cppn (2013). https://github.com/neale/CPPN
20. Robert, J., Webbie, M., et al.: Pydub (2018). http://pydub.com/
21. Sharma, H., Clark, C.: London: Singing dragon; 2012. Ayurvedic Healing.[Google Scholar]
22. Touvron, H., et al.: Llama 2: open foundation and fine-tuned chat models (2023)
23. Wang, X., Mo, X., Fan, M., Lee, L.H., Shi, B.E., Hui, P.: Reducing stress and anxiety in the metaverse: a systematic review of meditation, mindfulness and virtual reality (2022)

On the Impact of Directed Mutation Applied to Evolutionary 4-Part Harmony Models

Elia Pacioni$^{(\boxtimes)}$ and Francisco Fernández De Vega

Universidad de Extremadura, Av. Santa Teresa de Jornet, 38, 06800 Mérida, Spain
{eliapacioni,fcofdez}@unex.es
https://www.unex.es

Abstract. This paper analyzes the difficulty of finding solutions for the 4-part harmony problem, both from the point of view of the size of the search space and the time required to run the algorithm. These considerations led to improved running time through parallelization, precalculation of the fitness function, and directed mutation, which reduces the time to solution. Moreover, we show how combining these techniques allows us to extend the improvements to different harmony models, including new synthetic ones that may be proposed.

Keywords: 4-part harmonization · Directed mutation · Anticipated fitness evaluation · Harmonic models

1 Introduction

Harmonization for soprano, alto, tenor, and bass (SATB) is a commonly used technique in choral music. This method attained its apex during the Baroque era with J.S. Bach, who extensively utilized and augmented this style of harmonization in his musical works. Hence, the establishment of essentials for choral composition occurred at this time. However, scholars have continued to refine SATB harmonization over time, adapting the rules to suit evolving musical styles. These principles encompass aspects such as the progression of chords, the prohibition of certain dissonant combinations, melodic movements, and the balance of sound between different voices.

Lopez-Rincon et al. (2018) describe a taxonomy of AI methods used in algorithmic music composition, including heuristic methods like evolutionary algorithms and dynamic programming; deep learning methods like convolutional and recurrent neural networks; stochastic methods; and symbolic AI-based methods such as agents, declarative programming, and grammatical representation [1].

Liu and Ting's (2017) study analyzes AI methods used in music composition and identifies the most appropriate techniques for each task. Notably, neural networks exhibit the highest proficiency for imitative systems, whereas Markov models excel at predicting musical notes based on previous ones. Genetic algorithms (GAs) are very good at generating chord progressions [2]. Over the years,

C. Johnson et al. (Eds.): EvoMUSART 2024, LNCS 14633, pp. 311–325, 2024.
https://doi.org/10.1007/978-3-031-56992-0_20

numerous studies have been conducted, but the issue still needs to be solved, as satisfactory results using accurate scores have yet to be achieved. Simulations that do not reflect reality have almost always used a restricted and simplified set of problems. Hence, this area of research continues to pose a significant challenge even now.

This paper builds upon prior research published in [3], which presented the first version of the evolutionary algorithm and described various evolutionary approaches. This new research aims to enhance the evolutionary process further and increase the number of rules considered, making the problem more complex. Therefore, our primary objectives are to decrease the execution time by implementing parallelization and precalculating fitness values, develop novel harmonic models, and introduce a directed mutation operator to steer the algorithm towards potentially superior solutions. All the improvements and new developments obtained in this area are implemented in the Sharpmony application, which uses evolutionary algorithms to help music students learn harmony [4].

The paper's structure is as follows: Sect. 2 presents the literature review, analyzing the techniques used to address the problem and the solutions proposed. Section 3 describes the methodology applied to conduct the work, defining the approaches used and the theoretical concepts on which they are based. Section 4 presents the experiments performed and reviews the results obtained. Finally, Sect. 5 contains our conclusions.

2   4-Part Harmonization and Evolutionary Algorithms

2.1   SATB Harmonization Problem

Computer-assisted musical composition is an area of interest across various computer science domains. Specifically, numerous GA approaches have been proposed, although many are restricted in scope to simplified versions of the actual problem. Over the years, multiple contributions have been made to constructing correct harmonization for a given melody. For instance, Horner and Goldberg's work [5] disregarded essential details, while McIntyre's early contributions in the 1990s [6] suggested that the problem becomes manageable if a melody and pitch are provided, leading to the completion of the remaining three voices. However, the process of harmonizing from 5 to 9 notes required the efforts of thousands of individuals over hundreds of generations. It should be noted that this exercise is a simplified representation and cannot serve as a basis for any conclusive claims that the problem has been resolved.

More recently, Kaliakatsos et al. in 2014 [7] stated that the traditional rules are somewhat contradictory, so probabilities must be used to select the rules to apply. In this case, the authors prioritize the generation of chord progressions over completing the remaining voices.

Finally, in 2017, new approaches were presented that achieved results of similar quality to those achieved by music students at the beginning of their studies, with only 10 errors in the entire 32-chord musical score. However, only 11 rules were considered, providing ample room for improvement. Notably, this approach has gained support from harmony teachers at professional music conservatories, who continually contribute to the development of new rules. This makes the system look more like the real problem of 4-part composition but also makes it harder to solve with an evolutionary algorithm since the number of available solutions in the search spaces becomes smaller as more rules are added [3].

The approaches pursued thus far are of interest, but the 4-part harmonization problem remains unresolved.

2.2 Anticipating Fitness Evaluation

Pre-computing or anticipating the evaluation of individuals or fragments of individuals can be an efficient practice to reduce computing time and energy consumption during the execution of an EA. The literature predominantly employs three techniques: fitness approximation, surrogate models, or fitness inheritance. The use of these methods increases execution speed but also brings in an element of error or uncertainty connected to the estimated fitness value, which ultimately fails to represent the actual value accurately.

The technique of approximating the fitness function can be performed in multiple ways. Jin provides an extensive overview of approximation techniques and their integration into GAs [8]. He divides the approximations into problem approximations and function approximations. Problem approximation aims to substitute the initial problem statement with a slightly modified one that is easier to solve. In contrast, function approximation strives to generate an alternative expression that can replace the fitness function.

Shi introduces the concept of inheritance, where a child's fitness value is determined by a weighted combination of its parents' values [9]. Additionally, Shi discusses various statistical approximation methods, including polynomial models [10], Krigin models, and support vector machines (SVMs). Bajer emphasizes the significance of using surrogate models, such as radial basis function networks, and notes that this technique reduces the number of evaluations required by roughly 70%. Surrogate models may also adopt a Gaussian approach [11] or rely on neural networks [12].

2.3 Directed Mutation

Finally, directed mutation is a method that follows the concept of induced mutation in biological systems [13]. Here, the mutation operator focuses on specific chromosomes flagged during evolution due to their weaknesses or issues.

Bhandari et al. proposed this approach in 1994 [14]. They presented a novel operator to enhance the exploration of the solution space and expedite the algorithm's convergence.

One of the most recent studies showcases the Adaptive Directed Mutation (ADM) operator, demonstrating its superior efficiency compared to five conventional mutation operators. This improved the convergence, accuracy, and reliability of the tested algorithm [15].

Additionally, the directed and adaptive mutation can adapt to changes in the problem, facilitating improved fitness values, even in the face of complex issues. However, a high rate of targeted mutation may cause the solution to converge towards a local optimum. Thus, an attentive parameter configuration of the GA is necessary.

A distinct approach is suggested, in which direct mutation is implemented by determining the mutation likelihood of each gene composing the chromosome while considering its impact on enhancing the individual. The mutation probability is then adjusted based on the results achieved through the applied mutation. [16]

Recently Carvalho et. al proposed a similar approach in the context of genetic programming (GP) and, in particular, evolutionary grammar (GE) [17].

3   Methodology: Directed Mutation and Precalculation of Fitness Values

Before showing how the techniques previously described have been applied to improve solutions to the problem we address, we present here an analysis of the problem that allows us to understand the size and structure of the search space.

As described above, in the first version of the algorithm, 11 harmony rules were considered. However, the standard harmony rules that students must learn are many more, and the current version of the fitness function is made up of almost 50 different rules and exceptional cases. To the best of our knowledge, this is the first time such a number of rules have been considered for an EA to address the problem.

Making the set of rules as close as possible to those learned by students dramatically increases the computational time required for an EA to find a solution as the number of available solutions decreases, and the search process becomes harder.

To estimate the solution space, we consider all the chords available in a given key: to do this, we must take into account the fact that in a major key, e.g., C major, we usually work with the diatonic notes C, D, E, F, G, A and B, and each of them, according to its position in the scale, represents the scale degrees: I, II, III... VII. For each of these allowed degrees, and taking into account that each chord requires four notes since we are working with four voices, according to the basic rules of harmony and voice arrangement, we have computed a total of 2855 possible valid chord configurations, including secondary dominants and chords with properly configured seventh and ninth. This number represents all possible chord configurations: keep in mind that the notes of each chord can be distributed among the four voices; moreover, two different octaves are possible in each voice, and some of the rules establish what is valid for a chord configuration. Figure 1 shows a small set of chords dispositions available for the first degree in C Major.

Some rules establish a relationship among the notes in a single chord, while others consider the notes in consecutive chords; finally, some rules require an

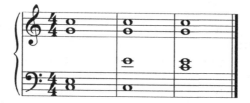

Fig. 1. Examples of first degree chord dispositions available in C major, among more than 2800 available ones.

analysis of a given chord, its preceding chord, and the following chord; therefore, three chords must be analyzed simultaneously.

When two chords are analyzed, and given that the order of the chords within the pair is relevant to the calculation of the error, (the number of errors occurring from a given chord disposition A to another chord disposition B is different from the number of errors occurring from B to A), we perform the Cartesian product, obtaining $2855^2 = 8\,151\,025$ pairs of possible chord dispositions in a single key. On the other hand, when three consecutive chords must be analyzed, we obtain $2855^3 = 23\,271\,176\,375$ combinations available.

These values correspond to the space of possibilities in a single key, but we must remember that we have 12 possible major keys: C, D-flat, D, E-flat, E .... B flat, B; and as many minor keys.

It is possible to move from a key to another easily: musicians usually apply the process of transposing when a score must be played by an instrument in a different key. Transposing is an easy process that consists of adding or subtracting a given musical interval to every note (minor third, major third....). Therefore, we could consider a single key, computing chord progressions, and the same could be applied to the other ones using a transposing process. This is useful when no modulation process is present, and we know that all of the chords in a sequence belong to the same key. When that is not the case, things quickly deteriorate: every chord disposition from a key may be followed by any other chord disposition from any different available key. Yet, in this paper, we do not consider modulations.

To summarize, concerning the experiments proposed in this article, we will focus only on errors between pairs of chords in C major, leaving minor keys and errors affecting three consecutive chords for future work.

Given the complexity of the problem and the size of the solutions space, we consider the application of parallelization and precalculation of fitness function as a way to address the problem we face in a reasonable time.

3.1   Precalculation of Fitness Values

A technique for anticipating the fitness calculation is proposed here. This choice is based on the study of the cost of the fitness function and its impact on the problem.

Specifically, each time an individual is evaluated, it is necessary, among other things, to compare each consecutive pair of chords to calculate the number of errors between them. Although other generic calculations are required, the one described above is the most time-consuming for calculating the fitness function.

Currently, error calculation between a single pair of chords takes between 20 and 25 s, significantly contributing to the total time required by the fitness function and, therefore, to the entire algorithm to find a solution.

The idea is to replace the error computation between a pair of chords with a precomputed database containing the errors previously computed for all possible pairs of chords, a total of 8 151 025 for a given key.

To use this method, we first need to compute and store every pair of chords and the total number of errors found; we use a relational database to do this. Given the nature of the problem, the information to be stored can be represented by a single entity: chord1, chord2, and the total number of errors found. Therefore, since not all chords have the same length in characters, a varchar stores this information, while an integer type can be used for errors. Assuming an average length of 45 characters for each chord, considering the weight of 1 byte for each character face and 4 bytes for an integer, we get $(45 * 1 * 2 + 4) * 8\,151\,025 = 766\,196\,350$ bytes, equivalent to about $766.20MB$. This is a reasonable size for a relational database, which shows that this solution can be viable.

The populated database must be optimized for maximum performance in the search mode since the evolutionary algorithm does not need write access, only read access. Therefore, the database is configured to be read-only, thus optimizing access time.

Read access through indexes has a computational cost on the order of $log(n)$, where n is the number of records in the table. On the other hand, using indexes penalizes write access since the index tree must be updated every time the database is written, generating an overhead. However, since no write operations are required once the database is ready, this disadvantage is not a problem for the evolutionary algorithm.

From a volume perspective, the index adds an amount of data proportional to the number of rows in the table and the size of the index keys. In the case of MariaDB with the InnoDB storage engine, it is possible to calculate the size of the index as follows: the index is created on the chord1 and chord2 columns, which have a total size of 90 kb, to which an overhead of 6 kb must be added for managing the index, for a total of 96 kb, that multiplied by the number of rows reaches a size of 746 MB; to this last value must be added the overhead due to the weight of the internal nodes to store pointers to the child nodes. A final estimate of the overhead caused by the index structure is about 900 MB, bringing the database to more than double its initial size.

3.2   Parallelization

The application of parallel computing is justified only in high computational loads. Otherwise, the price to be paid for the overhead introduced by parallelism may outweigh the benefits obtained.

As described above, we must compute all the errors corresponding to every possible pair of chords stored in the database. Considering that the average time for error calculation between a pair of chords using the sequential fitness function available is 22.5 s, the total time required to compute the whole database is 183 398 062.5 s, equivalent to about 5.8 years. Although this is an estimation considering a sequential approach with a single computer, this can be significantly improved using parallel approaches and a cluster. Moreover, parallel techniques have been applied to the main steps of the Evolutionary Algorithm. We show below how the algorithm's running time improves when both the precalculation of the database and parallel techniques are included in the main Ea in charge of finding 4-part harmony solutions for a given melody.

3.3   Creation of New Harmonic Models

Once the database is available, we will not need the rules anymore until they are changed. We could use the database with errors as the fitness function component to evolve different 4-part harmony for a given melody. Moreover, given that we can recompute the database when some rules are changed, we could also understand that different databases will embody different harmony rules, allowing us to work with different music styles and periods: classicism, impressionism, romanticism ... even jazz music. Not only the DB might embody any available harmony model, but also synthetic models could be tested.

This approach could lead to composing innovative scores that explore new sonorities or describe new compositional practices. These can be subjected to analysis by users to define the degree of satisfaction and consequently accept or not accept the proposed new harmonic model. Although we are not yet considering that possibility in this research stage, we will use a synthetic model to check if any model is usable and allows an EA to converge to solutions within that model.

To build this database, an analysis of the database that is being created was carried out using the rules of the standard harmonic model. Currently, errors have been calculated among about 130000 chord pairs, which corresponds to about 1.6% of the total chord pairs to be calculated. However, it is interesting to study the distribution of errors, and we have found that error-free chord pairs correspond to 8% of the total already calculated. We have thus decided to evenly distribute the chord pairs by the number of errors, from 0 to 9, with a 10% rate, when creating the complete set of the synthetic harmony models to be tested.

3.4   Directed Mutation

Once the database with the synthetic model is ready, the EA can be run using the number of errors between every pair of chords to compute the fitness value for every individual.

Although standard mutation operators may be applied, we explore the directed mutation technique here. The idea is to employ the info provided by the fitness function when a whole score, namely a series of chords, is analyzed: given that we compute errors for each of the consecutive pair of chords, we will know not only the total number of errors in a given individual but also the position of every pair of chords that features errors. This way, instead of randomly mutating a chord contained in the score, a chord that will generate errors may be selected for mutation among the available list of positions of chords with errors. Thus, Creating a list containing only positions with errors is equivalent to assigning the same mutation probability to positions with errors: $\frac{100}{N_{list\_items}}$ and 0% to error-free positions.

A high rate of guided mutation can be crucial, accelerating the algorithm's convergence to optimal solutions. This approach increases the likelihood that mutations will make significant improvements rather than simply random variations. A directed mutation can strategically guide the algorithm's evolution by leveraging specific information about errors or inefficiencies in the system. This type of mutation, if well calibrated, has the potential to explore the solution space more effectively, reducing the number of generations required to reach a result deemed satisfactory and increasing the quality of final solutions.

On the other hand, this approach could push the solution toward local minimums, so it is also necessary to consider intermediate solutions that overcome this problem and maintain a good level of genetic variability.

3.5   The Problem

For an exhaustive comprehension of the problem, it is necessary to know how the problem is encoded: an individual is represented as a series of chords corresponding to the four voices: one of the voices is provided (the melody, for instance), and the other three must be evolved. Thus, the chromosome encodes four notes per chord, together with the degree corresponding to every chord, that is required for a proper evaluation. Although this may be enough to launch the evolutionary algorithm, sometimes, students try first to decide the appropriate degree for each of the notes provided in the melody. Once the degrees have been established, the chord disposition must be selected for every degree. We can emulate this process by encoding an initial chromosome with the chosen degree and running a previous EA to evolve the best possible progression. Then, the second EA, which decides which notes are assigned to each voice, can be run. Both possibilities are evaluated below.

To test the validity of our method, we chose a melody as a reference that is complex enough to pose a genuine challenge, thus avoiding the overly simple problems often discussed in previous studies. The score includes 8 bars and

29 notes, considerably more extensive than those commonly examined in such research. Importantly, this is an exercise assigned to students at a music conservatory. Figure 2 illustrates the melody used as a reference. The key used for the proposed score is C Major.

Fig. 2. Melody used by the evolutive algorithm.

3.6 Tests Performed

In the tests performed, different configurations of the algorithm are tried, keeping some common features fixed, particularly the number of individuals fixed at 8 and the number of generations at 100. The chromosome comprises the four voices, the soprano voice -provided as the melody- is kept fixed while the other three must evolve.

Concerning the progressions to be applied, two approaches are tested: the first with 50 generations to evolve a proper chord progression, and the second with no evolution of the chord progresion. This adds an additional level of difficulty to the evolutionary algorithm.

In addition, to measure the goodness of the new mutation operator, tests are carried out with directed mutation, with random mutation, and finally with a percentage of both directed and random mutation, set at 50%.

The parameters used to test the algorithm, which are the same in all runs, particularly regarding mutation operator applied are described below. On the other hand, the crossover probability is 50% and selection type roulette.

The other parameters will be specified by analyzing the different configurations proposed for each run.

4   Experiments and Results

As described in the previous section, different configurations of the algorithm were adopted with the aim of improving the quality of the results. This section describes the experiments performed and the results obtained applying fitness function precalculation, parallelism and directed mutation.

4.1   Execution Time

Error calculation time between two consecutive chords was checked first by performing 100 tests for both configurations and calculating the average. The results, being a comparison between sequential and parallel, are analyzed by means of the speedup measure as follows:

$$S_n = \frac{T_s}{T_p}$$

where T_s represents the time consumed by the function performed sequentially, while T_p is the time employed by the parallel version.

When the speedup is computed using the values obtained, as described in Table 1 we obtain a speedup of 22,500. Considering that at the time of evaluating an individual it is necessary to calculate errors between pairs of successful chords, this result gives an idea of the improvement that the use of parallelism and precalculation of the fitness value can bring. Therefore, this technique actually helps to speed up the evaluation of an individual and consequently the execution of the entire GA.

Table 1. Error calculation time between two consecutive chords, in seconds.

Sequential	Parallel + DB
22,5	0,001

However, it is necessary to consider that this operation is not the only computationally expensive one. Therefore, although the speedup value is extremely high in the analysis of errors between chord pairs, the GA will not obtain such a high overall speedup. The operations for managing the structure of an exercise and the calculations needed to reconstruct errors that require information previous to the treated chord represent complex operations that are particularly impactful on the performance of the algorithm. However, these operations are fundamental to running a GA that can work on real problems.

Figure 3 presents a comparison of the execution time of the entire algorithm in the sequential, parallel and parallel versions with the database. In all tests, the configuration defined in the previous section is used. The result of the comparison is clearly visible and shows that between the sequential and parallel versions there is a significant improvement from 776 min to 532, consequently the speedup is about 1.5. Taking the parallel version with the database into consideration, we reduce the computation time required to 42 min, resulting in an overall speedup of 12.8.

This demonstrated how judicious use of parallelism can bring considerable advantage to the algorithm and, how the use of precomputation of fitness values drastically impacts the performance of the system, proving to be an extremely useful technique.

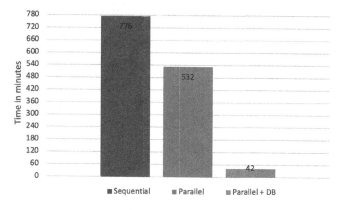

Fig. 3. Comparison of algorithm execution times between sequential and parallel version with database, in minutes.

4.2 Directed Mutation

The general configuration of the algorithm has been introduced in Sect. 3.5 and 3.6; We include in the Table 2 the parameters used to test every configuration of the GA, which changes according to the experiment run. For each configuration, 10 runs were performed, and the average of the fitness values was employed for comparison purposes.

Table 2. Genetic algorithm configurations used for experiments

Exp	Evolving Chord Progression	Mutation
EV0	YES	100% DIRECTED
EV1	YES	100% RANDOM
EV2	YES	50% DIR. 50% RANDOM
EV3	NO	100% DIRECTED
EV4	NO	100 % RANDOM
EV5	NO	50% DIR. 50% RANDOM

Figure 4 shows the results where on the x-axis we find the generations and on the y-axis the average fitness values over the 10 runs. We can see that most of the solutions have the same trend and at generation 100 are in a fitness range between 29 and 33, except for the experiment *EV0*, which obtains significantly better values than the others. Therefore, we can say that direct mutation together with a previous evolution of chord degrees, improves the solution obtained, allowing the algorithm to converge faster and obtain better solutions.

Moreover, a distinction can be made between experiments with 50 generations of progressions and those with only one generation. Only in the case of

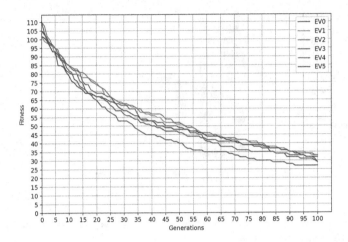

Fig. 4. Plots the average fitness values for different GA configurations.

using direct mutation there are significant improvements in using a larger number of generations of progressions, while in the other cases it does not seem so relevant.

4.3   Longer Runs

Although the time required for each run is significant, some long runs were performed to see if the GA could progress to better solutions. Given the importance of directed mutation, we performed two experiments with a larger number of individuals. In the first case 20 individuals, in the second case 8 individuals, giving a maximum of 1740 individuals evaluated in the evolutionary process. In all cases, 50 generations of progression are used.

Figure 5 compares the number of individuals evaluated with the best fitness value obtained. In this case, the experiment with 8 individuals gives significantly better results than the one with 20, showing a faster convergence to optimal values. Therefore, it can be considered more convenient to work with a smaller number of individuals for longer generations.

The GA found a solution with only 13 errors in the version with a population of 8 individuals, a quality level similar to other ones previously described in the literature.

In summary, all implementations tested to solve the problem have not yet reached the limit of their capabilities. The best solution with 13 errors is slightly lower than that of the previous version of the algorithm. However, it is necessary to consider that the new implementation applied a different set of rules, defined through the synthetic database. Further improvements could be achieved by studying more thoroughly the parameters employed by the GA and testing additional genetic operators.

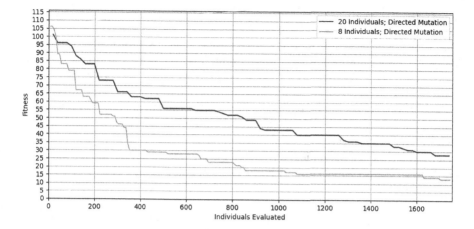

Fig. 5. Comparing total number of individuals evaluated and best fitness. Directed mutation applied & chords progression evolved first

4.4 Synthetic Harmonic Models

The last experiment conducted involves the creations of new harmonic models. Specifically, by changing the percentage of error-free chords within the database we are varying the harmonic model used to evolve the musical sheet. In particular, four cases are compared: 1%, 10%, 20% of error-free chord pairs, respectively.

Although short runs have been counducted, they provide clues on what we can expect. Figure 6 shows the GA progress in all cases. The examples shown were conducted with a single run of 4 individuals and 10 generations, since the only purpose is to demonstrate that by changing the database the algorithm continues to work, and this means that the EA seems to be able to cope with any harmonic model we may consider.

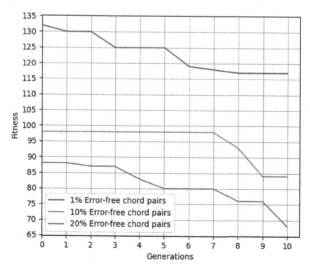

Fig. 6. Tested synthetic harmony models

5 Conclusions

This paper advances the solution of the 4-part harmony problem using GAs. Many previous approaches have used simple and unrealistic problems. In this paper, more than 22 rules were considered, and 25 exceptional cases within the rules, so over 50 different harmony rules were included, provided by a professional team of music conservatory teachers. The analysis of this model has allowed to build a number of synthetic models that are useful to test how an EA can cope with different music styles.

Multiple tests were applied, and results were examined to evaluate the use of parallelism, the anticipation of fitness value calculation, and the directed mutation operator.

On the one hand, the anticipation of the fitness calculation has not only allowed shrinking computing time but also shows that new harmonic models can be easily applied by building the database embodying errors detected between every pair of chords. We tested the idea using several synthetic harmony model that were built considering the percentage of errors found with the standard SATB classical harmony model. Results show the EA can cope with models tested, and this allow us to foresee the potential to exploit different harmony models. Secondly, the directed mutation operator, which applies mutation to positions where some errors are detected, provides results better than the standard mutation that operates randomly along the chromosome. We have improved speedup up to 12.8 with the new parallel approach.

In the future, some concepts of adaptive mutation could also be incorporated into the oriented mutation by variating the mutation probability of different chords featuring errors.

Acknowledgements. We acknowledge support from Spanish Ministry of Economy and Competitiveness under projects PID2020-115570GB-C21 funded by MCIN/AEI/10.13039/501100011033. Junta de Extremadura under project GR15068.

References

1. Lopez-Rincon, O., Starostenko, O., Ayala-San Martin, G.: Algoritmic music Composition based on artificial intelligence: a survey. In: International Conference on Electronics, Communications and Computers (CONIELECOMP), pp. 187–193 (2018). https://doi.org/10.1109/CONIELECOMP.2018.8327197
2. Liu, C.H., Ting, C.K.: Computational intelligence in music composition: a survey. IEEE Trans. Emerg. Top. Comput. Intell. 1(2), 2–15 (2017)
3. de Vega, F.F.: Revisiting the 4-part harmonization problem with GAs: a critical review and proposals for improving. In: IEEE Congress on Evolutionary Computation (CEC), Donostia, Spain, pp. 1271–1278 (2017). https://doi.org/10.1109/CEC. 2017.7969451
4. Fernandez De Vega, F., Alvarado, J., Sanchez, A., Serrano, M., Pacioni, E.: Evolutionary algorithms: a new hope for the future of music teaching. In: Proceedings of the Companion Conference on Genetic and Evolutionary Computation (GECCO '23 Companion), pp. 65–66. Association for Computing Machinery, New York, NY, USA (2023). https://doi.org/10.1145/3583133.3596945

5. Horner, A., Goldberg, D.E.: Genetic algorithms and computer-assisted music composition. In: International Conference on Mathematics and Computing, pp. 437–441 (1991)

6. McIntyre, R.A.: Bach in a box: the evolution of four part baroque harmony using the genetic algorithm. In: Evolutionary Computation: IEEE World Congress on Computational Intelligence. Proceedings of the First IEEE Conference 1994, pp. 852–857 (1994). https://doi.org/10.1109/ICEC.1994.349943

7. Kaliakatsos-Papakostas, M., Cambouropoulos, E.: Probabilistic harmonization with fixed intermediate chord constraints. In: ICMC (2014). https://doi.org/10.13140/2.1.3079.5526

8. Jin, Y.: A comprehensive survey of fitness approximation in evolutionary computation. Soft Comput. **9**(1), 3–12 (2005). https://doi.org/10.1007/s00500-003-0328-5

9. Shi, L., Rasheed, K.: A Survey of Fitness Approximation Methods Applied in Evolutionary Algorithms. Computational Intel. in Expensive Opti. Prob., ALO, 2 (2010)

10. Hosder, S., et al.: Polynomial response surface approximations for the multidisciplinary design optimization of a high speed civil transport. Optim. Eng. **2**, 431–452 (2001)

11. Bueche, D., Schraudolph, N.N., Koumoutsakos, P.: Accelerating evolutionary algorithms with gaussian process fitness function models. IEEE Trans. Syst. Man Cybern. Part C **2**(35), 183–194 (2005). https://doi.org/10.1109/tsmcc.2004.841917

12. Ulmer, H., Streichert, F., Zell, A.: Model-assisted evolution strategies. Knowl. Inc. Evolutionary Comput. 333–355 (2005)

13. Melamed, D., et al.: De novo mutation rates at the single-mutation resolution in a human HBB gene-region associated with adaptation and genetic disease. Genome Res. (2022). https://doi.org/10.1101/gr.276103.121

14. Bhandari, D., Pal, N.R., Pal, S.K.: Directed mutation in genetic algorithms. Inf. Sci. **79**(3–4), 251–270 (1994). https://doi.org/10.1016/0020-0255(94)90123-6

15. Tang, P.-H., Tseng, M.-H.: Adaptive directed mutation for real-coded genetic algorithms. Appl. Soft Comput. **13**(1), 600–614 (2013). https://doi.org/10.1016/j.asoc.2012.08.035

16. Puerta, B.R., Barrancas, F.D., Chavez, F., de Vega, F.F.: Un análisis preliminar de nuevos modelos de mutación dirigida en algoritmos genéticos. In: Conferencia de la Asociación Española para la Inteligencia Artificial (CAEPIA 2018), Granada, España

17. Carvalho, P., Magane, J., Lourenco, N., Machado, P.: Context matters: adaptive mutation for grammars. In: Pappa, G., Giacobini, M., Vasicek, Z. (eds.) Genetic Programming. EuroGP 2023. LNCS, vol. 13986, pp. 117–132 . Springer, Cham (2023). https://doi.org/10.1007/978-3-031-29573-7_8

Evaluation Metrics for Automated Typographic Poster Generation

Sérgio M. Rebelo[1](✉)[iD], J. J. Merelo[2][iD], João Bicker[1][iD],
and Penousal Machado[1][iD]

[1] Department of Informatics Engineering, CISUC/LASI – Centre for Informatics and
Systems of the University of Coimbra, University of Coimbra, Coimbra, Portugal
{srebelo,bicker,machado}@dei.uc.pt
[2] Department of Computer Engineering, Automatics, and Robotics and CITIC,
University of Granada, Granada, Spain
jmerelo@ugr.es

Abstract. Computational Design approaches facilitate the generation
of typographic design, but evaluating these designs remains a challenging
task. In this paper, we propose a set of heuristic metrics for typographic
design evaluation, focusing on their legibility, which assesses the text
visibility, aesthetics, which evaluates the visual quality of the design,
and semantic features, which estimate how effectively the design conveys
the content semantics. We experiment with a constrained evolutionary
approach for generating typographic posters, incorporating the proposed
evaluation metrics with varied setups, and treating the legibility metrics
as constraints. We also integrate emotion recognition to identify text
semantics automatically and analyse the performance of the approach
and the visual characteristics outputs.

Keywords: Computational Creativity · Design Measures ·
Evolutionary Design · Graphic Design · Layout · Poster Design

1 Introduction

Computational Design has revealed a significant potential to transform practices
in the field of Graphic Design (GD), including Typography and Layout domains
[7,18]. This potential involves automating typesetting tasks (*e.g.* applying text
styles, creating tables, *etc.*) and facilitating design exploration. Nonetheless,
assessing the outcomes of these computational processes remains a challenging
task, given that their evaluation relies on subjective design factors.

This paper is based upon work from a scholarship supported by SPECIES, by the
Foundation for Science and Technology, I.P./MCTES through national funds (PID-
DAC), within the scope of CISUC R&D Unit - UIDB/00326/2020 or project code
UIDP/00326/2020, and by Ministerio español de Economía y Competitividad (Span-
ish Ministry of Competitivity and Economy) under project PID2020-115570GB-C22
(DemocratAI::UGR).

In this paper, we present a set of ten heuristic metrics for evaluating typographic designs. These metrics consist of the application of design rules that enable automated assessment of various characteristics of typographic designs, especially posters. Our motivation behind developing these metrics is to streamline and facilitate computational typesetting processes, ultimately leading to faster and more efficient GD practices. The proposed set of metrics encompasses the evaluation of designs in terms of their legibility, aesthetic features, and coverage of content semantics. Legibility metrics determine whether all text content is adequately displayed and readable within the design, including the evaluation of *text legibility* and *grid appropriateness*. Aesthetic metrics assess the visual quality of the designs, examining aspects like *alignment*, *balance*, *justification*, *regularity*, *typeface pairing*, and *negative space fraction*. Semantic metrics focus on assessing how the composed text effectively conveys the semantic meaning of the content in terms of *layout* and *typography*.

We study the practical application of these metrics in guiding the creation of typographic poster designs by developing an Evolutionary Computation (EC) approach. In this context, we considered that the primary goal of posters is to fully display their content. This way, we explored a constrained evolutionary methodology where legibility metrics are treated as constraints that the generated outputs must satisfy, while other metrics define the objective value of designs. This approach is inspired by the workflow of traditional typography design processes, as conducted in nineteenth-century print houses. Back then, typographers employed an algorithmic method for typesetting content to fill all the space in a matrix. They use condensed typefaces for lengthy sentences and extended typefaces for shorter ones, while also emphasising the most significant parts of the content typographically [12].

Furthermore, we designed a user interface to support the developed approach, enabling users to input text content and specify the desired visual features of the outputs. To facilitate the generative process, we develop procedures to automatise the text division and to recognise the more semantically significant parts of the content using emotional recognition. The development of this approach follows an agile science methodology, structured around potential user cases and scenarios for the application of the proposed metrics in poster design [13]. The code repository for this project is accessible at github.com/sergiomrebelo/evo-poster. Supplementary materials are available at cdv.dei.uc.pt/projects/evoposter (websites visited: 8 November 2023).

We conducted experiments to examine the influence of the proposed evaluation metrics on the evolutionary generation of typographic poster designs. These experiments involved the legibility-constrained evolution of typographic posters using input texts of varying lengths and emotional content. The experiments consisted of three stages, each focusing on the evolution of posters based on either semantic metrics, aesthetic metrics, or a combination of both. The results demonstrate that the proposed metrics effectively guide an evolutionary process, producing finished and legible designs from a variety of text inputs while considering both aesthetics and semantics.

The primary contribution of this paper is the set of metrics for evaluating typographic designs. Other prominent contributions include (I) a functional constrained evolution approach for creating typographic posters; (II) a multi-purpose, domain-driven, and easily understandable representation of poster designs; (III) an investigation into weighted objective function strategies for evaluated designs based on aesthetics and semantics; (IV) an exploration of integrating computational design into a poster design field.

The remainder of the paper is organised as follows. The Sect. 2 provides a review of the related work, specifically focusing on design metrics. The Sect. 3 comprehensively describes the metrics. The Sect. 4 explains our experimental approach, the setup, and the results. The Sect. 5 summarises our contributions and outlines potential directions for future research.

2 Related Work

Heuristic metrics and measures for visual assessment are frequently used in the computational generation of visuals, especially in the context of EC. These metrics have been explored to overcome the limitations of cooperative evolutionary approaches, like Interactive Evolutionary Computation strategies, allowing the incorporation of subjective human-related data into the evolutionary design processes. However, they can lead to user fatigue and inconsistent evaluations [11].

Typography and Layout metrics are employed for either generative or optimisation objectives. Geigel and Loui [5] employed a set of design measures, including page balance, spacing, and emphasis, to automatically evolve page layouts for photography albums. Harrington et al. [6] proposed a non-linear layout evaluation measure, combining a set of heuristic metrics for document design, such as alignment, regularity, separation, balance, white-space fraction, white-space flow, proportion, uniformity, and page security. Building upon these measures, Purvis et al. [15] and Rebelo et al. [16] developed generative evolutionary approaches, adapting the proposed measures for the characteristic of the outputs. Lok et al. [8] introduced a technique for computing visual balance using lightness weight maps from images. Balinsky et al. [1] presented measures for determining page alignment and regularity, based on the extraction and quantification of "alignment lines." Bylinskii et al. [3] and Xie et al [21] developed Machine Learning (ML) models which unveiled some potential for auto-completion and layout retrieval. More recently, Lopes et al. [9] developed a pixel-based approach to assess the balance of design.

The reviewed related work primarily focuses on visual attributes like balance, regularity, and alignment while overlooking typographic attributes. Nevertheless, these typographic attributes are an essential consideration for graphic designers during their creative processes. Factors like typography pairing and content emphasis significantly contribute to the distinctiveness and effectiveness of the conveyed message. Recent ML approaches show a promising manner to evaluate the designs. However, their effectiveness in poster design scenarios is constrained by the limited availability of poster layout data.

3   Metrics

We introduced a set of ten metrics to evaluate typographic designs, specifically posters that feature short text messages. These metrics aim to address the lack of consideration of typography and representation of content semantics in current work on computational assessment of visual designs.

The computation of these metrics assumes that designs consist of text boxes, each representing a single line of text. These text boxes are characterised by attributes such as size, alignment, and font. They are organised sequentially within a one-column grid. In addition, typefaces used must provide data related to the category, *e.g.* serif, mono-space, sans-serif, *etc.* Certain metrics require quantifying the emotional charge of each text line. An optimal poster layout is also considered, which describes the expected distribution of each text box in terms of percentages of the poster's height based on the emotional analysis.

The metrics are evaluated on a scale ranging from 0 (poor score) to 1 (perfect score), and they can be divided into three evaluation objectives: legibility; aesthetics; and semantics. They have been implemented as a standalone module, and the source code is available in the project repository.

Legibility metrics evaluate whether the text content on a poster is fully visible in the design. This objective includes two distinct metrics: (I) the *text legibility* and (II) the *grid appropriateness*.

The *text legibility* metric assesses whether all text content is fully visible within the text boxes that compose the poster. It is calculated by the arithmetic mean of the legibility score of all text boxes. The legibility score of each text box is the variance between the width of the text box's content, as rendered, and the available width of the design (*i.e.* the width of the poster without the horizontal margins). This score is then normalised, gradually prejudicing the cases when the text exceeds the available width. A score of 0 indicates that the text width doubles the width of the container, while a value of 1 indicates that the text fits within the space.

The *grid appropriateness* metric assesses whether the grid used in the design is suitable. It does so by comparing the size of the grid (both width and height, including the margins) with the size of the poster. This metric can only have two values. A value of 1 indicates that the grid entirely fits within the design dimensions, while a value of 0 suggests that it does not.

Aesthetic metrics are used to evaluate the visual and typographic features of a design. These metrics include (III) *alignment*, (IV) *balance*, (V) *justification*, (VI) *regularity*, (VII) *typeface pairing*, and (VIII) *negative space fraction*. We defined these metrics based on works of Harrington et al. [6] and typographic principles outlined by Lupton [10] and Bringhurst [2].

The *alignment* metric assesses the consistency of the horizontal alignment of text. The estimation involves two main steps. First, it computes the variance in text width between neighbouring text boxes. Then, it calculates the arithmetic mean of these variances (d), using the non-linear $A/(A+d)$ where A is a constant that controls how fast values fall away from 1 as the distance between entries increases. Second, the metric checks if the line alignment of the text box is uniform. This is determined by the division of 1 (the expected line alignments on the design) by the count of different line alignments identified on the posters. The overall alignment score is a weighted average, with width variance contributing 80% and text alignment contributing 20%. This score is subsequently normalised within a range from 0 (high width variance and different line alignments) to 1 (low width variance and consistent use of a single line alignment).

The estimation of *balance* metric initially involves the calculation of the visual weights and the balance centres of each text box which compose it. The visual weight of a text box (vw) is defined by multiplying its area by its optical density. The optical density (oD) is calculated by considering the relative Luminance, which is computed on the average pixel values for the red (r), green (g), and blue (b) colour channels within the text box, using the formula:

$$oD = \log_{10}\left(\frac{1}{n}\sum_{n}^{t=1} 0.2126 \times r_t + 0.7152 \times g_t + 0.0722 \times b_t\right) \qquad (1)$$

The balance centre of a text box (x_t, y_t) is determined based on its line alignment (x_t) and the overall vertical alignment of the poster. For instance, if the text box is left-aligned and vertically aligned to the top, its balance centre (x_t, y_t) is the upper-left corner. When the vertical alignment is set to centre, the vertical balance centre position is visually adjusted moving one-twelfth towards the top. Next, it calculates the centre of the visual weight of the entire design (w_x, w_y), by considering all text boxes (t) within it as:

$$w_x = \left(\frac{\sum_{t=1}^{n} x_t \times vw_t}{\sum_{t=1}^{n} vw_t}\right) \text{ and } w_y = \left(\frac{\sum_{t=1}^{n} y_t \times vw_t}{\sum_{t=1}^{n} vw_t}\right) \qquad (2)$$

Subsequently, it estimates the expected balance centre of the design (c_x, c_y) employing the same method as used for text boxes, considering the alignment of the first text box (c_x) and the same global vertical alignment (c_y), albeit taking into consideration the full poster size. Finally, the overall balance (B) score is determined considering the calculated current and expected balance centres and poster sizes as follows:

$$B = 1 - \left[\left(\left(\frac{w_x - c_x}{width}\right)^2 + \left(\frac{w_y - c_y}{height}\right)^2\right)/2\right]^{\frac{1}{2}} \qquad (3)$$

The *justification* metric evaluates whether the text fully occupies the available space. This metric is inspired by the traditional aesthetics of nineteenth-century letterpress posters where text content was traditionally justified within the available space, ensuring that text occupies all the available areas when possible. The calculation of this metric is similar to the *text legibility*. The overall justification score is determined as the arithmetic mean of the justification scores of all text boxes. The justification score of a text box is calculated by considering the variance between the text width of each text box and the available space. The variance value is normalised penalising both designs with text overflow and with excessive white space. The penalty is lower when text fully fits within the poster, being divided by a factor. A lower variance results in a higher justification score. A justification score of 1 indicates zero width variance, while if the variance doubles the design width the justification is set to 0.

The *regularity* metric evaluates how regular is text box heights in the design. It measures the distances (d) between the vertical positions of the top edges in neighbouring text boxes using a non-linear function $A/(A + d)$, where A is a constant that determines the rate at which values decrease as the distance between entries increases, similar to *alignment*. The overall regularity score is the arithmetic mean of the value of all neighbouring pairs is normalised within a range between 0 (low regularity) and 1 (high regularity).

The *typeface pairing* metric evaluates the compatibility of the typefaces used on the poster, considering their categories on typographic classification. It begins to create a unique list of categories from the typefaces used in the design. The overall typographic pairing score is determined by the count of categories found in the design, normalised between 1 (if there is only one category) and 0 (if all the used typefaces are from different categories).

The *negative space fraction* metric assesses the appropriateness of the percentage of background colour in the design. It calculates the current percentage of the poster occupied by background colours and computes the deviation from an optimal value chosen by the user. The differences are then normalised to a range between 0 (twice the optimal background percentage) and 1 (optimal background percentage).

Semantic metrics assess whether the placement of the typography in the designs conveys the semantic meaning of the content. This analysis encompasses the evaluation of the semantic significance of the IX *layout*, and X *typography*.

The *semantic significance of the layout* metric evaluates whether text lines which contain more emotional charge are appropriately highlighted in the layout. The greater the emotional charge on a line, the more height this line of text should be typeset on the design. This calculation is done at the line (or text box) level. So, for each text box, it calculates the distance between its height and the height in the optimal layout, represented as a percentage of the poster's total height. This metric can operate in two modes, "Fixed" and "Relative". In the "Fixed" mode, it considers that content must fulfil the total poster available height. In "Relative" mode, it only considers the current height of the composition, ignoring the empty space. By default, it operates in "Fixed" mode. The overall score for the semantic significance of the layout is the arithmetic mean of the normalised distance of all text boxes, ranging from 0 (significant differences between the actual and optimal layout) to 1 (perfect match between the optimal and current layout).

The *semantic significance of typography* metric evaluates if the most emotional parts of the content are typographically emphasised. This involves calculating variations in weight, stretch, and type design across fonts used in the text boxes within the design. The score for the variable typographic features (weight and stretch) is computed as the mean of the distances between the current and expected values. These distances are normalised between 0 (maximum distance) and 1 (no distance). To determine the expected values for each text box, the range of values used in the design is established. Then, for each text box, an expected distance is assigned based on the recognised emotional levels in the content. The text box with the highest emotional context recognised is assigned the maximum distance value within the range, while the one with lesser emotional charge receives no distance. Ultimately, for each text box, the distance between the current value and the value of a text box with no expected difference is calculated, and the current distance is determined by comparing the results with the assigned expected distance. The type design score is calculated by counting the number of typefaces that deviate from the expected ones, normalised based on the total number of text boxes. A score of 0 indicates that the output aligns with type design expectations, whereas a score of 1 signifies that it does not meet any expectations. To calculate the expected typefaces, typefaces are initially assigned to emotional levels based on their first use within each emotional level. Each typeface is uniquely associated with one emotional level. If a typeface is already assigned to another level, the typeface for this level is set as undefined. The overall score of the *semantic significance of typography* is the highest score among the three types of variations. To prevent excessive emphasis on the same content, the score is penalised when multiple features receive evaluations exceeding a high threshold. In such cases, the score is divided by the number of features that exceeded this threshold.

4   Experiments

We conduct evolutionary experiments using the proposed metrics and a Genetic Algorithm to automatically evolve typographic posters using different texts. We experiment with a constrained evolution approach employing a stochastic ranking method [19] to fitness assignment the designs. Legibility metrics serve as constraints that the generated outputs must adhere to. Aesthetic and semantic metrics determine the design value of the outputs, through a weighted multi-criteria objective evaluation function. In addition, we implement a constraint penalty approach to facilitate the sorting and visualisation of population and elitism. The penalty is calculated as the inverted weighted arithmetic mean of the two legibility metrics.

We adopted a Domain-Driven Design [4] approach to define the representation of posters. Each genotype is encoded in the JSON format, with each representational attribute identified by a key term commonly used in the GD domain. The genotype comprises two parts: (I) text boxes data and characteristics, namely content, typeface, weight, stretch, size, and alignment; and (II) poster characteristics, specifying the size (width and height), margins (in percentages of size), and vertical alignment. The grid of the poster is inferred based on the text box size and the poster characteristics. One descriptive example of a genotype is provided in supplementary materials.

The generative process starts with the random initialisation of a poster population. For each line of text, a text box is assigned, and its characteristics are randomly determined from available choices. This process involves randomly selecting a typeface, along with the weight and stretch based on its available options. The font size is defined within a predefined range, while alignment is randomly set to left, centre, or right. The vertical alignment of the poster is also randomly defined as top, middle, or bottom.

Tournament selection is used to choose individuals for breeding based on their fitness, and variation operators (mutation and crossover) are applied to the selected individuals, with a certain probability, to generate new offspring. We use a uniform crossover method [20], which randomly determines which parent will pass down genetic material to the descendant. It flips the vertical alignment gene and all text boxes on the new individual to determine which parent will pass down its characteristics to the offspring. The text boxes are passed to the descendants with all of their characteristics.

The mutation variation operator determines if an attribute of the individual in new offspring will undergo change, based on a certain probability. The mutation method employed varies depending on the selected attribute. When poster vertical alignment or the alignment, weight, and stretch attributes of a text box are selected for mutation, the current value is modified by one of the available options. If the typeface attribute of the text box is selected, it replaces the current typeface with one of the available options, and it adjusts the weight and stretch attributes if needed, selecting values close to the current ones. In the case of the font size of the text box, it adjusts the size based on random values within a specific range, typically between -5 and 5.

We employ an elitist process, where the best individuals from the current generation are combined with the new offspring. To select these individuals, it sorts the population using a penalty constraint approach.

Before the evolution, the system emotionally analyses input content to determine the parts that must be emphasised on the layout. The analysis starts with the preprocessing of the text by (I) handling contracted word forms, (II) replacing abbreviations and slang expressions with formal equivalents, (III) substituting emojis with their actual meanings, (IV) changing negations to antonyms, (V) eliminating stop words and URLs, and (VI) tokenising the text. Subsequently, it conducts a lexicon-based analysis using a word-emotion association lexicon [14]. Based on the result, emotional scores are assigned to individual words. The emotional scores for each line are calculated by summing the scores of words on it.

We designed a user interface that allows users to input the text and configure the desired visual features of the outputs and the evolutionary approach. Furthermore, new typefaces, options, and default evolution parameters through a dedicated configuration file. Typefaces are incorporated into the system in the OpenType Variable Font format. Users can input content either divided into lines or choose to let the system automatically divide the text. Text division is performed using a Sentence Boundary Detection algorithm [17] to split the text into sentences and then divide the longer sentences based on a random factor within a predetermined line size range. Screenshots of the developed interface are available in the supplementary material.

Table 1. Metrics weights in the respective evaluation objective.

Aesthetic metrics		Semantic metrics	
Alignment	10%	Layout	50%
Regularity	10%	Typography	50%
Balance	20%		
Negative Space Fraction	20%		
Justification	30%		
Typeface Pairing	10%		

Table 2. Experimental Parameters.

Type	Parameter	Value
Evolutionary	Generations	400
	Population size	30
	Elite size	1
	Crossover probability	90%
	Mutation probability	10%
	Tournament size	10
Posters' features	Poster size (px)	141×100
	Default margins (ltrb)	5% 5% 5% 5%
	Min colour contrast	2.5
Aesthetic metrics	Alignment/regularity A constant	10
	Justification factor	3
	Optimal negative space factor	50%
Semantic metrics	Significance of layout mode	Fixed
	Significance of typography threshold	0.2

4.1 Experimental Setup

The conducted experiments are divided into three different stages: (S1) evolving only considering semantic objective; (S2) evolving only considering aesthetic objective; and (S3) evolving considering both objectives, with equal weight on objective evaluation. The score of each evaluation objective is determined as the weighted arithmetic mean of the score of metrics that compose it. The weight of each metric on the objective function is defined empirically, with the goal of creating poster designs inspired by the aesthetics of nineteenth-century letter-press posters, which was one of the main motivations for the development of this work. Table 1 displays the weight assigned to each metric in both the semantic and aesthetic groups.

The conducted experiments occur on the client side of a Chromium web browser. We loaded 8 OpenType Variable Font, all of which have weight and

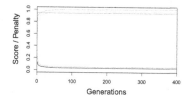

Fig. 1. Progression of constraint penalty and legibility metrics across generations. Visualised results present the average of three stages, totalling 1,800 runs. Black lines depict penalty constraint values, while pink and blue lines show the scores of *text legibility* and *grid appropriateness* metrics, respectively. Fittest individuals are depicted with solid lines, and the average population with shaded lines. (Color figure online)

stretch axes available. The experimental parameters for these experiments were empirically defined and are summarised in Table 2. In the three evaluation stages, we generated typographic posters for 20 different texts in three languages (English, Portuguese, and French). These texts expressed various purposes, emotional contexts, and lengths. On average, each input text contained 39.55 characters, divided into 7.15 words, with a maximum of 83 characters and a minimum of 17. The average text line length was 4.15 lines. The selected texts, apart from being neutral, were classified by representing emotions such as anticipation, disgust, joy, sadness, surprise, and trust. Further description of the text and the typographic available options are provided in the supplementary material.

4.2   Experimental Results

The experimental results demonstrate that the proposed metrics and approach enable the evolution of legible designs for various types of text inputs. Analysing the progression of constraint penalty in the population (see Fig. 1), we observed that the proposed approach leads to a faster reduction in the number of individuals that do not comply with the legibility constraints, either in the best individuals or the population average. Individuals evaluated with maximum fitness are present in the population but do not are legible. A detailed analysis of *grid appropriateness* metric reveals that the proposed approach can consistently generate and maintain individuals that use valid grids. On the other hand, the evolution of *text legibility* is inversely proportional to the constraint penalty value. The results also reveal that the fittest individuals at the end of evolution are always individuals with all text legible.

The evolution based on **semantic** metrics (S1) reveals that the proposed semantic metrics can guide the evolution of posters towards designs where the more emotional parts of the content are highlighted. The experimental results (see Fig. 2) unveil that the best individual in the population achieves high fitness values as well as high scores in both metrics. However, the metrics values stabilise after a dozen generations, primarily due to the legibility constraints. These constraints restrict the semantic evolution of certain individuals, as they compromise legibility. In this sense, some individuals with high semantic scores are

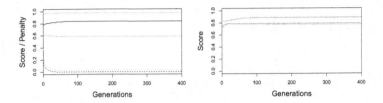

Fig. 2. Evolutionary progress in the S1 experiment. The results are averages from 600 runs, considering varied text inputs. The chart on the left illustrates best individual fitness (solid lines), constraint penalty (dashed lines) and maximum fitness (dotted). On the right, the chart shows score metrics related to the significance of semantics in *layout* (green lines) and *typography* (orange lines). Solid lines represent the average of the fittest individuals, while shaded lines represent the average of the entire population. (Color figure online)

also not fully legible designs. It is also noted that the evolution of the semantic significance of layout is slightly faster than the typography.

The generated outputs exhibit certain parts of the content that are emphasised in terms of layout, typography, or both (see Fig. 3a). This emphasis is primarily directed toward the more emotional sections of the content. However, when the emphasis affects both layout and typography, it can sometimes become excessive, leading to a visually heavy impact, particularly in designs with fewer emotional variations. Additionally, certain design issues were observed in the outputs, including non-uniform alignment, unbalanced layout, or excessive use of different typefaces.

The outcomes of the aesthetics evolution (S2) indicate that it is possible to evolve designs complying with the proposed aesthetic metrics. Nonetheless, these results also reveal the challenge of evolving designs that fulfil all the metrics simultaneously. The performance of metrics depends on the text content, and the necessity for legibility limits the range of aesthetic design possibilities.

Experimental results (see Fig. 4) reveal that global fitness increases faster in the initial generations but slows down as the constraint penalty decreases. We noted that some individuals in the population achieve higher than the best individuals in the population, but they are not legible posters. When examining the score of each metric individually, *regularity*, *alignment*, and *typographic pairing* progressively increase over the generation both in the fittest individuals and population average. This suggests that enhancing these metrics does not compromise the legibility of the posters. On the other hand, *balance*, *justification*, and *negative space fraction* maintain relatively stable scores over the generations, with slight decreases in scores on the best individuals, while legibility and other aesthetic metrics increase. The *justification* score of the best individual was even, on average, below the population average. Further experimentation is necessary to understand how other metrics and experimental parameters impact their evaluation. However, high scores on these metrics result in the creation of invalid individuals, even in more advanced generations, who are subsequently

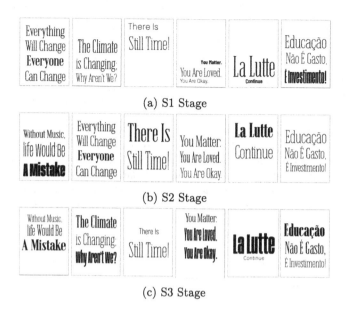

(a) S1 Stage

(b) S2 Stage

(c) S3 Stage

Fig. 3. Examples of designs evolved during the three experimental stages. More results can be found in the supplementary material folder.

removed from the population. Due to this, legibility metrics had worse scores than in the S1 stage.

The visual analysis of the outputs (see Fig. 3b) unveils that, even though they are created with the same input texts, the generated posters are more balanced and regular in typographical terms compared with S1 outputs. Nonetheless, we noticed that in some outputs less meaningful parts of the content are typographically highlighted (e.g. conjunctions), making the content more difficult to read.

The results of the evolution combining both aesthetic and semantic metrics (S3) indicate that optimising all the metrics simultaneously is a challenge, being the design possibilities influenced by the text content and the recognised emotional charge. The experimental results (see Fig. 5) exhibit an evolutionary pattern akin to the two other experiments: faster evolution in early generations followed by gradual stabilisation as the constraint penalty values decrease. However, we noted that there are individuals with higher fitness values in the population, yet they are not legible.

The analysis of the evolution of semantic metrics separately reveals that their combined evaluation with aesthetic metrics presents challenges to evolving semantic metrics, being constrained by the other aesthetic metrics. In terms of the *significance of layout*, this experiment indicates that it is restricted by aesthetic metrics, showing a relatively flat trajectory, with the fittest individuals and the population average achieving similar values. The evolution of the *significance of typography*, although exhibiting some improvement in the early generations, it faster stabilises and remains below the scores achieved in the S1 experiment. The

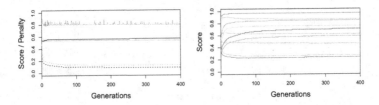

Fig. 4. Evolutionary progress in the S2 experiment. The results are averages from 600 runs, considering multiple text inputs. The chart on the left illustrates best individual fitness (solid lines), constraint penalty (dashed lines) and maximum fitness (dotted). On the right, the chart shows score metrics related to aesthetics, including *alignment* (navy), *regularity* (green), *justification* (violet), *typography pairing* (pink), *balance* (turquoise), and *negative space fraction* (orange). Solid lines represent the average of the fittest individuals, while shaded lines represent the average of the entire population. (Color figure online)

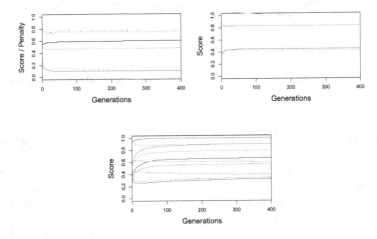

Fig. 5. Progression of evolution in the S3 experiment. The results are averages from 600 runs, considering various text inputs. The top-left chart presents the progression of the best individual fitness (solid lines), constraint penalty (dashed lines) and maximum fitness (dotted). The top-right and bottom charts illustrated the scores of metrics related to semantics and aesthetics, respectively. Semantic metrics include the significance of semantics in *layout* (green) and *typography* (orange). Aesthetic metrics include *alignment* (navy), *regularity* (green), *justification* (violet), *typography pairing* (pink), *balance* (turquoise), and *negative space fraction* (orange). Solid lines represent the average of the fittest individuals, while shaded lines represent the average of the entire population. (Color figure online)

results of aesthetic metrics resemble the results of S2. *Alignment, regularity,* and *typographic pairing* progressively increase over the generations, while *balance, justification,* and *negative space fraction* maintain relatively stable scores, seemingly constrained in their progress by legibility metrics. We observed a slight evolution of *justification* scores, which appear to be related to the inclusion of semantic metrics and the exploration of layouts that are not highly valued by other aesthetic metrics but, from a semantic perspective, are considered superior and enable improved content justification.

The empirical visual analysis of the outputs (see Fig. 3c) reveals an increase in aesthetic features value present, compared to S1. Simultaneously, it is possible to observe the emphasis of parts with more emotional weight, whether through layout or font selection, in comparison to S2. However, this emphasis in certain designs should be more pronounced in some posters.

5 Conclusions and Future Work

Computational approaches in GD unveil potential, but evaluating outcomes remains challenging due to reliance on subjective factors. We introduced ten heuristic metrics for autonomously evaluating typographic designs, addressing both visual and semantic aspects. These metrics are categorised into three evaluation objectives: Legibility metrics assess text display effectiveness, including *text legibility* and *grid appropriateness*. Aesthetic metrics evaluate visual quality, considering *alignment, balance, justification, regularity, typeface pairing*, and *negative space fraction*. Semantic metrics focus on typographic content representation for *layout* and *typography*. We applied these metrics in a constrained evolution approach to generate typographic poster designs for various texts, with legibility metrics as constraints and other metrics in a weighted multi-criteria objective evaluation function. Our preliminary experiments covered three stages, evaluating posters using semantic, aesthetic, and both metrics.

The main motivation behind developing these metrics is to streamline computational typesetting processes. Despite their preliminary nature, evaluation experiments reveal their effectiveness in guiding typographic poster design generation for diverse text contents, highlighting their potential for automation in GD practices. However, evolving multiple metrics simultaneously, in this constrained setting, restricts metrics' progression, necessitating further research to understand the individual metrics' progression and their interaction with constraints. Future work also includes less constrained evolutionary experiments with different parameters. Moreover, we intend to conduct further evaluation studies with human participants, aiming to assess the visual impact of the proposed metrics in output designs.

References

1. Balinsky, H.Y., Wiley, A.J., Roberts, M.C.: Aesthetic measure of alignment and regularity. In: Borghoff, U.M., Chidlovskii, B. (eds.) DocEng '09: Proceedings of the 9th ACM Symposium on Document Engineering, pp. 56–65. ACM, September 2009. https://doi.org/10.1145/1600193.1600207
2. Bringhurst, R.: The Elements of Typographic Style, 2nd edn. Hartley & Marks Publishers, Seattle (1997)

3. Bylinskii, Z., et al.: Learning visual importance for graphic designs and data visualizations. In: Gajos, K., Mankoff, J., Harrison, C. (eds.) Proceedings of the 30th Annual ACM Symposium on User Interface Software and Technology, pp. 57–69. ACM, October 2017. https://doi.org/10.1145/3126594.3126653

4. Evans, E.: Domain-Driven Design Reference: Definitions and Pattern Summaries. self-publishing (2015). https://www.domainlanguage.com/ddd/reference. Accessed 18 Jan 2024

5. Geigel, J., Loui, A.C.P.: Automatic page layout using genetic algorithms for electronic albuming. In: Beretta, G.B., Schettini, R. (eds.) Internet Imaging II, vol. 4311, pp. 79–90. SPIE (2000). https://doi.org/10.1117/12.411879

6. Harrington, S.J., Naveda, J.F., Jones, R.P., Roetling, P., Thakkar, N.: Aesthetic measures for automated document layout. In: Munson, E.V., Vion-Dury, J.Y. (eds.) DocEng '04: Proceedings of the 2004 ACM Symposium on Document Engineering, pp. 109–111. ACM, October 2004. https://doi.org/10.1145/1030397.1030419

7. Levin, G., Brain, T.: Code as Creative Medium: A Handbook for Computational Art and Design. The MIT Press, Cambridge (2021)

8. Lok, S., Feiner, S., Ngai, G.: Evaluation of visual balance for automated layout. In: Vanderdonckt, J., Nunes, N.J., Rich, C. (eds.) Proceedings of the 9th international conference on Intelligent User Interfaces, pp. 101–108. ACM, January 2004. https://doi.org/10.1145/964442.964462

9. Lopes, D., Correia, J., Machado, P.: Towards the automatic evaluation of visual balance for graphic design posters. In: Pease, A., Cunha, J.M., Ackerman, M., Brown, D.G. (eds.) Proceedings of the 14th International Conference on Computational Creativity, ICCC 2023, Waterloo, 19–23 June 2023, pp. 192–199. Association for Computational Creativity, July 2023

10. Lupton, E.: Thinking with Type: A Critical Guide for Designers, Writers, Editors, & Students, 2nd edn. Princeton Architectural Press, New York (2014)

11. Machado, P., Romero, J., Manaris, B.: Experiments in computational aesthetics. In: Romero, J., Machado, P. (eds.) The Art of Artificial Evolution. Natural Computing Series, LNCS, pp. 381–415. Springer, Berlin, Heidelberg (2008). https://doi.org/10.1007/978-3-540-72877-1_18

12. Meggs, P.B., Purvis, A.W.: Meggs' History of Graphic Design, 6th edn. John Wiley & Sons, Inc., Hoboken (2016)

13. Merelo, J.J.: Agile (data) science: a (draft) manifesto, July 2022. arXiv preprint arXiv:2104.12545. https://doi.org/10.48550/arXiv.2104.12545. Accessed 18 Jan 2024

14. Mohammad, S., Turney, P.: Emotions evoked by common words and phrases: using mechanical Turk to create an emotion lexicon. In: Inkpen, D., Strapparava, C. (eds.) Proceedings of the NAACL HLT 2010 Workshop on Computational Approaches to Analysis and Generation of Emotion in Text, pp. 26–34. ACL, June 2010

15. Purvis, L., Harrington, S., O'Sullivan, B., Freuder, E.C.: Creating personalized documents: an optimization approach. In: Proceedings of the 2003 ACM Symposium on Document Engineering, pp. 68–77. ACM, August 2003. https://doi.org/10.1145/958220.958234

16. Rebelo, S.M., Martins, T., Bicker, J., Machado, P.: Exploring automatic fitness evaluation for evolutionary typesetting. In: Sas, C., Maiden, N.A.M., Bailey, B.P., Latulipe, C., Do, E.Y.L. (eds.) C&C '21: Creativity and Cognition Virtual Event Italy, 22–23 June 2021. ACM, June 2021. https://doi.org/10.1145/3450741.3465247, (Article no. 12)

17. Reynar, J.C., Ratnaparkhi, A.: A maximum entropy approach to identifying sentence boundaries. In: Grishman, R. (ed.) Proceedings of the Fifth Conference on Applied Natural Language Processing, pp. 16–19. ACL, March 1997. https://doi.org/10.3115/974557.974561

18. Richardson, A.: Data-driven Graphic Design: Creative Coding for Visual Communication. Bloomsbury Publishing Plc, London (2016)

19. Runarsson, T.P., Yao, X.: Stochastic ranking for constrained evolutionary optimization. IEEE Trans. Evol. Comput. 4, 284–294 (2000). https://doi.org/10.1109/4235.873238

20. Syswerda, G.: Uniform crossover in genetic algorithms. In: Schaffer, J.D. (ed.) Proceedings of the 3rd International Conference on Genetic Algorithms, pp. 2–9. Morgan Kaufmann Publishers Inc., Cambridge, June 1989

21. Xie, Y., Huang, D., Wang, J., Lin, C.Y.: CANVASEMB: learning layout representation with large-scale pre-training for graphic design. In: MM'21: Proceedings of the 29th ACM International Conference on Multimedia, pp. 4100–4108. ACM, October 2021. https://doi.org/10.1145/3474085.3475541

Enough is Enough: Learning to Stop in Generative Systems

Colin Roitt[✉][iD], Simon Hickinbotham[iD], and Andy M. Tyrrell[iD]

School of PET, University of York, York, UK
{colin.roitt,simon.hickinbotham,andy.tyrrell}@york.ac.uk

Abstract. Gene regulatory networks (GRNs) have been used to drive artificial generative systems. These systems must begin and then stop generation, or growth, akin to their biological counterpart. In nature, this process is controlled automatically as an organism reaches its mature form; in evolved generative systems, this is more typically controlled by hardcoded limits, which can be difficult to determine. Removing parameters from the evolutionary process and allowing stopping to occur naturally within an evolved system would allow for more natural and regulated growth. This paper illustrates that, within the appropriate context, the introduction of memory components into GRNs allows a stopping criterion to emerge. A Long Short-Term Memory style network was implemented as a GRN for an Evo-Devo generative system and was tested on one simple (single point target) and two more complex problems (structured and unstructured point clouds). The novel LSTM-GRN performed well in simple tasks to optimise stopping conditions, but struggled to manage more complex environments. This early work in self-regulating growth will allow for further research in more complex systems to allow the removal of hyperparameters and allowing the evolutionary system to stop dynamically and prevent organisms overshooting the optimal.

Keywords: Generative Design · Self-regulation · EvoDevo

1 Introduction

Evolutionary generative systems can be extremely powerful and interesting tools for generating a variety of different designs, such as 2D images [22] or 3D worlds [2]. As with many evolutionary systems, the more complex the generative system, the more parameterisation can increase, leading to parameter tuning problems. One major parameter that must be considered is when a generative system should terminate [10].

Artificial growth in generative systems can vary greatly in implementation, but all must come to a point where they stop growing. This is typically achieved with some hardcoded limit, often even built into the mechanism that generates phenotypes, or as a limit set by a user [11,13,14]. However, to take full advantage of the evolutionary emergence of properties, it may prove helpful to give

C. Johnson et al. (Eds.): EvoMUSART 2024, LNCS 14633, pp. 342–356, 2024.
https://doi.org/10.1007/978-3-031-56992-0_22

more control to the evolved generative systems. Indeed, it could be considered detrimental to the principle of evolutionary systems, with hard-coding aspects of a solution reducing exploration of the algorithm. In an ideal evolutionary system, stopping would be determined dynamically based on the state of the system. The question then becomes to what extent evolved systems are capable of stopping at the right time within the context of what they design. Too many steps and compute time is wasted at best, or at worst the organism overshoots the optimal; too few, and it is unable to reach it.

In biology, the systems that control gene expression, and therefore morphogenesis (how organisms grow) are Gene Regulatory Networks (GRNs) - the interface between evolutionary and developmental processes. Two biological systems, for the large part of the twentieth century, that were considered separate later became considered together as Evolutionary Development (EvoDevo) [1,16]. Three primary characteristics of GRNs have been identified as heterochrony, spatial patterning, and interactions between genes and gene products [3].

The impact of these three characteristics is a strong indirect encoding of complex features; GRNs control the process of growth, not the final organism. In nature, a relatively small amount of DNA encodes the biology of a remarkably varied and complex organism. Indirect encoding of a genotype in a phenotype is only possible by allowing interactions between genes, thus creating a GRN, which has been shown to be powerful and key in the artificial domain [1,8].

GRNs can be emulated within generative systems through computational structures such as feedforward neural networks and have been successfully implemented for multi-objective problems in [11]. That contribution used a fixed number of development steps and did not investigate the controlled stopping of growth. The development of neural networks into some recurrent structure with a memory may aid in the emergence of halting decisions.

The ability to stop requires context not only in the current environment but also on the historical actions made, thus requiring a construct of memory. However, it has been shown that it is problematic for evolutionary systems to develop these internal constructs, as it often requires a number of evolutionary steps that are not immediately effective in increasing the fitness score [17].

Long short-term memory networks have long been used for sequence learning as a form of recurrent neural network [12]. These systems have persistent memory that is carried over recurrent iterations, allowing them to remember and forget information within their architecture. However, the artificial evolution of LSTMs is not as well understood as the more traditional training methodologies based on backpropagation, for example. Where LSTMs have been used in evolved systems, they have not been used in the context of generative systems and often do not evolve the network directly [15,20,21].

Through the implementation of the recurrent Long Short-Term Memory (LSTM) networks, the work outlined in this paper will examine how evolutionary systems make use of memory during generative tasks and, crucially, reveal if and how stopping criteria emerge; questions that become applicable to both the aesthetic- or physical-property domains of generative applications.

The complications involved in evolving memory have been approached in several ways, such as including *helper objectives* that promote memory in a system and by pre-training an LSTM with an information maximisation objective [20]. The topological evolution of LSTM has also been seen in [21]; producing a genetic programming style tree that then builds the topology of the LSTM cell - something that has been investigated outside of evolutionary spaces [15].

LSTMs have been shown to work with sequence data well; in particular, some examples of this in an artistic generative case include generating rap lyrics [19] and typefaces [18], both examples incorporating features the network has learnt from a known corpus. With particular focus on song lyrics, an evolved system that could generate lyrics to a contextually appropriate length rather than being hardcoded could prove an interesting use case for the system outlined in this paper.

The motivation is to understand how these ideas of memory affect the evolution of developmental systems, on a simple level, and into more complex systems. This background suggests that an LSTM style memory mechanism might aide the development of stopping processes in developmental growth - Sect. 2 details the methodology of this paper in an EvoDevo system.

2   Methodology

To evaluate recurrency in GRNs, this paper proposes a simple LSTM-GRN that addresses the issue of halting generative developmental processes. Of the key properties of GRNs outlined above, the inclusion of recurrency is intended to facilitate the interaction of genes and their products in an artificial space, where proteins and diffusion gradients are not inherent properties of an artificial system, something that biological systems heavily leverage. The use of the system state allows for more complex interactions over time.

An LSTM was chosen above a recurrent neural network as it already has a framework for handling memory inbuilt. This new LSTM-GRN is evaluated through a range of experiments. The following sections present the GRN model and then the experimental methodology and setup of the environment in which these tests were performed.

2.1   LSTM-Gene Regulatory Network

The LSTM-GRN, as shown in Fig. 1, has three major parts: the LSTM cell, the feedforward layer, and the activation function. LSTM cells are commonly placed either before or after a perceptron layer. In the current design, the fully connected feedforward perceptron layer follows the LSTM cell to process information that is captured temporally by the LSTM memory [20].

LSTM Cell. This part of the LSTM-GRN is a simple construction of the basic LSTM unit outlined in [12]. The unit itself contains a forget gate, an update gate, and an output with an activation function.

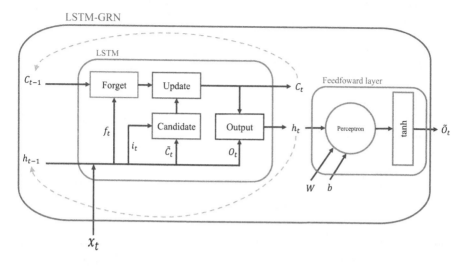

Fig. 1. Overview of the LSTM-GRN architecture shown for an N-dimensional input vector. Visible are the three components: the LSTM cell (the large box, left), the feedforward layer (the circle, middle) and the activation function (the small box, right)

The LSTM cell works by having two streams of long-term and short-term memory flow through it. Short-term memory and the new data input are used, along with weights, to determine how *much* long-term memory to forget. The short-term memory and input are then calculated with additional weights to determine candidate information; this is used to update the long-term memory C_t, which is one output of the cell, fed back in at the next memory input stage, and maintained by the forget and update actions. Finally, the long- and short-term memory are used with more weights to calculate the final output (h_t), which is also fed back into the new short-term memory. The output is then passed on to the feedforward section of the LSTM-GRN constructed of a perceptron and activation function, completing one full pass through the LSTM-GRN at any given development step.

At any given time in the devo loop t, the inputs are defined as the long-term and short-term relationship from the previous loop, C_{t-1} and h_t-1, respectively, and the current input state x_t. The weights and biases used in each equation are given as W_f and b_f, W_i and b_i, W_c and b_c and W_o and b_o for forget, input, candidate, and output stages, respectively.

The forget gate, f_t, is defined by Eq. 1 which will act on C_{t-1} to give C_t.

$$f_t = \sigma(W_f \cdot [h_{t-1}, x_t] + b_f) \tag{1}$$

The update step requires candidate memory update values and some information from the input, and those are given as:

$$i_t = \sigma(W_i \cdot [h_{t-1}, x_t] + b_i) \tag{2}$$

$$\tilde{C}_t = \tanh(W_c \cdot [h_{t-1}, x_t] + b_c) \tag{3}$$

The final update step is given by:

$$C_t = f_t \times C_{t-1} + i_t \times \tilde{C}_t \tag{4}$$

The last step is to calculate the cell output, which is also used as the short-term input for the next iteration, h_t.

$$O_t = \sigma(W_o \cdot [h_{t-1}, x_t] + b_o) \tag{5}$$

$$h_t = O_t \times \tanh(C_t) \tag{6}$$

Feedforward and Activation. The final two components of the LSTM-GRN are a single feedforward layer with an activation function, which here is the *tanh* function. In principle, this can allow the output of the LSTM cell to handle temporal data through its memory gates; in this way, the flow of information is embedded with temporal information before being handed on to the feedforward perceptron layer, which then functions as a more classical neural network unaware of any recurrency. This is given simply as:

$$\tilde{O}_t = \tanh(h_t \cdot W + b) \tag{7}$$

The *tanh* function is a construct of the LSTM-GRN output requirement such that the values can be normalised to a range between -1 and $+1$. This function is also present to introduce non-linearity into the output of the system.

The network described here cannot be trained by a data set because there is no appropriate training set for the task outlined in Sect. 3. Therefore, evolution holds a key solution to producing a GRN that behaves appropriately.

2.2   Evo-Devo Loop

Evo-devo presents a strategy to take advantage of indirect encoding of gene regulatory networks in an evolutionary space. This process combines two straightforward approaches to both the evolution of a genome and the development of an organism through the LSTM-GRN.

The process behind Evo-devo can be considered simply as two nested loops. The evolution of the LSTM-GRN takes place in the outer loop, while the inner loop acts as a fitness function for the evolutionary selection process. The mapping of genotype to phenotype occurs within this inner devo loop. The simple evolved genotype is interpreted as the LSTM-GRN; this controls the complex growth of the phenotype. The Devo-loop implementation is described in Algorithm 1.

For initial experiments to consider the basic functionality of this new method, the LSTM-GRN has a fixed topology and is represented in the search space as an array of values, each encoding a specific weight or bias value used in the model,

---

Algorithm 1: Fitness function for a given individual

Data: *genome*
Result: $fitness = [distance, steps]$
$t \leftarrow TARGET$;
$curr\_loc \leftarrow (0,0)$;
$S, D, H \leftarrow None, None, None$;
$n \leftarrow 0$;
while $n \neq MAX\_DEVO\_STEPS$ & $H < 0$ **do**
   | $n \leftarrow n + 1$;
   | $S, D, H \leftarrow GRN(genome, [curr\_loc_x, curr\_loc_y, H])$;
   | **if** $H < 0$ **then**
   | | $step = |S| \cdot 10$;
   | | $direction = |D| \cdot 10$;
   | | $x_i = step \cdot \cos \,(direction)$;
   | | $y_i = step \cdot \sin \,(direction)$;
   | | $curr\_loc = (x_i, x_y)$;
   | **else**
   | | END DEVO
   | **end**
end
$distance \leftarrow |curr\_loc - t|$;
$return[distance, n]$;

W_f and b_f, W_i and b_i, W_c and b_c and W_o and b_o. This keeps the functions used for genetic operations simple and available in most libraries; in this case, the library chosen was the DEAP library for Python [9]. The evolutionary component is a standard genetic algorithm (GA); however, the inclusion of the devo steps introduces an indirect encoding that can evolve to generate complex behaviour - this is extended by the memory component in the LSTM-GRN.

Evolution is driven by a number of genetic operators that are crucial to achieving good convergence and achieving it in acceptable time. Simulated binary crossover [7] (probability: 0.5, η: 0.4) [6] was chosen for this task, as it is capable of solving a number of issues not present in a binary value crossover, but for a genome constructed of a series of real numbers. Similarly, a Gaussian mutation function (individual probability: 0.3, gene probability: 0.2, μ: 0, σ: 0.3) was used to randomly mutate individuals, allowing for frequent small mutations in the real-coded genome, but also some infrequent larger mutations. Finally, the selection algorithm used was NSGA-II, a multiobjective algorithm that has become a common standard in evolutionary space [5].

The output of this evolvable generative network is then used to drive the morphological development of two structures outlined in Sect. 3. Source code made available at https://github.com/ColinRoitt/LSTM-GRN

3   Experiments

In order to show that the LSTM can work within this context, this paper first presents a simple "single point" problem where the organism must reach a point in an artificial environment. Then to extend that to allow for a more visible devo process a "point cloud" based problem is presented. Finally, the idea is extended to that of structural growth to see if the organism is able to evolve the appropriate strategies to reach a task-specific point arrangement.

The approaches presented here are inspired by biological forms, specifically the natural and efficient growth exhibited in an experiment with the slime mould *Physarum polycephalum* [23]. Wasted energy, while an abstract concept within these experiments, is represented by optimised growth strategies; learnt behaviours that react efficiently to an environment.

In the first experiment, a single point is used as a target for a single structure to grow from its starting point to the target. The second experiment is a more complex problem: An organism is placed in the middle of a field of targets randomly placed in a band at a specified radius. The organism can then, over each devo step, take a number of actions: step forward, change the angle, or branch. Thus, it is possible for an LSTM-GRN to develop a structure to cover the target area, efficiently collecting many targets.

4   Results

The aim here is to first validate the basic principle that a stopping criteria can be met by a recurrent system and then go on to evaluate these established abilities in a more complex problem.

4.1   Single-Point Target

The organism begins as a point in the bottom left of the arena shown in Fig. 2. Through each Devo step, the LSTM-GRN outputs a learnt step size and direction in which to grow. Fitness, measured in the final Devo step, is given by the distance between the final point reached and the *food* and the number of steps taken. In this series of experiments, the position of the target is fixed at the top of the arena, and the organism is blind to its location; the only feedback that reaches the organism of its quality is after the devo process is complete. Throughout the evolution of the organism, it must evolve the correct series of steps to efficiently reach the target and, crucially, when to stop growing so as not to expend unnecessary energy.

The evolutionary process was given 100 generations and a maximum of 100 development steps. The final and best individual terminated its development cycle after 7 steps and achieved a final distance from the target of 0.29 units. Individuals from three different generations produced through an evolutionary run are shown in Fig. 2. Starting from a single point with little growth to a long indirect path to the target, before finally finding the shortest path and refining

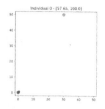 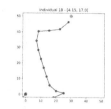 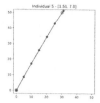

(a) Gen: 1, steps - 100, dis-   (b) Gen: 40, steps - 17, dis-   (c) Gen: 70, steps - 7, distance - 57.65   tance - 4.15   tance - 1.51

Fig. 2. An example of some of the visually selected best individuals evolved over 3 of the 100 generations. The organism can be seen to go through a number of stages of evolution; first finding the target and then optimising its path. This experiment was carried out 7 times and achieved similar results in each run.

the step size to hit the centre of the target around generation 70. This illustrates that the LSTM-GRN presented here is capable of balancing both the pathfinding and termination criteria to encode a solution.

These early results suggest that it is possible to evolve a GRN that can perform a stopping operation in an appropriate manner. However, it is important to note that this is a relatively simple and, crucially, static test; The target is in a fixed location, so location information can be encoded directly into the genome and *saved* via evolution not development. To provide a more challenging problem, the LSTM-GRN should be able to interpret spatial input based on its environment, which will be addressed with more complex issues in the next section.

4.2   Point Cloud

In this series of experiments, more targets are placed in the environment and the organism must be able to sense its surroundings and follow routes efficiently to collect the *food*.

Although the organism's ability remains the same as in the previous experiment, it now has the additional ability to push and pop its current location to and from a stack in memory, allowing for branching growth. Another ability it is given is a look-ahead vector pointing to its nearest food source. This is a critical piece of information as the points are no longer fixed between evaluations and are distributed randomly within a specified radius.

Importantly, the organism retains the ability to terminate its growth, reducing the number of steps it has to take. An upper bound of 1,000 devo steps was imposed to make the experiments tractable, this is compared to the previous 100 devo steps.

The results presented in Fig. 3 show solutions taken from the Pareto front of the final population. The phenotype does present reasonable coverage of the environment and notably skips around the gap in the middle of the space. However, the key feature that must be investigated is the stopping of the growth process at an appropriate point.

It is apparent that although an acceptable path has been discovered through the environment, the organism has not terminated its growth. However, there are a number of potential reasons why this may be the case. There are many more targets in the arena that it may try to collect; this increase in the *score* fitness may be winning out in the selection process over any organisms that halt before this is done. This is a feature that can be seen in a number of results.

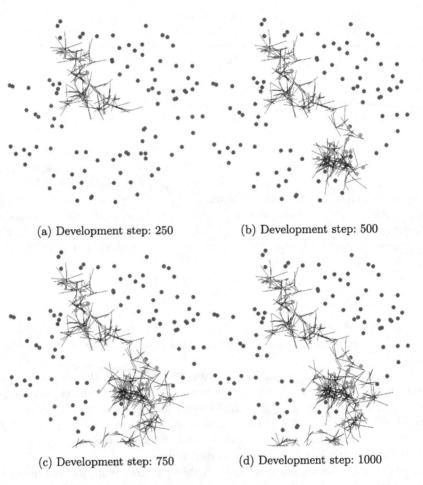

(a) Development step: 250   (b) Development step: 500

(c) Development step: 750   (d) Development step: 1000

Fig. 3. An example of branches developing when directional distance information is given to the network about the nearest point (unconstrained distance). Growth starts in the centre of the arena (orange X), blue dots are targets, and orange dots are captured targets (Color figure online)

The final Pareto front (Fig. 4) from another run of this experiment yielded a typical distinct split across the solution space, and although it is clear there are changes across generations (Fig. 5), it is not yet clear how much of the solution space is reachable with the current configuration of the LSTM-GRN.

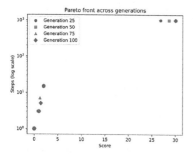

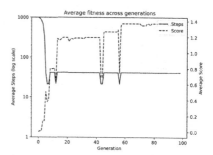

Fig. 4. The solutions generated in this run occupy tight clusters suggesting difficult local minima to escape. The final generation is marked with the red diamond. (Color figure online)

Fig. 5. Average of both fitness measures across 100 generations - steps (minimised) and score (maximised). The average step stabilises around generation 20 as the populations become fixed at the maximum or minimum number of steps. The score shows good improvement over the generations.

A range of phenotypes emerge in the final population. A consistent trait through the populations of this task is the large circles that are present in Fig. 6a. These structures are large and repetitive, but they cover a significant area and collect a good number of targets, as highlighted in orange. This pattern emerges frequently when no termination action is activated in the majority of the population.

Figure 6b is similar to the experiment run for Fig. 3, however, in this iteration of the experiment, the LSTM-GRN has a limited distance in which it can see the direction of its nearest target. This results in large swooping arcs compared with the unconstrained look-ahead of quite sharp turns. This is likely due to it often lacking an understanding of its environment, and as such the best strategies to maximise score involve finding ways to cover the space as much as possible.

4.3   Structured Cloud

The final experiment replaces the random field of targets with deliberate structures. In this, the targets are in the form of a capital letter Z and a straight line. There are several reasons why these layouts were chosen. Letters of the alphabet have been used in the past as a reference point for generative systems [4,18]. However, mainly it gives some complexity to the task of finding the shortest

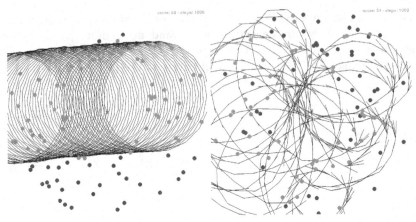

(a) Large circles commonly evolved a strategy to maximise score in a landscape few networks perform well minimising the number of steps.

(b) A branching organism developed when directional distance information given to the GRN is constrained to within 300 units.

Fig. 6. Two common results from the point cloud test

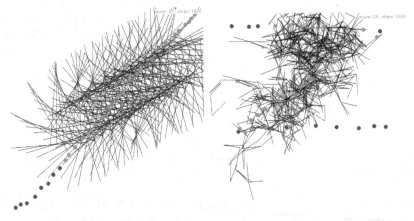

(a) The organism explores a straight line from a centre point out in both directions - it explores more effectively towards the top right, neglecting the bottom left.

(b) In order to cover the area of the letter Z, the organism behaves similarly to the examples seen when presented with a point cloud.

Fig. 7. Two examples of structure being latent within the solution space.

path, similar to solving mazes. Within the straight line, a very clear optimal route is developed, with the letter Z that also provides the challenge of tight turns for exploration.

Figure 7a is the first example of a response to a deliberate structure in the environment. As before, the organism starts in the middle of the arena, and the expected structure would be for the branching to occur towards the top right and bottom left. The actual phenotype that evolves is not significantly different from the expected result. However, it does take a lot of steps exploring either side of the line rather than taking advantage of the proximity of the nearby targets.

The quality of the exploration strategy of the organism is more difficult to evaluate qualitatively here due to the tight structure of the targets. The organism clearly does not adhere rigidly to the intended structure. However, the area below the bottom line of the letter Z (Fig. 7b) was briefly explored; then abandoned and wasteful areas of exploration, such as the acute top angle of the structure, are quite close to each other, so it would not be unreasonable to search in space for more densely packed targets. Also note that the LSTM-GRN was unable to stop earlier than the termination criteria allowed.

5 Discussion

The LSTM-GRN presented here can certainly evolve to stop, as shown in the first experiment; however, as the complexity grows, it becomes harder to interrogate small changes in efficacy. It may simply be a case of expanding the evolutionary search strategy in selection or mutation or perhaps allowing some change to the topology of the GRN. It is not unreasonable to expect this incongruity, as it has been pointed out that the search space for evolving memory is *"deceptively large"* [20]. It is also likely that it is simply challenging to escape the local minima of the fitness space.

An important distinction to make is the difference in complexity between the two experiments outlined here. The success of the more simple experiment in halting the devo process does suggest that this problem becomes quite challenging to search over, with many local minima seen throughout the solution space.

There are a number of explanations that resolve the discrepancy presented between these experiments. One such explanation is that the phenotypically controlled organism never found all the points in the environment, and, given the current experimental set-up, this may be a requirement before beggining to minimise the step fitness. That is to say, there must first be a path to a maximum score fitness before the evolutionary strategy is able to optimise the route thus minimising the steps fitness - as seen in the first experiment.

Another limitation of this fitness strategy is that the collection of targets is a binary measure. Using some proximity such that almost every mutation in development process has some immediate feedback on quality may help guide the search.

There is also an argument to suggest that the more complex problem is simply poorly suited to this type of solution. If this were to be used, for example, in the context of designing supports for a bridge, there may be some 'have targets to grow to, but also some fixed boundary definitions. And fitness may need to be based on other characteristics required by that of a bridge, such as physical properties.

The question also arises of when the LSTM-GRN is learning to stop or optimising to stop similar to questions on generalisability. Using the steps taken as a fitness measure may lean into this idea that it's simply an optimisation process, rather than the LSTM-GRN learning to dynamically grow to its environment. There is a great deal of nuance between these two ideas, and while they are not the same thing, they are closely coupled. The inclusion of proxy fitness measures may aid this, for example, including some amount of energy an organism can use up and replenish within the environment.

6 Future Work

Many more avenues for exploration are possible and this paper only scratches the surface of deep integration of recurrent systems into Evo-Devo environments and generative designs stopping their growth.

Within the current framework, it would be valuable to adjust the fitness function in order to guide the evolution process in a more granular way. For example, including an amount of reward for proximity to a target would resolve the issue with the binary scoring system. This makes small changes to the organism's growth more likely to receive either positive or negative feedback, allowing these small mutations time to be taken advantage of through generations.

This problem of learning to stop or optimising to stop can be a challenge to seperate; the simplest way would be to remove any idea of duration from the fitness function. In this way, the system is no longer optimising for stopping, it would have to learn when a stopping condition was met. One premise is the idea of giving the organism some amount of energy. It can move around and collect targets and gain more energy, or it can terminate with its current energy level. To encourage termination, a strong penalty can be included for an organism running out of energy or '*dying*'. Removing direct references to stopping early in the fitness function now removes one-half of the challenge of preventing a GRN simply optimising for stopping. Then allowing the GRN to decide when to stop based on energy levels and the environment. Upon a search of the literature, there seems to be little discussion of these artificial ideas of energy within the context of Evo-Devo algorithms.

7 Conclusion

While it is clear that including reccurency does allow termination criteria to be handled internally by a GRN in more simple environments, it remains to be seen whether this is true for much more complex generative systems. There is clear evidence for the efficacy of the LSTM-GRN in handling generative tasks, but it will become increasingly important to set a benchmark for how stopping is handled across a number of GRN constructs.

It becomes clear through consideration of the results here that optimising to stop and learning when to stop are similar but slightly different tasks. It remains to be seen what the best way to ensure a system is learning when to stop, but it is apparent a development in this area will significantly aide the understanding of stopping in generative systems as a whole.

Acknowledgements. The authors acknowledge the support of a School-funded PhD studentship and the support of the EPSRC project RIED EP/V007335/1.

References

1. Banzhaf, W.: On the dynamics of an artificial regulatory network. In: Banzhaf, W., Ziegler, J., Christaller, T., Dittrich, P., Kim, J.T. (eds.) ECAL 2003. LNCS, pp. 217–227. Springer, Heidelberg (2003). https://doi.org/10.1007/978-3-540-39432-7_24
2. Broughton, T., Tan, A., Coates, P.S.: The use of genetic programming in exploring 3D design worlds. In: Junge, R. (ed.) CAAD futures 1997, pp. 885–915. Springer, Dordrecht (1997). https://doi.org/10.1007/978-94-011-5576-2_68
3. Davidson, E.H.: Genomic Regulatory Systems. In Development and Evolution, Elsevier (2001)
4. Dawkins, R.: The evolution of evolvability. On growth, form and computers, pp. 239–255 (2003)
5. Deb, K., Pratap, A., Agarwal, S., Meyarivan, T.: A fast and elitist multiobjective genetic algorithm: NSGA-II. IEEE Trans. Evol. Comput. **6**(2), 182–197 (2002)
6. Deb, K., Agrawal, R.B., et al.: Simulated binary crossover for continuous search space. Complex Syst. **9**, 115–148 (1994)
7. Deb, K., Sindhya, K., Okabe, T.: Self-adaptive simulated binary crossover for real-parameter optimization. In: Proceedings of the 9th Annual Conference on Genetic and Evolutionary Computation, pp. 1187–1194. GECCO 2007, Association for Computing Machinery, New York, NY, USA (2007)
8. Eggenberger, P.: Evolving morphologies of simulated 3d organisms based on def-erential gene expression. In: Harvey, I., Husbands, P. (eds.) Proceedings of the 4th European Conference on Artificial Life, pp. 205–213. Springer (1997)
9. Fortin, F.A., De Rainville, F.M., Gardner, M.A., Parizeau, M., Gagné, C.: DEAP: evolutionary algorithms made easy. J. Mach. Learn. Res. **13**, 2171–2175 (2012)
10. Frazer, J.: Chapter 9 - creative design and the generative evolutionary paradigm. In: Bentley, P.J., Corne, D.W. (eds.) Creative Evolutionary Systems, pp. 253–274. Morgan Kaufmann, San Francisco (2002)
11. Hickinbotham, S., Dubey, R., Friel, I., Colligan, A., Price, M., Tyrrell, A.: Evolving design modifiers. In: 2022 IEEE Symposium Series on Computational Intelligence (SSCI), pp. 1052–1058 (2022)
12. Hochreiter, S., Schmidhuber, J.: Long short-term memory. Neural Comput. **9**(8), 1735–1780 (1997)
13. Hornby, G.S., Lipson, H., Pollack, J.B.: Evolution of generative design systems for modular physical robots. In: Proceedings 2001 ICRA. IEEE International Confer-ence on Robotics and Automation (Cat. No.01CH37164), vol. 4, pp. 4146–4151. IEEE (2001)

14. Hornby, G.S., Pollack, J.B.: Body-brain co-evolution using l-systems as a generative encoding. In: Proceedings of the 3rd Annual Conference on Genetic and Evolutionary Computation, pp. 868–875. GECCO'01, Morgan Kaufmann Publishers Inc., San Francisco, CA, USA (2001)
15. Jozefowicz, R., Zaremba, W., Sutskever, I.: An empirical exploration of recurrent network architectures. In: Proceedings of the 32nd International Conference on International Conference on Machine Learning, vol. 37, pp. 2342–2350. ICML 2015, JMLR.org (2015)
16. Lohmann, I.: The birth of evo-devo. Nat. Rev. Mol. Cell Biol. **24**(5), 311 (2023)
17. Ollion, C., Pinville, T., Doncieux, S.: With a little help from selection pressures: evolution of memory in robot controllers. In: Artificial Life 13. MIT press (2012)
18. Phon-Amnuaisuk, S., Salleh, N.D.H.M., Woo, S.-L.: Pixel-based LSTM generative model. In: Omar, S., Haji Suhaili, W.S., Phon-Amnuaisuk, S. (eds.) CIIS 2018. AISC, vol. 888, pp. 203–212. Springer, Cham (2019). https://doi.org/10.1007/978-3-030-03302-6_18
19. Potash, P., Romanov, A., Rumshisky, A.: GhostWriter: using an LSTM for automatic rap lyric generation. In: Màrquez, L., Callison-Burch, C., Su, J. (eds.) Proceedings of the 2015 Conference on Empirical Methods in Natural Language Processing, pp. 1919–1924. Association for Computational Linguistics, Lisbon, Portugal (2015)
20. Rawal, A., Miikkulainen, R.: Evolving deep LSTM-based memory networks using an information maximization objective. In: Proceedings of the Genetic and Evolutionary Computation Conference 2016. ACM, New York, NY, USA (2016)
21. Rawal, A., Miikkulainen, R.: From nodes to networks: evolving recurrent neural networks (2018)
22. Secretan, J., et al.: Picbreeder: a case study in collaborative evolutionary exploration of design space. Evol. Comput. **19**(3), 373–403 (2011)
23. Tero, A., et al.: Rules for biologically inspired adaptive network design. Science **327**(5964), 439–442 (2010)

Generating Emotional Music Based on Improved C-RNN-GAN

Xuneng Shi and Craig Vear(✉)

University of Nottingham, Nottingham, UK
`craig.vear@nottingham.ac.uk`

Abstract. This study introduces an emotion-based music generation model built upon the foundation of C-RNN-GAN, incorporating conditional GAN, and utilizing emotion labels to create diverse emotional music. Two evaluation methods were employed in this study to assess the quality and emotional expression of the generated music. Objective evaluation utilized metric calculations, comparing the generated music to the music in the EMOPIA database, including factors like note range, chord count, and chord consistency. Additionally, subjective assessment involved inviting 20 listeners to hear a set of both real and generated music. Listeners were asked to distinguish between real and generated music and evaluate emotional expression and harmony. The results indicate that the model-generated music successfully conveys a variety of emotions and approaches the quality of human-composed music.

Keywords: AI music composition · deep learning · C-RNN-GAN · music emotion · controlled music generation

1 Introduction

Music is a unique universal language that transcends cultural and linguistic barriers, fostering global artistic exchanges [27]. With the continuous advancement of machine learning technology, deep learning techniques have been successfully applied to generate sequence-dependent music data [13]. Thanks to the creation of various large music databases, researchers have produced high-quality music, to the point where it is almost indistinguishable from compositions crafted by human ears. An example of such techniques include the C-RNN-GAN [20]. Olof Mogren introduced the C-RNN-GAN architecture, which employs adversarial training to model the joint probability of an entire sequence, enabling the generation of data sequences [20]. This model excels in handling various musical attributes, including pitch, duration, frequency, intensity, and continuous sequence data related to time. It generates pitches and rhythms that closely resemble human-composed music.

In the research described in this paper, we aimed to construct a music model capable of generating music with rich emotional expressions. The emotional

C. Johnson et al. (Eds.): EvoMUSART 2024, LNCS 14633, pp. 357–372, 2024.
https://doi.org/10.1007/978-3-031-56992-0_23

labels for this model are based on the Russell emotional model [22,25], categorizing emotions into four basic types according to arousal and valence (i.e., the positivity or negativity of emotions). Given the successful application of the C-RNN-GAN model in music generation [20], this work adopts this model as its foundation and extends it. We introduced a system for generating emotional music using the C-RNN-GAN model. According to the literature review, we believe this is the first instance where the C-RNN-GAN model is combined with additional emotional labels for generating emotional music sequences, setting the goal for this model to enable ordinary people, even those without a background in music, to create music corresponding to specified emotions. Our new insights include: 1. During data preprocessing, specific attention was given to extracting fundamental note information, along with the inclusion of rhythm data, to enhance the expression of musical emotions. 2. When constructing the generator, a comprehensive CNN and RNN structure was employed, with the addition of emotional labels as input, increasing the complexity and personalization of music generation. 3. An alternating training strategy was implemented. This approach not only alleviates mode collapse but also prevents the discriminator from updating too rapidly, ensuring the generator fully learns from the music data in the database. 4. Through these enhancements, high-quality and stable music generation was achieved, validated by a combination of objective and subjective assessment methods, confirming the practical applicability of the generated music.

2   Literature Review

2.1   Generative Adversarial Networks and C-RNN-GAN

Generative Adversarial Networks (GANs) are models introduced in the course of deep learning development. GAN models have a clear architecture, comprising two neural networks: a generator and a discriminator, which learn from and interact with each other during the data generation process. The generator takes random noise and processes it through the network to produce results that resemble real data. The discriminator assesses the generated data to distinguish between real and generated data, computing errors to obtain a loss function for updating model parameters. Training a GAN model is akin to a zero-sum game [4,30]. The objective of a GAN is to train the generator G and the discriminator D such that the generator G can produce samples resembling the real data distribution pdata(x).

Traditional GAN models almost universally face a common challenge: they are primarily built using fully connected layers, which can often be challenging when dealing with high-dimensional structured data. This challenge is particularly pronounced when handling time-series data due to the strong continuity and sequential characteristics of such data. To address this issue, Kaleb Smith introduced a novel Time Series Conditional GAN (TSGAN) [28]. Unlike traditional GANs, TSGAN not only allows users to employ additional conditional information to guide the generation process but also features a dual GAN architecture.

The first GAN in TSGAN is responsible for converting randomly generated latent vectors into spectrograms, while the second GAN utilizes these spectrograms to generate time series data. This architectural design enables TSGAN to more effectively capture the diversity and complexity of time series data, making it more efficient in simulating and generating time series.

Recurrent Neural Networks (RNNs) and Long Short-Term Memory networks (LSTMs) are particularly effective in handling time series data because they can remember and utilize past information, capturing sequential dependencies [6, 19]. Therefore, in a project conducted by Shi, she incorporated an LSTM network within the generator to receive and process input time series data. However, when generating time series using GAN models, adversarial training between the generator and discriminator is required, placing strict demands on the model's loss function and training process, which makes training complex. This can lead to issues such as mode collapse, gradient vanishing, and training imbalances. Researchers Soumith Chintala and León Bottou made a series of improvements in 2017 to address these challenges [2]. They enhanced the stability and performance of Wasserstein GAN by adopting the 'Wasserstein distance' as the loss function and introducing a gradient penalty mechanism. These improvements helped prevent gradient vanishing problems, allowing the model to learn more effectively. These enhancements strengthened the performance of GAN models in various application scenarios.

RSGAN (Relativistic GAN), proposed by Alexia Jolicoeur-Martineau [11], represents another direction of improvement in the loss function of traditional Generative Adversarial Networks (GANs) by introducing the concept of relativistic adversarial training [11]. Unlike traditional GAN models, RSGAN places a greater emphasis on the relative probability between generated and real samples. It considers the interactions and relative rankings of multiple samples within a broader context. Such a design enables the model to more effectively balance the competitive relationship between the generator and discriminator during the training process, thereby improving the overall stability of the model and the quality of generated samples.

In the field of intelligent generation, the C-RNN-GAN model leverages the strengths of Recurrent Neural Networks (RNN) and Convolutional Neural Networks (CNN), along with the powerful generation capabilities of Generative Adversarial Networks (GAN) [20]. Olof Mogren and his team were the first to demonstrate the versatility of the C-RNN-GAN model framework, paving the way for subsequent research [20, 35]. Their work showcased the integration of RNNs, CNNs, and GANs in handling complex generation tasks, such as music generation. However, despite the considerable versatility exhibited by their model, there are still some limitations, especially in specific applications like music generation [29].

In the research described in this paper, the envisioned model adopts the C-RNN-GAN generator architecture, integrating convolutional layers with LSTM layers to comprehensively and efficiently address the complex requirements of music generation. To further optimize the C-RNN-GAN model, the C-RNN processes emotion labels by transforming them into specific latent vector representations, which are then provided to the generator as conditional inputs. Subsequently, the generator receives the transformed input vectors and random noise. These samples undergo upsampling through deconvolution layers, gradually generating music segments consistent with the emotion labels. Then, the discriminator evaluates the consistency between the generated music segments and their respective emotion labels. Integrating emotion labels into the C-RNN-GAN model adds a new dimension to the expression of the generated music. It allows the model to evoke specific emotions in the audience, resonate with human experiences, and enhance the emotional depth of the compositions. This innovation makes a significant contribution to the AI-driven music generation field, providing a nuanced approach to creating emotionally resonant music. By customizing the generated music to match specific emotions, this work expands the capabilities of the C-RNN-GAN model, with the aim of bridging the gap between machine-generated music and human-created music.

2.2   Music Generation: Symbolic Music Generation and Generate Music with a Given Sentiment

Innovation in recurrent neural networks (RNNs) and convolutional neural networks (CNNs) continues to drive rapid advancements in the field of AI music [3,17,31], resulting in various variants of RNNs and CNNs that further enrich this research domain [1,14,36]. Presently, due to the adept handling of inherent temporal dependencies and complex structures in music, RNNs dominate the music generation field. However, when dealing with long time sequences, the "memory" capacity of RNNs weakens as the number of time steps increases, leading to the issue of vanishing gradients. This problem results in extremely small gradients during training, effectively halting network learning. Vyas, N et al. introduced the LSTM (Long Short-Term Memory) model, which maintains a constant error flow by integrating specialized units, effectively "remembering" past information [26]. The gate mechanisms in LSTM can efficiently manage long sequence data, providing faster learning and higher task success rates compared to other RNN algorithms, especially in handling complex tasks with extended time lags. Gleb Rogozinsky and his colleagues advocate the use of a variational autoencoder (VAE) to enhance flexibility and address issues in traditional RNNs in music generation [23]. Unlike LSTM, VAE focuses on transcending the treatment of time series data. It enhances the application of distributed genetic algorithms through the use of a variational autoencoder (VAE), encoding input data into a continuous latent space, allowing for smooth transitions within the latent space and the generation of novel music samples.

In the past two years, emotion-based music databases have been applied to music generation models. Krause & Layla [16] was the first individual to achieve emotion-based music composition. She improved upon the LSTM model by introducing the MLSTM model. This model optimized the weights of specific neurons responsible for emotional signals, allowing control over the emotional classification of music [16]. Jacek and his team created their own dataset using an encoder-decoder seq2seq architecture and developed a model to calculate and classify emotions based on the chords present in each piece of music [5].

While the aforementioned research can identify emotions within given music segments, research on generating music based on specified emotions remains limited. Currently, most studies focus on classifying existing music or transferring emotions [7,12,33]. These methods primarily involve recombining and rearranging original musical notes rather than generating them. Music comprises not only the combination of notes but also elements like melody, harmony, and rhythm, making it a significant challenge to ensure that these elements align structurally with the target emotion. Traditional training methods such as MLE are subject to exposure bias. Therefore, Pedro L. T. Neves and their team have proposed a noteworthy approach, utilizing the Transformer architecture in a GAN training model for generating music that responds to specific emotions. Their research not only marks significant progress in music generation but also achieves remarkable results in enhancing the quality of generated music. Through the clever application of the Transformer architecture, this model excels in capturing the long-term dependencies and abstract features within music, providing powerful modeling capabilities for autonomously generating music with specific emotional responses. [21]. Pedro L. T. Neves employed the EMOPIA dataset for handling music samples and emotion label inputs [8].

Despite significant progress in existing models within the music generation field, research on music generation based on specific emotion labels is still relatively nascent [18,32]. Understanding and implementing music features associated with different emotion labels remains a primary challenge for the models. Currently, there might not be a large-scale, diverse, and labeled music dataset created specifically for this purpose. Creating music datasets with various emotion labels would contribute to the training and validation of these models. Current data processing in research involves segmenting music into smaller sections for model input, although most emotions are conveyed through the development of an entire track, rather than a short segment [15,24]. Considering the overall music structure is crucial for generating emotionally rich music.

3   Methodology

3.1   Data

Data Selection and Data Processing. In the research described in this paper, a MIDI format music dataset was used as the training database for the model. MIDI format is the standard for music data exchange. In addition to note information, it includes instrument selection, control information, and other rich content that can be transmitted and played on various devices, containing complete information about music compositions.

This study utilized the EMOPIA database as the training dataset [8]. The database consists of 387 songs annotated by professional musicians, covering four basic emotion labels. Furthermore, 1087 music segments were randomly selected from these 387 songs, reorganized and extended to ensure diversity, consistent length, and avoidance of repetition. This provided rich and distinct data for studying the relationship between music and emotion.

In contrast to other databases that input all information from MIDI files, this study processed music files from the database using a method similar to natural language processing. Instead of using the entire MIDI file as input, representative features were extracted. This not only ensured the capture of essential features but also sped up the model's training and inference processes. Furthermore, it helped to mitigate the impact of irrelevant or noisy data on the model's performance. Shi selected pitch, duration, and velocity in the music as the input features for the model, as these musical attributes are primary criteria for assessing music quality and emotion.

First, notes and chord pitches from the music files in the dataset were extracted, and an integer mapping was created for the notes. Additionally, each music segment was associated with its emotion label for later retrieval and processing. Rhythm in the music was specifically extracted and used as additional input for the generator. The rhythmic information of the music is represented as a time series, where each time step corresponds to a bar, a beat. This is then encoded with discrete values (e.g., 0 for silence, 1 for a note). Rhythm not only determines the temporal structure of the music but is also can be closely related to the emotion, style, and expression of the music. After the basic data extraction, the data underwent preprocessing. The extracted data was cleaned to eliminate interference from noise and invalid data. The extracted 1087 long music segments were cut into shorter segments, and rhythm and emotion labels corresponding to each segment were retrieved. To ensure consistency in data range and distribution among different samples, data standardization and normalization were applied.

A linear scaling method was employed to re-adjust the data to a range between -1 and 1. This standardization aids in improved performance of the data within certain neural network models. Standardized data can expedite the convergence speed of algorithms and enhance their performance. Finally, all processed note sequences and rhythm sequences were transformed into numpy arrays suitable for machine learning, while emotion labels were converted into one-hot encoding.

To further enhance the model's robustness and reliability, a series of error handling mechanisms were embedded in the code. These mechanisms played a crucial role in the process of retrieving emotion labels from the CSV file, ensuring that the data processing workflow can still function correctly when faced with inconsistencies or errors. To ensure that the model is effective and has generalization capabilities in practical applications, the dataset was split into 80% for training, ensuring the exposure of the model to a variety of samples to capture key features, and 20% for evaluating the model's performance to check for overfitting. After data preprocessing, a total of 233,303 bars were available for training. For each bar, this study set the width (temporal resolution) to 100 (i.e. BPM = 100) to model common time patterns like whole notes, sixteenth notes, and even (the smallest) thirty-second notes. Additionally, the height was set to 84 to cover the pitch range from C1 to C8, with 12 keys per octave (C, D, E, F, G, A, B, and the black keys C#, D#, F#, G#, A#). Consequently, the data tensor size for each bar was 100 (time steps) × 84 (notes).

3.2 Model

Currently, state-of-the-art models applied to music generation include progressive transformer models, bidirectional LSTM models, among others [10]. These models have made groundbreaking advancements in the quality of music generation. They are capable of generating diverse music while overcoming issues such as pattern collapse and incoherent musical segments. However, what makes C-RNN-GAN more appealing is its higher adaptability. Olof Mogren [20] conducted multiple experiments with C-RNN-GAN, successfully generating continuous music data and demonstrating better stability during the training process compared to other GAN variants. Therefore, in this study, C-RNN-GAN was chosen as the base model, and efforts were focused on further improving it.

Construction of Generator and Discriminator. To construct a model capable of generating music sequences with given emotion labels, we attempted to combine C-RNN-GAN with conditional GAN. Initially, the basic generator and discriminator architecture of C-RNN-GAN were modified to incorporate two neural networks with different depths. In GANs, the discriminator can sometimes overpower the generator, resulting in the so-called gradient vanishing problem. As a result, the discriminator was set to consist of only two dense layers: one with 256 neurons and another with 512 neurons, each followed by a LeakyReLU activation function. Subsequently, two Dropout layers were added to the network to drop some nodes during training to prevent overfitting.

The discriminator is designed as a binary classifier with the primary task of distinguishing between real music sequences and sequences generated by the generator. The discriminator's input consists of music sequences along with their corresponding emotion labels. The emotion labels are embedded and duplicated to match the shape of the sequences and then concatenated with them. This combined input is processed through a series of dense layers with leaky rectified linear unit (LeakyReLU) activations.

The discriminator has two outputs: "Validity" Output: This includes a dense layer with a single node and an Sigmoid activation function, which outputs a value representing the authenticity (real or fake) of the input sequence. If the validity value is greater than 0.5, the discriminator considers the sequence to be real; if it's less than 0.5, the discriminator identifies the sequence as generated by a machine. "Emotion Label" Output: This includes a dense layer with "self.num_classes" nodes and employs a softmax activation function to output a probability distribution representing the emotion label of the input sequence. The reason for using the Sigmoid activation function in the "Validity" output is its typical use in binary classification problems, transforming any real value into a range between 0 and 1. The Sigmoid function has its maximum derivative (slope) when the input is close to 0, and as the input moves away from 0, its derivative gradually approaches 0.

On the other hand, the output of the "emotional_label" uses one-hot encoding. The Softmax function is typically used in multi-class classification problems. Its purpose is to transform a vector with K-dimensional real values into a K-dimensional probability distribution vector. For a given input vector x = [x1, x2, x3, ... , xk], the corresponding Softmax function output y = [y1, y2, ... , yk], where for each element, the numerator is the exponential of the input element, and the denominator is the sum of the exponentials of all input elements. This ensures that all elements of the output y are between 0 and 1, and their sum is 1, making them interpretable as a probability distribution.

The generator's network structure is more complex. It includes two 1D convolutional layers (Conv1D) to capture local features, followed by two LSTM layers (one unidirectional and one bidirectional) to capture temporal dependencies. Such a design allows for the full use of different levels and directions of information in a sequence modelling task, thus improving the performance and generalisation of the model. Subsequently, the model comprises four densely connected fully connected layers, each followed by a LeakyReLU activation function and a BatchNormalization layer to ensure stable training of the network. In training mode, a Dropout layer follows the convolutional layers. The processed emotional label, corresponding rhythm, and noise input are fed into the network with a shape of (self.seq_length, self.latent_dim). The emotional label, as an additional input, initially passes through a fully connected layer and is then repeated self.seq_length times. The structure of the generator's network is described in Fig. 1.

Fig. 1. The picture shows the processing flow of each layer of the network in the model.

Next, the noise input, emotional label, and rhythm input are concatenated along the final axis. These data are then processed through a series of neural network layers in the generator, including densely connected layers with Leaky Rectified Linear Unit (LeakyReLU) activation and batch normalization. The generator's output is a 2D tensor representing the generated music sequence, where one dimension is activated using the hyperbolic tangent (tanh) function to keep values within the range of −1 to 1, while the other dimension remains linear. Subsequently, the generator and discriminator are combined into a comprehensive model for training. During training, it uses noise and emotional labels to generate sequences, which are then validated by the discriminator. The model aims to minimize the difference between generated sequences and real sequences and correctly label the emotions of the generated sequences. Different learning rates are employed for the discriminator and the combined model, with an Adam optimizer using a beta_1 value of 0.9. The utilized loss functions include binary cross-entropy for sequence authenticity and categorical cross-entropy for emotional labels. Hyperparameters may be adjusted in subsequent tests based on the result analysis.

3.3   Training and Generation Process

Model Training. After the aforementioned data and model preparation stages, a multi-epoch training loop was conducted. At the beginning of each training session, the model first establishes some ground truths: it marks real sequences as 'real' (represented by one) and sequences generated by the generator as 'fake' (represented by zero)[1].

Within the loop iterating through epochs, batch_size random integers ranging from 0 to len(self.sequences_tra) are chosen as indices to retrieve corresponding real music sequences, emotional labels, and rhythms from the database for training input. The discriminator is trained with sequences generated by the input, labeled as fake, and real sequences retrieved from the database, labeled as real. This approach is repeated in each training epoch, and the values of the generator and discriminator from the previous iteration are updated via backpropagation, improving the performance of the generator in the

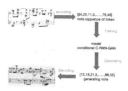

Fig. 2. Figure showing the process of encoding and decoding a sequence of notes into machine-readable data.

next iteration. Most intriguingly, even though the generator knows it is creating fake music, during the training process, we instruct it that these music sequences should be considered real. This strategy compels the generator to strive for creating increasingly realistic music. What's more, in order to cope with the problem of pattern collapse in GAN models, we uses an alternating training strategy when training the model. Every three times the generator is trained, the discriminator is trained, which not only avoids pattern collapse, but also prevents the database music information that the generator cannot adequately learn because the discriminator data is updated too quickly.

Model Generation. After training, we now have a model capable of generating music that matches a specific emotion. The authenticity, complexity, and creativity of the music are achieved through the mutual competition and learning between the generator and the discriminator. The generation process is somewhat similar to the training process, but this time, we rely solely on the generator without the need for the discriminator's assistance.

Given an emotion and random noise, the generator produces a piece of music. Generate a midi file of the same size as the input. The generated music sequence is scaled back to its original MIDI range and converted into a note sequence—transforming the MIDI file into a format recognizable by a computer (Fig. 2). For each note/chord:

- If it's a rest note ("R"), create a rest note and randomly set its duration.
- If it's a chord (e.g., "C4.E4.G4"), split the chord into individual notes and choose a random rhythm. Each note gets a volume and is assigned to a voice in the sequence.

---

[1] We refer to human-composed music as 'real' and music generated by the models as 'fake' from now on.

– If it's a single note, set a random rhythm and volume for the note and add it to the part.
– Update the note offset based on the duration of the note so that the next note knows where to begin.
– Finally, save the MIDI stream as a file, converting the predicted notes/chords and their corresponding rhythms into a playable MIDI file.

4   Results and Evaluation

4.1   Data Driven Evaluation

During the model learning process, an analysis of the loss rate change and the music segments generated at different epochs revealed that the model achieved optimal performance after 500 epochs of training. Therefore, it was decided to use this model trained for 500 epochs to generate music segments. The generated music is typically smooth and exhibits artistic aesthetic qualities.

From the charts of generator and discriminator loss rates (Fig. 3), it can be observed that the generator's loss rate decreases rapidly after 100 rounds of training, which is a positive sign. This indicates that the generator is getting better at deceiving the discriminator, and the generated data is becoming closer to the distribution of real data. The specific cross-training strategy we employed, training the generator three times and then the discriminator once, clearly had an impact. In contrast to the base C-RNN-GAN model, where it takes 250 training rounds to see a decrease in generator loss, our improved model made significant progress after just 100 rounds of training. This not only improves the training speed but also demonstrates the effectiveness of the generator in capturing data early.

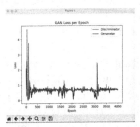

Fig. 3. Figure shows Generator and discriminator loss rate.

Evaluating the generated music is a highly complex and challenging task. This study employs both subjective and objective evaluation methods. This dual approach ensures that the generated music meets certain technical standards and is also artistically and emotionally appealing. This multi-faceted, multi-level evaluation strategy provides a more comprehensive and in-depth understanding, further optimizing the music generation model to align with aesthetic and emotional needs.

The goal of this study is to assess that the generated music sounds natural, harmonious, and emotionally rich. Therefore, rhythmic consistency, pitch range, and average chord count were chosen as objective evaluation metrics. Automatic analysis using muspy calculation tools and algorithms was employed to ensure that the generated music meets specific technical standards. By comparing the computed metrics of generated music to those of real music, we can evaluate the authenticity of the generated music using these parameters: *Pitch Consistency*: Measures the consistency of pitch in music relative to the key of C major, helping to understand the choice of tonality and harmony. *Pitch Range*: The difference between the highest and lowest pitches in the music, revealing the dynamic range and expressiveness of the music. *Pitch Range*: The range of pitches that the music can cover within a specific timeframe. *Average Chord Count*: The number of notes played at the same time on the same time scale, reflecting the complexity and texture of the music.

These metrics collectively provide us with a comprehensive and objective perspective to help evaluate the quality and characteristics of the music.

From the results above, we can observe that the gap in rhythmic consistency between real music from the database and the generated music has narrowed to within 1. Specifically, the rhythmic consistency of real music is approximately 3.47, while the rhythmic consistency of the generated music is 3.06. In terms of average chord count, the two are very close, differing by only 0.1. These data indicate that the generated music has made significant progress in simulating real music. However, we also noticed an interesting phenomenon: the pitch range value of real music is approximately 53, while the pitch range of generated music is around 72. This suggests that the latter has a broader pitch range, possibly indicating greater complexity or variability in terms of musicality, expressiveness, or emotional depth. These findings highlight the progress made by the generated music in terms of rhythmic consistency while also indicating its potential for greater complexity and variability in the pitch range, which may contribute to its musicality, expressiveness, or emotional depth. However, this needed testing at a human level of experience.

4.2   Human Subjectivity Evaluation

Subjective evaluation is necessary as objective assessments cannot comprehensively capture the emotional and affective aspects of music, given that a piece of music's appeal is related to the emotions and cultural backgrounds of the listeners. Subjective evaluation involves feedback from human listeners, including overall satisfaction, emotional experience, and willingness to listen again. Since everyone's musical background and aesthetics are unique, collecting diverse feedback is essential for ensuring the effectiveness and comprehensiveness of the study. To achieve this, 20 participants from different countries, age groups, and genders, with 60% female and 40% male, were selected for the survey. The majority of the sample were under the age of 30.

During the testing process, participants provided basic information, such as age, country, gender, and musical preferences. They listened to 10 music samples, each approximately 10 s long. These samples included music generated by the model representing different emotions (happiness, sadness, calmness, anger) and real music clips. The sources of the samples were kept confidential to eliminate potential biases. This design aimed to elicit the most authentic responses from the participants. After listening to each sample, participants were required to answer three questions: First, they categorized the sample into one of the four basic emotions. Then, they evaluated the music's harmony using a rating scale from 0 to 10. Finally, they assessed whether the music sounded like it was generated by a machine. These questions were designed to assess performance in emotion classification, musical harmony, and authenticity. Here is a detailed breakdown of participants' evaluations for each music audio in various aspects, including structural assessment, emotion classification, and whether they perceived the music to be generated by a machine [9]. The average scores for these evaluations are shown in Fig. 4.

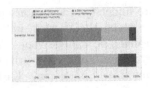

Fig. 4. Figure shows comparison chart of database music and generated music harmony.

The results show that the majority of users considered real music to have high harmony. Model-generated music performed well but primarily fell within the "moderately harmonious" range. In contrast, the proportion of real music rated as "extremely harmonious" is about twice that of model-generated music. Model-generated music was rarely rated as "disharmonious", while real music was almost never rated as "disharmonious". This suggests that both can achieve a certain level of harmony. We found that 65% of the test participants had an accuracy rate of less than 50% when trying to differentiate between model-generated music and real music from the EMOPIA database. This implies that they could not accurately distinguish the source of the music. To gain a deeper understanding, we interviewed those test participants with an accuracy rate of over 75%.

The results showed that even within the high-accuracy group, some individuals admitted that they were guessing when making judgments. From the results in Fig. 5, we can see that when judging music with happy and sad emotions, the audience generally has a high accuracy rate. The accuracy rates are 87% and 73%, respectively. However, music with calm and anger emotions is more challenging to distinguish compared to music with distinct emotions like happiness and sadness, with accuracy rates below 50%.

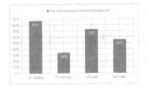

Fig. 5. Figure show how accurately different emotions are correctly distinguished in music.

In the objective evaluation, we found that the generated music is highly similar to real music in terms of average chord count and note consistency. This indicates that our model has successfully captured core musical attributes, including compositional principles and harmonic structures. From a technical perspective, this suggests that our model is highly accurate in terms of rhythm, structure, and melody, comparable to real music. From the audience's perspective, this high degree of consistency provides generated music with an auditory experience almost indistinguishable from human-composed music, increasing its value in practical applications.

5   Discussion

While our model has achieved good results in the above analyses, there are several issues we cannot ignore. The cross-training strategy we adopted helps maintain the dynamic balance between the generator and discriminator, accelerates the learning of the generator, and alleviates gradient vanishing. However, it is not a perfect solution, as a fixed training frequency may not perform optimally in some cases. Additionally, it also increases the model's sensitivity to other hyperparameters such as learning rates and batch sizes. Future research can be improved in various ways. Firstly, more flexible training strategies can be employed, adjusting the training frequency of the generator and discriminator based on performance metrics to adapt to different contexts and datasets. Introducing multi-objective optimization can also enhance overall performance, including generating quality, diversity, and interpretability. Additionally, reducing computational complexity and improving training efficiency is an important direction. A deeper understanding of the theoretical foundations of cross-training strategies can help design more efficient training methods.

Differences in the pitch range may stem from various reasons. Real music is typically created by experienced musicians who can convey profound emotions and themes within a limited pitch range, which is challenging for models to emulate. Furthermore, the model's complexity may be insufficient to mimic all the nuances of real music. While the model exhibits consistency in composition, rhythm, and melody, there is still room for improvement in pitch range details. The gap between the music generated by the model and the music in the database in evaluation results may be influenced by multiple factors. The music in the database is created by experienced musicians, meticulously designed, and the music generated by the model, although possessing some degree of harmony, may not match the complexity and intricacy of professional music compositions. Expanding the model to generate multi-track music is an important future task, as it can enhance emotional expression and increase the rhythm and harmony of the music.

In the user testing for this project, most of the information collected came from the under-30 crowd. User testing that focuses on the under-30 age group may give rise to the potential bias of limited representation. Due to the limited age range, we may have overlooked music preferences, emotional experiences, and cultural differences across age groups. To improve the comprehensiveness of the evaluation, we should broaden the audience to include users of different ages, cultures, and social backgrounds, and conduct stratified analyses to gain a deeper understanding of the feedback from each group. In addition, a deeper investigation into music preferences and emotional experiences, as well as the introduction of cross-cultural elements, will ensure that the assessment results are more representative and universal. Such improvements will contribute to a more comprehensive understanding of user feedback on the generated music system, improving its applicability and user experience.

6   Conclusion

In this study, we improved the C-RNN-GAN model by incorporating conditional GAN, creating a more powerful GAN model capable of deeply understanding the nuances of emotion and generating diverse music. Shi introduced emotion labels as additional inputs to establish a closer connection between emotion and music composition. This ensures that the generated music is not only structurally coherent but also emotionally consistent and accurate.

Cleverly chosen loss functions for the generator and discriminator enhanced the stability of the model's music generation, reducing the limitations of traditional GAN models. Through an alternating training strategy, the generator quickly learned to understand both musical note information and emotional hierarchies. The combination of subjective and objective analyses confirmed the effectiveness of our alternating training approach. The generated musical notes closely resemble those in the real music database, providing strong support for the practicality of the model. It also ensures that the generated music meets not only music theory standards but also individual emotional assessment needs. Through this comprehensive evaluation approach, we gain a more comprehensive understanding of the extent to which the generated music conveys emotions. This helps enhance the authenticity and emotional resonance of the model's creative output.

Furthermore, this work also points out several possible trends for future research. Expanding into multi-modal data generation involves integrating visual and auditory information into a comprehensive content creation approach [34]. This means that in the future, not only can music matching specific emotions be generated, but corresponding visual elements, such as images or videos, can be generated simultaneously. This will provide a richer and multi-layered sensory experience.

Audio Examples from the Research. https://sites.google.com/view/ai-musi-c/.

References

1. Alam, M., Samad, M.D., Vidyaratne, L., Glandon, A., Iftekharuddin, K.M.: Survey on deep neural networks in speech and vision systems. Neurocomputing **417**, 302–321 (2020)
2. Arjovsky, M., Chintala, S., Bottou, L.: Wasserstein generative adversarial networks. In: International Conference on Machine Learning, pp. 214–223. PMLR (2017)
3. Chen, K., Zhang, W., Dubnov, S., Xia, G., Li, W.: The effect of explicit structure encoding of deep neural networks for symbolic music generation. In: 2019 International Workshop on Multilayer Music Representation and Processing (MMRP), pp. 77–84. IEEE (2019)
4. Daskalakis, C., Ilyas, A., Syrgkanis, V., Zeng, H.: Training GANs with optimism. arXiv preprint arXiv:1711.00141 (2017)
5. Grekow, J., Dimitrova-Grekow, T.: Monophonic music generation with a given emotion using conditional variational autoencoder. IEEE Access **9**, 129088–129101 (2021)
6. Hadjeres, G., Nielsen, F.: Anticipation-RNN: enforcing unary constraints in sequence generation, with application to interactive music generation. Neural Comput. Appl. **32**(4), 995–1005 (2020)

7. Han, B.J., Rho, S., Jun, S., Hwang, E.: Music emotion classification and context-based music recommendation. Multimed. Tools Appl. **47**, 433–460 (2010)
8. Hung, H.T., Ching, J., Doh, S., Kim, N., Nam, J., Yang, Y.H.: EMOPIA: a multimodal pop piano dataset for emotion recognition and emotion-based music generation. arXiv preprint arXiv:2108.01374 (2021)
9. Imasato, N., Miyazawa, K., Duncan, C., Nagai, T.: Using a language model to generate music in its symbolic domain while controlling its perceived emotion. IEEE Access **11**, 52412–52428 (2023). https://doi.org/10.1109/ACCESS.2023.3280603
10. Ji, S., Luo, J., Yang, X.: A comprehensive survey on deep music generation: multi-level representations, algorithms, evaluations, and future directions. arXiv preprint arXiv:2011.06801 (2020)
11. Jolicoeur-Martineau, A.: The relativistic discriminator: a key element missing from standard GAN. arXiv preprint arXiv:1807.00734 (2018)
12. Juslin, P.N., Västfjäll, D.: Emotional responses to music: the need to consider underlying mechanisms. Behav. Brain Sci. **31**(5), 559–575 (2008)
13. Kamilaris, A., Prenafeta-Boldú, F.X.: Deep learning in agriculture: a survey. Comput. Electron. Agric. **147**, 70–90 (2018)
14. Khan, A., Sohail, A., Zahoora, U., Qureshi, A.S.: A survey of the recent architectures of deep convolutional neural networks. Artif. Intell. Rev. **53**, 5455–5516 (2020)
15. Kim, J., André, E.: Emotion recognition based on physiological changes in music listening. IEEE Trans. Pattern Anal. Mach. Intell. **30**(12), 2067–2083 (2008)
16. Krause, B., Lu, L., Murray, I., Renals, S.: Multiplicative LSTM for sequence modelling. arXiv preprint arXiv:1609.07959 (2016)
17. Kumar, V.B., Padmaveni, K.: A review on evolution of automatic music generation using machine learning techniques. In: AIP Conference Proceedings, vol. 2794. AIP Publishing (2023)
18. Liebman, E., Stone, P.: Artificial musical intelligence: a survey. arXiv preprint arXiv:2006.10553 (2020)
19. Mangal, S., Modak, R., Joshi, P.: LSTM based music generation system. arXiv preprint arXiv:1908.01080 (2019)
20. Mogren, O.: C-RNN-GAN: continuous recurrent neural networks with adversarial training. arXiv preprint arXiv:1611.09904 (2016)
21. Neves, P., Fornari, J., Florindo, J.: Generating music with sentiment using transformer-GANs. arXiv preprint arXiv:2212.11134 (2022)
22. Posner, J., Russell, J.A., Peterson, B.S.: The circumplex model of affect: an integrative approach to affective neuroscience, cognitive development, and psychopathology. Dev. Psychopathol. **17**(3), 715–734 (2005)
23. Rogozinsky, G., Shchekochikhin, A.: On VAE latent space vectors distributed evolution driven music generation. In: Proceedings of the 11th Majorov International Conference on Software Engineering and Computer Systems. MICSECS (2019)
24. Rubin, S., Agrawala, M.: Generating emotionally relevant musical scores for audio stories. In: Proceedings of the 27th annual ACM Symposium on User Interface Software and Technology, pp. 439–448 (2014)
25. Russell, J.A.: Measures of emotion. In: The Measurement of Emotions, pp. 83–111. Elsevier (1989)
26. Shah, F., Naik, T., Vyas, N.: LSTM based music generation. In: 2019 International Conference on Machine Learning and Data Engineering (iCMLDE), pp. 48–53. IEEE (2019)
27. Small, C.: Musicking: The Meanings of Performing and Listening. Wesleyan University Press (1998)

28. Smith, K.E., Smith, A.O.: Conditional GAN for timeseries generation. arXiv preprint arXiv:2006.16477 (2020)
29. Tsai, T.J.: Towards linking the lakh and IMSLP datasets. In: ICASSP 2020-2020 IEEE International Conference on Acoustics, Speech and Signal Processing (ICASSP), pp. 546–550. IEEE (2020)
30. Xie, J.: A novel method of music generation based on three different recurrent neural networks. In: Journal of Physics: Conference Series, vol. 1549, p. 042034. IOP Publishing (2020)
31. Yang, L.C., Chou, S.Y., Yang, Y.H.: MidiNet: a convolutional generative adversarial network for symbolic-domain music generation. arXiv preprint arXiv:1703.10847 (2017)
32. Yang, Y.H., Chen, H.H.: Machine recognition of music emotion: a review. ACM Trans. Intell. Syst. Technol. (TIST) 3(3), 1–30 (2012)
33. Zentner, M., Grandjean, D., Scherer, K.R.: Emotions evoked by the sound of music: characterization, classification, and measurement. Emotion 8(4), 494 (2008)
34. Zhang, H., Xie, L., Qi, K.: Implement music generation with GAN: a systematic review. In: 2021 International Conference on Computer Engineering and Application (ICCEA), pp. 352–355. IEEE (2021)
35. Zimmermann, J.B.: Jamendo: une plate-forme de musique libre en ligne. Entretien avec laurent kratz, pdg de jamendo. Terminal. Technol. l'inf. Cult. Soc. (102) (2008)
36. Zou, Q., Jiang, H., Dai, Q., Yue, Y., Chen, L., Wang, Q.: Robust lane detection from continuous driving scenes using deep neural networks. IEEE Trans. Veh. Technol. 69(1), 41–54 (2019)

Building an Embodied Musicking Dataset for Co-creative Music-Making

Craig Vear[1]([✉]), Fabrizio Poltronieri[1], Balandino DiDonato[2], Yawen Zhang[1],
Johann Benerradi[1], Simon Hutchinson[3], Paul Turowski[4], Jethro Shell[5],
and Hossein Malekmohamadi[5]

[1] University of Nottingham, Nottingham, UK
`craig.vear@nottingham.ac.uk`
[2] Edinburgh Napier University, Edinburgh, UK
[3] New Haven University, Newhaven, USA
[4] Liverpool University, Liverpool, UK
[5] De Montfort University, Edinburgh, UK

Abstract. In this paper, we present our findings of the design, development and deployment of a proof-of-concept dataset that captures some of the physiological, musicological, and psychological aspects of embodied musicking. After outlining the conceptual elements of this research, we explain the design of the dataset and the process of capturing the data. We then introduce two tests we used to evaluate the dataset: a) using data science techniques and b) a practice-based application in an AI-robot digital score. The results from these tests are conflicting: from a data science perspective the dataset could be considered questionable, but when applied to a real-world musicking situation performers reported it was transformative and felt to be 'co-creative. We discuss this duality and pose some important questions for future study. However, we feel that the datatset contains a set of relationships that are useful to explore in the creation of music.

Keywords: dataset · music performance · embodied AI

1 Introduction

To a musician, making music is about more than organising sound in space. Depending on the musician and the context, music-making can be a deeply felt embodied experience, one that shifts their understanding of self and their relationships with others. These bonds built on trust are augmented through creative engagement and play. Crucially, these relationships occur within music-making. Similar bonds are created when we listen to music, but, from the perspective of the performer, there is a two-way interplay that can lead to moments of creativity and novelty which are surprising, unexplainable, and meaningful to the individual.

This paper outlines a research project called the *Embodied Musicking Dataset*. Its aim was to build a dataset useful in training artificial intelligence (AI) for

C. Johnson et al. (Eds.): EvoMUSART 2024, LNCS 14633, pp. 373–388, 2024.
https://doi.org/10.1007/978-3-031-56992-0_24

co-creative and meaningful real-time music-making. It is important to stress that this is a proof of concept, as no precedent existed to guide the team. The concept of musicking [24] was used as a lodestone with which to build the foundational concept of embodied musicking. Introduced in more detail below, the premise is that 'musicking establishes relationships at its location, and therein lies its meaning.' [24]. From this foundational concept, we designed a dataset to capture certain musicological, physiological and psychological aspects of a musician's performance. In an attempt to build a gold-standard dataset, we formalised the process and used a single backing track that would act as a strict matrix for the dataset. We also designed a self-labelling process. 10 musicians each performing 2–5 solos, were employed, and the dataset was built.

In this paper, the authors evaluate the dataset from two perspectives: 1) a data science 'deep dive' seeking evidence of correlations or causation between the features across participants, and 2) a practice-based musical experiment that tested meaning-making in the co-creation of music with AI trained on the dataset. Our findings show that the data science techniques unearthed very little evidence of correlation or causation. The authors of this paper would even admit that, from this perspective, it is questionable. And yet, when applied to the practice-based experiments *inside* music-making, the experience for the musicians was transformational and the AI was perceived to be operating on a level of co-creativity expected from human musicians.

2   Related Work

2.1   Embodied Musicking

When musicians perform, they do not simply output sound into the world but engage in an embodied experience of *becoming* the sound they create in the flow of music-making [4,13,24,26]. Relationships between musicians and music flow are particularly evident in improvisational music like jazz, free, indeterminacy, or live electronics. Performers interact with represented ideas (scores and charts) while negotiating a constantly evolving state (real-time idea generation and sound invention). They act as autonomous agents (turn-taking) or simultaneous co-creators (joint playing). (e.g., [12,17]). Social and musical interactions create meaning in a participatory way inside the flow. Participation requires expressive alignment with these elements at deeper psychological and physiological levels than simply making a sound. [17] state, 'In order to share the act of producing and perceiving sound and movement, we need to examine the *embodied- inter(en)acted phenomenological experience* of music-making'. For the purposes of this research, we define the concept of *embodied- inter(en)acted phenomenological experience* [17] of music-making as simply *embodied musicking* using the following existing concepts:

Musicking. The composer and music theorist Christopher Small describes the embodiment of music as musicking. He defines it as:

to music is to take part, which can happen 'in any capacity, in a musical performance, whether by performing, by listening, by rehearsing or practising, by providing material for performance (what we call composing) [24].

Small stresses that 'the act of musicking establishes in the place where it is happening a set of relationships, and it is in those relationships that the meaning of the act lies'. Simon Emmerson clarified Small's principle of 'meaning' to infer the 'what you mean to me', (this subtle shift circumvents the significant issues of value and who is doing the evaluation of meaning) [4]. Therefore, meaning (or the what-you-mean-to-me) is to be found by examining the relationships within, across, through and emergent of the creative acts of musicking and the materials of these acts, e.g. *people, sound, space, and time.*

Embodied Music Cognition. As defined by Nijs et al. [21], within music performance the 'embodied interaction with the music implies the corporeal attunement of the musician to the sonic event that results from the performance' [21]. They further depict the embodied experience of participating in musicking as a direct, engaged approach, relying on perceiving the musical environment and skillfully coping with challenges (affordances and constraints) arising from the complex musical interactions [21]. For Nijs et al. [21], and as adopted by this research project, the optimal embodied experience (flow) occurs when the:

musician is completely immersed in the created musical reality (presence) and enjoys himself through the playfulness of the performance. Therefore, direct perception of the musical environment, skill-based playing and flow experience can be conceived of as the basic components of embodied interaction and communication pattern. [21]

Flow Theory. Csíkszentmihályi's Flow Theory (1975) supports Small's and Nijs's et al. argument that the acts of doing in music are to be considered an immersive and embodied experience. Csíkszentmihályi defined flow as 'the state in which people are so involved in an activity that nothing else seems to matter'. [2] He discusses how this deep engrossment in the here-and-now of action can occur in physical pursuits (athletes entering the zone) and in 'interactions with symbolic systems such as mathematics and computer languages' (such as concentration in computer game puzzles and video game immersion), [23] both of which describe parameters of immersion within live music performance with AI. Whilst Privette & Bundrick (1991) [22] focus on the 'intrinsically enjoyable experience' of flow (emotion being another quantifiable parameter of music performance), Csíkszentmihályi & LeFevre (1989) [3] and Massimini & Carli (1988) [18] argue for characterisation of flow as a balance of 'challenges and skills' proportionately beyond normal levels found in wakefulness. In short, through the act of musicking, musicians become embodied in the music through a sense of incorporation within their environment (the soundworld), shared effort, and a loss of awareness of their day-to-day wakefulness and corporeal self-consciousness.

2.2 Musicking Datasets

Most music datasets focus on two aspects of musicking: 1) the physical properties of sound (pitch, onset, timbral construction, dynamic warp from strict count etc.) or 2) the mechanics of music organisation (harmonic progression, melodic line prediction, long and short-term time-based feature extraction) [1,6,7,9,11,14,15,20,25]. However, there is no precedent for the design and development of a music dataset of embodied musicking. Recent works which built high-quality datasets that did have a direct influence on our project include the University of Rochester Multimodal Music Performance (URMP) dataset by Li et al. (2018) [15][1]. This is a gold-standard dataset for multi-modal musician analysis. It is primarily aimed at audio-visual analysis of music performance. However, it does not dig deep into the embodied nature of musicking. Also, the recording process of this dataset prioritised the quality of the recorded media, which, if adopted here, would have had a negative influence on the quality of the embodied performance of the musicians. *GrooVAE* (Google Magenta project) *Learning to Groove with Inverse Sequence Transformations* [8] is another gold-standard music performance dataset that influenced this project heavily. The focus is on extracting the features of drummers' groove in order to train a neural net to introduce a sense of humanisation to matrix-composed music such as midi-players and loop generators.

3 Methodology (for Building the Dataset)

3.1 Visual Model

The design model for our embodied musicking dataset (EMD), indeed our whole proposition, is based on capturing the multi-dimensional interrelationships of Embodied Musicking (Fig. 1). We have conceptualised these relationships as an interconnected matrix of the main components extracted from the theoretical proposition above. This concept was then used to identify the most significant human parameters to capture to create a dataset that adhered to the governing objectives of the project.

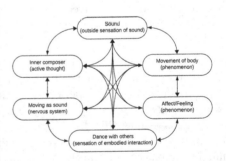

Fig. 1. Visual Model for Embodied Musicking

Subsequently, we determined which sensors to use and the most efficient way of designing the dataset-capturing environment (described below).

---

[1] http://www2.ece.rochester.edu/projects/air/publications/li2018creating.pdf.

3.2 Primary Considerations

We needed to merge individual data from each of the 10 musicians, into a single dataset, focusing on a single instrument, style, source (score), high musical proficiency, and linking performances with a single musical backing as the ground-truth temporal track, all within the funding scope. The properties we arrived at were: *Instrument*: Upright piano; *Style*: Jazz improvisation; *Proficiency*: professional jazz musicians; *Source*: "How Deep is the Ocean" by Irving Berlin; *Temporal ground-truth (backing)*: 2 different versions of a pre-recorded human bass player.

The backing track was a crucial element in binding the dataset into a consistent temporal framework and ensuring that elements across the different recordings and musicians were bound into a cohesive matrix. To this end, a bass player recorded two different versions as if he were working in a piano-bass duet. Both recordings were five choruses of the tune in the format: head, improv 1, improv 2, improv 3, head recap. There was a two-bar count in, an option for a click track (2 & 4) and no outro (i.e. the final chorus simply stopped at the start of the first bar of the next chorus).

The choice of 'How Deep is the Ocean' as the source score was determined through consultation with several professional pianists in advance of the recording. It was agreed that this track is a well-known standard and would, therefore, not require a great deal of rehearsal. Furthermore, its harmonic construction offered the consulted musicians enough interest to sustain repeated improvisations without the risk of repeated themes and material and facilitating in-the-flow engagement.

3.3 Dataset Design, Apparatus and Setup

The dataset was designed to encapsulate the parameters outlined in *the visual model* (Fig. 1). The sub-division consisted of:

Part 1: Physical-world Music: Backing track audio and associated score organisation (mono). Audio recording of the piano (mono). Video of hands and fingers (embedded with the audio track). A Windows laptop was used to capture all physical-world music data. An HD webcam was used to capture the hands and fingers of the pianist. The audio was captured using a USB microphone plugged into the laptop.

Part 2: Embodied Musicking: Electroencephalogram EEG [from BrainBit]. Electro-Dermal Activity EDA (arousal from Bitalino). Body tracking (using the Cubemos Skeleton SDK from the Intel Real Sense depth tracking camera). We used a Bitalino (r)evolution Bluetooth board[2] to capture EDA data from each musician. This streamed directly into the laptop for synchronous capture. This produced a single value. The BrainBit Headband[3] is a smart 4 channel EEG

---

[2] https://plux.info/kits/36-bitalino-revolution-board-ble-810121002.html.
[3] https://brainbit.com/.

sensors associated with T3 and T4 temporal lobe regions and O1 and O2 occipital lobe regions. With an Intel RealSense Depth Camera D455 we captured 12 body skeleton points. The camera is positioned in front of the musician and close enough to record both facial expressions and upper torso movement. After each recording, the musicians sit down and listen back to their recording and log their sense of depth of flow through that performance.

Part 3: Flow Protocol Analysis (post-recording): Self-flow-evaluation as a sliding scale. A single axis slider was used to log the depth of subjective/perceived musicking. This produced a single value and was used as a label.

A bespoke ecosystem was designed using mostly open-source technologies to support the data capture, ensure the data sync across all the sensors and preserve the data integrity. A custom software named "Blue Haze" was developed in Python captures data with a 10 Hz sample rate and stores them in MongoDB[4] database using JSON-like documents. Blue Haze is an open-source project available on GitHub and customisable to other dataset-building projects[5].

3.4 Flow Baseline

To gather an understanding of musicians' flow baseline in piano playing, we ask them to complete a *Flow Short Scale* questionnaire before they attend the recording session. This *Flow Short Scale* method has been previously used in music studies. For example, Haug et al. (2020) [10] and Martin et al. (2008) [16] to a sample of musicians. We used the *Flow Short Scale* of 13 items published by Engeser (2012) [5]. These items give us information about flow experience, perceived importance, performance fluency and activity absorption. Additional items look at demand, skills, and the perceived fit of demands and skills.

3.5 Dataset Contributors

We employed 10 solo improvising jazz musicians to contribute to the dataset: 5 from the UK and 5 from the US in the area of New Haven, Connecticut. We aimed for a mixture of flavours of jazz styles from commercial, through fusion, to modern. The recordings took place during the first 2020 COVID lockdown, with each musician contributing between 3 and 5 recordings to the dataset. We were strict about how we conducted these sessions and ensured that all preventative measures were implemented. The full data set can be found here https://rdmc.nottingham.ac.uk/handle/internal/10518.

4   Analysis of the Dataset. Test 1: Data Science

Using data science techniques, we sought to discover hidden relationships, correlations, and causation in the dataset. However, our conclusion is that there is

---

[4] https://www.mongodb.com/.

[5] https://github.com/Creative-AI-Research-Group/embodiedMusickingDataset/tree/master/blue%20haze.

limited evidence of these from this classic perspective. This analysis presented here, is adapted from a larger Master's project by Yawen Zhang at the University of Nottingham.

4.1 Data Preprocessing

In the Data Preprocessing phase, our focus was on ensuring the quality and completeness of the data collected from various musicians. To address missing values, we applied interpolation-based methods, carefully considering each scenario to avoid introducing biases—especially when a missing value might indicate an actual physiological or psychological state, it was retained as N/A. For physiological data like EDA and EEG, susceptible to external disturbances and equipment errors, we meticulously used statistical techniques and box plots for outlier detection, particularly scrutinizing anomalies at the start and end of performances. Data synchronization was a critical step, aligning different data types (physiological, psychological, and audio) accurately over time. This was achieved using a common backing track audio file for each performance, which provided a consistent time framework and enabled precise synchronization of data across multiple dimensions.

4.2 Data Analysis

In our project, the comprehensive exploration of varied datasets, encompassing EDA, EEG, and skeletal data, was critical. The initial phase of exploratory data analysis (EDA) was aimed at grasping the dataset's fundamental characteristics, identifying key patterns, anomalies, and relationships. This involved employing statistical methods like mean, median, and variance calculations to discern central tendencies and dispersions, complemented by visual tools such as histograms, box plots, and scatter plots to visualize data distributions and relationships between variables.

Fig. 2. Correlation Heatmap

This foundational work was integral for assuring data integrity and setting the stage for advanced analyses. It involved scrutinizing data distribution and addressing minimal missing values, particularly in essential metrics like 'flow', using median imputation to avoid biases and preserve statistical test validity. Subsequent stages of analysis ventured beyond basic descriptions, engaging in advanced statistical techniques and multifaceted data evaluations. These included deep dives into individual cases, cross-file comparative analyses using methods like ANOVA, and focused analyses employing clustering techniques. Each of these steps contributed to a holistic understanding of the dataset, revealing intricate details and complex interdependencies within the data.

A key part of our analysis was the use of a heatmap to illustrate correlations between variables (Fig. 2). We found strong inter-correlations among EEG channels (T3, T4, O1, O2), which is consistent with expectations for measures of brain electrical activity. Interestingly, the 'flow' state did not show strong correlations with physiological data like EDA or EEG, hinting that musician self-reports might be less reliable or influenced by complex factors beyond basic physiology. The skeletal data, specifically the X and Y coordinates for body parts, demonstrated significant correlations, suggesting patterns of synchronous movement. The limited correlation between 'flow' and physiological data underscored the need for a deeper, more intricate analysis to unravel the underlying relationships. Nevertheless, the heatmap was instrumental in laying the groundwork for comprehending the dataset's complexity.

Deep Dive into a Single File. In this phase, we explored a single performer's data, laying the foundation for future multi-file analyses. Our choice to analyze musician "Jn3VvBWcnDESzN9gUTh3bN" is significant. This musician has high data integrity, a "flow" score close to the dataset average, and noticeable variations in performances. Data is mostly complete, with just 1 and 3 missing values in sync-delta and flow attributes, respectively. We identified outliers using the Interquartile Range (IQR) method, mainly concentrated in skeletal data.

Analysis of Physiological Data and Flow. The trends of EDA and flow exhibit some parallels; at certain points where EDA peaks are evident, as visualized in the provided graph (Fig. 4) fluctuations in the flow state are also apparent, and occasionally, there seems to be a temporal lag between the peaks of the two. Acknowledging that EDA is influenced by factors like ambient temperature, humidity, and individual skin conductivity, we note that its peaks do not always correlate directly with "flow" states, suggesting that while EDA may indicate transient physiological changes during a performance, it may not fully capture the musician's psychological state. This complexity underlines the plausible

Fig. 3. Moving Window Average Time Series for the Four EEG Channels

but intricate link between EDA and the musician's psychological state, a connection we aim to explore further in subsequent analyses.

EEG data was constrained by the sampling frequency of 7.68 Hz, calculated as the inverse of the average sampling interval (130.153 ms), which only enables study frequencies up to 3.9 Hz (Nyquist-Shannon sampling theorem) and makes it challenging to conduct meaningful frequency domain analysis of EEG. Furthermore, EEG data is often susceptible to noise and may also be influenced by non-cognitive factors such as muscle movements, eye blinks, or even electrical interference. A moving window average of 10 s was employed during the trend analysis assisting us in observing the primary signal trends more clearly. Upon observation, we discern that the different EEG channels (T3, T4, O1, O2) exhibit highly similar patterns throughout the entire time span. The high correlation among the four channels is evident even without a formal correlation analysis (Fig. 3).

In our analysis of the motion data, significant issues were encountered due to a substantial number of data points being zeros, negative values, or having low confidence levels, leading to a large portion of data being deemed unusable. To confront these data irregularities, we explored various ARIMA (Autoregressive Integrated Moving Average) models, adjusting the parameters multiple times in an attempt to find an optimal fit for our non-stationary and volatile dataset.

The graph reflects the state of the motion data with one of the parameter sets we tested, where (p, d, q) = (1, 1, 10). This particular model configuration was an effort to mitigate the impact of unreliable data points while seeking to uncover any latent trends. However, even after processing with the ARIMA model, the data did not yield distinct "flow" patterns, underscoring the need for advanced analytical methods. Our choice of ARIMA was driven by its potential to model non-stationary time series data, a characteristic of the complex and noisy signals we encountered.

Fig. 4. Time Series Plot with Marked EDA Peaks

From the comparison, the musician's overall movement, as represented by the nose's position, doesn't show a clear correlation with their flow state. The dynamics of the nose's position might be influenced by various factors, and its interplay with the musician's psychological state (as represented by flow) isn't straightforward. The implications and potential reasons for these observations would warrant deeper exploration in further analysis.

Cross-File Comparisons. In our cross-file comparisons, we analyzed the data from multiple sessions to discern any overarching patterns across different musicians and performances. This involved a comparative analysis of EDA and 'flow' data, looking for consistencies and discrepancies that could inform our understanding of physiological responses during musical performances. The analysis found that EDA and 'flow' trends across different sessions and chorus IDs. The visualization captures the EDA and 'flow' data points for each chorus, scaled and overlaid to facilitate comparison. Through this graphical analysis, we can observe the variability and potential correlations between EDA responses and the reported 'flow' states across multiple performance sessions. The image underscores the diversity of physiological reactions and psychological states experienced by musicians, highlighting the need for a nuanced interpretation of these complex datasets.

Focused Single-File Analysis. After initial explorations and analyses, we delved deeper into specific file analysis to understand musicians' physiological reactions during specific tasks. Continuing with the second-stage analysis, our goal was to analyze their physiological reactions (EDA data) across five performances (i.e., five choruses). We used KMeans clustering method, an unsupervised machine learning method that groups similar data points together with different cluster numbers. In the analysis, three clusters emerged as the optimal choice. Their respective sizes were 38, 30, and 13, leading us to hypothesize Cluster 0: Encompassing the majority of bars, this suggests these bars share statistical similarities in EDA. Cluster 1: Bars in this cluster have a different EDA reaction than those in Cluster 0. Cluster 2: The smallest cluster bars here might represent sections where the musician showcases specific emotions or technical prowess. These bars, being fewer, might represent special or challenging parts of the track. Based on the clustering results, bars in Cluster 0 might predominantly belong to one or two choruses, representing sections where the musician showcases technical and emotional peaks. In contrast, bars in Cluster 2 might be spread across all choruses, acting as the "baseline" throughout.

4.3   Conclusion of Analysis 1

Through this analysis, we have come to recognize that there might be some form of correlation between music and physiological activity. Although no explicit patterns were discerned, potential connections and trends were observable from the data. The flow state is a unique psychological and physiological state linked to heightened focus, skill-challenge balance, and time perception distortion. Our findings suggest that when musicians enter a flow state, their physiological indicators might manifest patterns consistent with this psychological state.

5   Analysis of the Dataset. Test 2: Inside Musicking

To question the dataset's relevancy of purpose, we needed to evaluate it within the domain that it originated: namely, inside musicking. To this end, we designed *Jess+* an intelligent digital score system that uses AI and a robotic arm to amplify and communicate the creativity of an inclusive ensemble. This project was conducted in the real-world with professional musicians and in a collaboration between Orchestras Live (a national producer creating inspiring orchestral experiences for communities across England) and Sinfonia Viva (a British orchestra based in Derby, England).

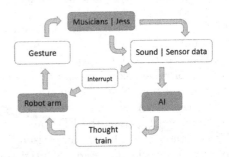

Fig. 5. Interaction Design

The digital score consisted of a robot arm, controlled by creative AI, which moved in a way much like a conductor or dancer might respond to music, in some experiments it drew on paper. The system consisted of a realtime feedback loop: live sound -> AI response -> movement -> human interpretation. The human ensemble consisted of a trio of musicians: Jess (disabled musician playing a digital instrument), Clare (violin) and Deidre (cello). The main goal was for the digital score to empower disabled musicians to engage in live music conversations despite disability-related barriers. Using a user-centered design and agile workflow, we created an AI-robot digital score with musician collaboration, following a closed-loop, realtime interaction design illustrated in Fig. 5. The design of *Jess+* is discussed in 2 separate papers, which are currently in production at the time of writing.

The Embodied Musicking Dataset was central in training a core part of the *Jess+* system. The AI Factory consisted of 7 convolutional encoder-decoder deep learning models with an hourglass-shaped architecture [19], trained to predict features from other features of the dataset including audio envelope, *core* position, EEG, EDA and flow, with the goal of this AI-stack to learn a representation of features with relation to each other in the context of musicking.

In realtime deployment, predicted data from deep learning networks was either fed to neighboring networks or used as raw data for deciding robot movements. These were predefined in advance and in consultation with the musicians. However, the AI was allowed to make critical decisions about speed, tempo, velocity, duration and interruption of each of these movements so as to surprise the human musicians. The team spent 5 months on the project, meeting the 3 musicians 5 times to test and develop the digital score. This culminated in a formal sharing with the partners using two types of mode: pens on paper (top photo in Fig. 6), and "dancing feather" (bottom photo in Fig. 6)[6].

5.1   Results

Through an extensive and iterative research process, we conducted a comprehensive qualitative investigation into musicians' reflections on using the digital score. Our findings revealed that the design decisions implemented in this case study enriched the musicians' experiences. They discussed how the AI and the robot were working with them inside musicking. And that its behaviours and interactions inspired relationships and bonding that they perceived to be co-creative. Though each musician had a unique connection with the robot, the disabled musician, Jess (who was also playing a digital instrument), felt a strong bond with the system during music-making, seeing *Jess+* as an extension of herself. She thought the extension's purpose was to visually represent the

Fig. 6. Final Performance

music, referring to it as a "friend" and "story-teller." Non-disabled musicians (playing cello and violin) recognized the Jess-system connection, labelling it a "creative accompanist".

---

[6] Here are some of the recordings from this sharing session: https://youtu.be/MB PQNmAXvXk https://youtu.be/7dQKIpjKJu4 https://youtu.be/sK4KAmv3ikw.

All musicians highlighted how they felt being in-the-loop with *Jess+* and how this transformed their own practices. Jess felt that the system allowed her to "express the emotions that she is sometimes not able to express through her current digital setup". She felt "extended through the system", which meant that she could "feel like she was able to express her feelings directly onto a score". This led to her wanting to "explore that part of me and I wanted - you know, I want my emotions that are in here to get expressed outwardly through that". For the non-disabled musicians, they reflected how

> The robot arm was liberating to improvise with as it was non-judgemental. At times, it united the three musicians' music, and at other times, it could also be independent from us (as we knew it would return to respond to what we later did). This, in turn, influenced the musicians to start or stop, to 'gel' together harmonically or feel the freedom to play outside harmonic or rhythmic frameworks. In my opinion, improvising with a human (especially someone new to you) carries psychological elements that could interfere with making music together, so the robot arm provided the opportunity for freedom of musical and emotional expression that would take much more time to establish and develop between humans".

5.2   Conclusion of Analysis 2

From an embodied and subjective perspective, our findings revealed that the musicians formed unexpected and distinct relationships with the creative-AI robot arm (*Jess+*). Although they viewed its role differently, they all acknowledged that it cultivated a set of relationships between the ensemble that a) augmented their creativity, and b) stimulated a sense of an inclusive ensemble without a hierarchy. They felt that an inclusive ensemble was formed involving disabled and non-disabled musicians *and* non-human musicians (*Jess+*). They also felt that they were in-the-loop with it *inside* musicking and that it had its own "voice". They were particularly taken by they way they felt that it listened to them and co-created with them, and significantly that it operated in a non-judgemental way. Discussed in much more detail in a forth-coming paper, they "viewed the system as an additional layer of creativity and felt empowered by its inclusive potential".

On the one hand, these insights shed new light on human-AI co-creativity and collaboration. On the other hand, however, the team does not know why the central AI, and its AI-factory design of deep learning models trained on the Embodied Musicking Dataset worked or was able to conjure such intense relationships as those experienced by these professional musicians. Simple good will or a lack of transparency were factors that were mitigated against through the research design and methodology. We suspect that it was to do with the musicking-behaviours embedded into the training data that were collected through improvised performance, and the musicking-focus behaviours embedded into the AI and symbolic algorithms.

6 Discussion

The experiences described by the professional musicians provide convincing evidence that something *musicking* was captured in this dataset. Classic data science techniques for extracting correlation and causation within the dataset, though, have been unable to point to why or how. This conflict brings forth more questions:

1. Can embodied musicking be represented in a dataset? We think this dataset provides some convincing evidence that it can. We are not certain that we were able to capture it in its entirety, but the approach and philosophy that we used to frame this proof-of-concept indicates a positive move forward. Musicking is more than just emitting notes into the air through time. There are emotions and affectual responses involved. There is also some brain activity involved, although we are not convinced that that is purely logical planning and reasoning (e.g. "I need to play this next", or "the next logical note to play is this"). We suspect that these decisions are more inherited and embedded into the whole mind-body system.

2. Should we be designing embodied musicking datasets using scientific principles? The EMD was built using a "gold-standard" approach. We had a single backing track, and focused on a single instrument, and invited professional musicians to contribute data using a standard system of instruments to capture their embodied musicking. Why? Surely, creativity and humanness are messy. The deep dive into the dataset through the classic data science techniques revealed very little to suggest that a "gold standard" approach was appropriate. There are some critical questions about the sample-rate at which the dataset was built (brain waves captured at 10 Hz are really only going to show limited information about mental activity), whether the EDA tracking of arousal was not standardised, and can the individual musician really be trusted to label their performance objectively (interesting to note that only recording they perceived to be "good" were allowed into the dataset, as such we have not data of a poor embodied performance). And yet, given these problematic concerns (from the perspective of data science), the AI that was implemented in *Jess+* transformed musicianship and creativity because the humans trusted, recognised and enjoyed the sensation of the AI.

This leads to our final question *3. what is really going on here?* Clearly, something is happening when we train neural nets using the EMD to co-create with human musicians, but what is it? Might it be the messy-ness of it that we recognise somehow as human? We mentioned in the introduction that, from a data science perspective, the EMD is questionable, and we are happy with that description. Perhaps there is some quality in its questionable-ness that we relate to: perhaps there is something in the way that the system was designed from an embodied perspective, and the AI-stack operated as a poetic random-number generator. Or perhaps the level of trust amongst the team and professional musicians convinced them that the AI was to be trusted, and they went along with it (although given the amount of discussion, challenge and debate in the practical sessions, we doubt this). Regardless, the next stages of our research need to lean into these questions.

7   Conclusion

In this paper, we shared findings from designing, developing, and deploying a proof-of-concept dataset aiming to capture physiological, musicological, and psychological aspects of embodied musicking. This dataset was then tested using a) classical data science techniques and b) practice-based techniques.

Data science revealed potential correlations between music and physiological activity. While no clear patterns emerged, we observed possible connections and trends in the data. Flow state has been verified as a unique psychological and physiological state associated with heightened focus, a balance between skills and challenges, and a distortion in the perception of time. Our findings suggest that when musicians enter a flow state, their physiological indicators might manifest patterns consistent with this psychological state. This offers a new view on understanding and promoting the flow state, implying that monitoring physiological indicators can identify and improve flow experiences. However, this testing also unveiled challenges. The low database sampling rate and skeletal data confidence issues hindered EEG and high-frequency data analysis. This calls for a more holistic approach in future research, incorporating a range of physiological, cognitive, and emotional measurement techniques.

In practice-based performance, the dataset trained a deep learning model and powered an intelligent digital score that extended the creativity of disabled and non-disabled musicians within an inclusive music ensemble. The professional musicians involved all acknowledged that the AI/robot was co-creative, that they felt in-the-loop with it and that it transformed their creativity. They recognised it as a co-creator, and a member of the ensemble, even a "friend". Their improvisations were open, enhancing and for all three, extending their techniques and confidence. In summary, we face a conflict: the dataset may seem flawed, yet we believe it holds something vital to musicking, a set of relationships that are somehow perceived to be from within musicking itself.

Acknowledgements. This dataset project was funded with a generous grants from the Human-Data Interaction Network funded through the Engineering and Physical Sciences Research Council grant number EP/R045178/1. This work on Jess+ funded by the ERC under the European Union's Horizon 2020 research and innovation programme (ERC-2020-COG - 101002086). Additional funding was received from TAS Hub and the Faculty of Arts at the University of Nottingham. We would like to address our sincere thanks for the support we received from Orchestras Live (https://www.orchestraslive.org.uk/) and Sinfonia Viva (https://www.sinfoniaviva.co.uk/), and of course our musicians Jess, Clare and Deirdre.

References

1. Cancino-Chacón, C.E., Grachten, M., Goebl, W., Widmer, G.: Computational models of expressive music performance: a comprehensive and critical review. Front. Digit. Humanit. **5** (2018). https://www.frontiersin.org/articles/10.3389/fdigh.2018.00025
2. Csikszentmihalyi, M.: Beyond boredom and anxiety: experiencing flow in work and play. San Fransisco, ca: Jossey-bass, p. 4 (1975)
3. Csikszentmihalyi, M., LeFevre, J.: Optimal experience in work and leisure. J. Pers. Soc. Psychol. **56**(5), 815 (1989)
4. Emmerson, S.: Living Electronic Music. Routledge, Abingdon-on-Thames (2017)
5. Engeser, S.: Theoretical integration and future lines of flow research. In: Engeser, S. (eds.) Advances in Flow Research, pp. 187–199. Springer, New York (2012). https://doi.org/10.1007/978-1-4614-2359-1_10
6. Friberg, A., Bresin, R., Sundberg, J.: Overview of the KTH rule system for musical performance. Adv. Cogn. Psychol. **2**, 145–161 (2006). https://doi.org/10.2478/v10053-008-0052-x
7. Friberg, A., Colombo, V., Frydén, L., Sundberg, J.: Generating musical performances with director musices. Comput. Music J. **24**(3), 23–29 (2000). https://doi.org/10.1162/014892600559407, https://direct.mit.edu/comj/article/24/3/23-29/93448
8. Gillick, J., Roberts, A., Engel, J., Eck, D., Bamman, D.: Learning to groove with inverse sequence transformations. In: International Conference on Machine Learning, pp. 2269–2279. PMLR (2019)
9. Giraldo, S., Ramírez, R.: A machine learning approach to ornamentation modeling and synthesis in jazz guitar. J. Math. Music **10**(2), 107–126 (2016). https://doi.org/10.1080/17459737.2016.1207814, https://www.tandfonline.com/doi/full/10.1080/17459737.2016.1207814
10. Haug, M., Camps, P., Umland, T., Voigt-Antons, J.N.: Assessing differences in flow state induced by an adaptive music learning software. In: 2020 Twelfth International Conference on Quality of Multimedia Experience (QoMEX), pp. 1–4. IEEE (2020)
11. Huang, C.Z.A., et al.: Music transformer: generating music with long-term structure. arXiv preprint arXiv:1809.04281 (2018)
12. Hytonen-Ng, E.: Experiencing Flow in Jazz Performance. Routledge, Abingdon-on-Thames (2016)
13. Leman, M.: Embodied Music Cognition and Mediation Technology. MIT Press, Cambridge (2007)
14. Lerch, A., Arthur, C., Pati, A., Gururani, S.: An interdisciplinary review of music performance analysis. arXiv preprint arXiv:2104.09018 (2021)
15. Li, B., Liu, X., Dinesh, K., Duan, Z., Sharma, G.: Creating a multitrack classical music performance dataset for multimodal music analysis: challenges, insights, and applications. IEEE Trans. Multimed. **21**(2), 522–535 (2018)
16. Martin, A.J., Jackson, S.A.: Brief approaches to assessing task absorption and enhanced subjective experience: examining short and core flow in diverse performance domains. Motiv. Emot. **32**, 141–157 (2008)
17. Martínez, I.C., Damesón, J., Pérez, J.B., Pereira Ghiena, A., Tanco, M.G., Alimenti Bel, D.: Participatory sense making in jazz performance: agents' expressive alignment. In: 25th Anniversary Conference of the European Society for the Cognitive Sciences of Music (Ghent, Bélgica, 31 de julio al 4 de agosto de 2017) (2017)

18. Massimini, F., Carli, M.: 16. the systematic assessment of flow in daily experience (1988)
19. Milletari, F., Navab, N., Ahmadi, S.A.: V-Net: fully convolutional neural networks for volumetric medical image segmentation. In: 2016 Fourth International Conference on 3D Vision (3DV), pp. 565–571. IEEE (2016)
20. Müller, M., Grosche, P., Wiering, F.: Automated analysis of performance variations in folk song recordings. In: Proceedings of the International Conference on Multimedia Information Retrieval, pp. 247–256, March 2010
21. Nijs, L., Lesaffre, M., Leman, M.: The musical instrument as a natural extension of the musician. In: Proceedings of the 5th Conference of Interdisciplinary Musicology, pp. 132–133. LAM-Institut jean Le Rond d'Alembert (2009)
22. Privette, G., Brundrick, C.M.: Peak experience, peak performance, and flow: correspondence of personal descriptions and theoretical constructs. J. Soc. Behav. Pers. **6**(5), 169 (1991)
23. Siekpe, J.S.: An examination of the multidimensionality of flow construct in a computer-mediated environment. J. Electron. Commer. Res. **6**(1), 31 (2005)
24. Small, C.: Musicking: The Meanings of Performing and Listening. Wesleyan University Press, Middletown (1998)
25. Todd, N.P.M.: The dynamics of dynamics: a model of musical expression. J. Acoust. Soc. Am. **91**(6), 3540–3550 (1992). https://doi.org/10.1121/1.402843, https://pubs.aip.org/asa/jasa/article/91/6/3540-3550/968369
26. Vear, C.: The Digital Score: Musicianship, Creativity and Innovation. Routledge, Abingdon-on-Thames (2019)

PatternPortrait: Draw Me Like One of Your Scribbles

Sabine Wieluch$^{(\boxtimes)}$ and Friedhelm Schwenker

Ulm University, 89069 Ulm, Germany
sabine.wieluch@uni-ulm.de
https://www.uni-ulm.de/in/neuroinformatik/

Abstract. This paper introduces a process for generating abstract portrait drawings from pictures. Their unique style is created by utilizing single freehand pattern sketches as references to generate unique patterns for shading. The method involves extracting facial and body features from images and transforming them into vector lines. A key aspect of the research is the development of a graph neural network architecture designed to learn sketch stroke representations in vector form, enabling the generation of diverse stroke variations. The combination of these two approaches creates joyful abstract drawings that are realized via a pen plotter. The presented process garnered positive feedback from an audience of approximately 280 participants.

Keywords: Generative AI · Sketch Data · Representation Learning

1 Introduction and Related Work

What exactly is a portrait? The most common association is a historical drawn painting of a person. Though portraits can be created through lot's of different types of media and art: sculptures, photography or film and also written portraits exist [9]. All these have in common that they want to capture essential features of the portrayed person like their appearance, personality, mood and sometimes also give an insight into the person's life story.

A portrait also always contains the artist's intent: this might be the truthful depiction of a person's life in a biography movie, emphasizing a person's beauty and wealth in classic historic paintings or exaggerating visual features in a caricature for entertainment. The missing intent is also why a quick photo snapshot usually is not considered a portrait.

For this paper we are aiming to create joyful abstract portrait sketches from images. These sketches should be drawn on paper by a drawing robot and also have a unique visual style.

Drawing robots have become essential tools for the digital art community, especially in the genre of generative art. Examples of well-known art projects and experiments have been created by Jon McCormack et al. [10], who built a swarm of small driving and drawing robots called "DrawBots". They utilize

C. Johnson et al. (Eds.): EvoMUSART 2024, LNCS 14633, pp. 389–400, 2024.
https://doi.org/10.1007/978-3-031-56992-0_25

evolutionary simulation to give each robot it's own aesthetic preference. Another artist is Sougwen Chung [3], who uses large industry robot arms to co-creatively paint large abstract paintings. Finally there is also "Sketchy" by Jarkman [1], which is a small Arduino-based drawing robot that can take pictures and draw very rough facial sketches using edge detection.

To give our portraits a unique aesthetic style, we aim to include generative patterns into the images as a form of shading. For this we draw single template sketches that act as reference for a neural net. This net is trained to learn the seen stroke shapes, recreate and also slightly alter them. These strokes are scattered in the darker areas of the image to create a unique shading style.

Freehand sketch representation learning, recognition and generation is a large and diverse research field [18], where we mainly focus on one-shot learning and derivate generation. A lot of different forms of data representation have been presented over the last years, though as we aim to use a drawing robot as output medium, research that mainly focuses on vector representation is of interest here. Sketch-RNN [5] for example uses a recursive neural net and interprets a sketch drawing as sequence of turtle-graphic-like moves drawn by a virtual pen. A more modern approach is our prior work StrokeCoder [16], which uses a transformer to learn freehand sketch data and is also able to create derivates, though only in a restricted context.

So for this paper we utilize graph neural nets in combination with graph convolution [8], which allows us to have a single vector representation of a full stroke and also benefit from the possibility of latent space exploration to generate stroke derivates.

2 Generating Sketch Portraits: System Overview and Constrains

"PatternPortrait" was born from the idea to create a drawing machine that would be capable of taking a snapshot of a person's face and then draw a quick, simplified portrait from that, injecting a unique style by shading with different patterns.

Our setup for this installation consists of a webcam mounted on a tripod to take pictures, an Axidraw V3 drawing robot in combination with a magnetic board to hold the paper in place. Both devices are connected to a notebook running the application. Images of this setup can be seen in Fig. 1.

Derived from the previously stated idea, we identified the following constrains for creating such a drawing bot installation:

1. The project's input is a regular pixel image but the output it drawn with a drawing robot (aka a pen plotter), therefore a transformation from one image type to the other is required.
2. The overall process of taking the picture, calculating the output image and the drawing process itself should not take longer than approximately 10 min to prevent boredom and frustration for the viewer.

(a) Used Devices   (b) Drawing in Process

Fig. 1. "PatternPortrait" setup: pictures are taken with a regular webcam in SD resolution. The drawings are created with an Axidraw V3 pen plotter.

3. The system should be capable of learning line styles from single template sketch drawings and arranging them into a simple pattern as shading. We aim for single template images, as we ourselves can not provide a dataset containing thousands of sketch drawings created by us. Also we promote the idea that AI systems should be transparent and ethical in their data usage, so we would like to construct systems that can work only on a single artist's data.

These three constrains will act as guidelines in the following sketch portrait generation process.

3 From Picture to Lines

The system's first task is to create lines from the initial input image. There are several approaches of creating artistic line drawings from images, for example through a reaction-diffusion [12] simulation, where parameters are changed depending on the pixel lightness or by calculating points of a weighted Voronoi net [14], which are then connected to a single line via approximating the Traveling Salesman Problem. Both examples can be seen as a form of dithering, as these approaches aim to also represent the original image's brightness distribution. We instead aim for an algorithm that only translates the main visible features and shapes, like people's facial or body features, into lines and so more closely resembles a human approach of drawing a face.

To achieve this translation from picture to lines, we first use canny edge detection [2] to retrieve a binary image that contains highlighted pixels belonging to a detected edge. In a next step, we need to identify pixels that visually belong to the same edge or line. This task is not trivial and several complex approaches [6,11] to solve this vectorization problem have been presented over the last years. Though for our artistic goal, we are not in need for a perfect exact result and therefore are using a very simple, greedy approach:

Every pixel is interpreted as a point $p \in P$ (with P being all found pixel points), located at the pixel's x and y coordinate. A random p is sampled from

P and it's closest neighbor point $p_{nbr} \in P$ is determined. If the distance between both points is smaller than a threshold $||p_{nbr} - p|| < d$, both points are connected to a path and removed from P.

Now this process is iteratively continued from both path ends: the next closest point $p_{nbr} \in P$ is detected and compared to the current path end point $||p_{nbr} - pend|| < d$. If the closest point to the path end is close enough, it is added to the path and so becomes the new end point and is also removed from P. This search is repeated until no point exists, that can be added to the path or P is empty. If this is true for both path ends, the path is considered finished and a new random point p is sampled from the remaining points in P until all points have been visited.

The so generated vector paths contain many unnecessary points which even make the path look jittery on close inspection. For this reason all extracted paths are simplified [13] via line fitting in a last step. An example processed picture can be seen in Fig. 2.

4   Generating Abstract Shading via Graph Neural Nets

After retrieving the vectorized canny edge detection lines, the only step missing is a form of shading. Here the goal is to search for dark areas in the original image and fill these areas with shapes generated from a single reference drawing.

To always include the full stroke in context, we decided to interpret each stroke as an undirected graph $G(V, E)$ which consists of n vertices v with $v \in V$ that are equally distributed along the stroke line. All vertices $v_1...v_n$ are ordered in stroke direction, meaning that v_1 represents the stroke start and v_n represents the stroke end. Also each vertex v has positional attributes $<\Delta x, \Delta y>$ that contain the positional delta to it's predecessor. Vertex v_1 acts as reference point and so receives the positional attributes $<0,0>$.

The graph also consists of $2(n-1)$ edges $e \in E$ where each vertex v_i is connected to the reference point v_1 and also to it's neighboring vertices v_{i-1} and v_{i+1} on the stroke line, if they exist. This graph structure allows for local neighborhood knowledge in each vertex v but also for global knowledge collected in vertex v_1 when graph convolution is used. A visual representation of the graph structure can be seen in Fig. 3.

The graph representation is then trained with a Graph Encoder, that shows similar structures to a variational autoencoder (VAE) [7]. The graph matrix of size $n \times 2$ (2 positional attributes Δx and Δy) is processed through two blocks of a graph convolution layer and a pooling layer. The graph convolution layer used is the same as described by Kipf et al. [8], which uses addition as the message passing aggregator. The pooling layer applies a pairwise one-dimensional max-pooling along the ordered vertices, which halves the matrix dimension. The result is then flattened and transformed into two vectors μ and σ via a fully connected layer. With help of the reparametrization trick by Kingma et al. [7], the latent vector z can be obtained from the distribution described by μ and σ^2. The decoding process of z is done via three fully connected layers and a final reshaping to receive the same dimensions on the output matrix as for the input.

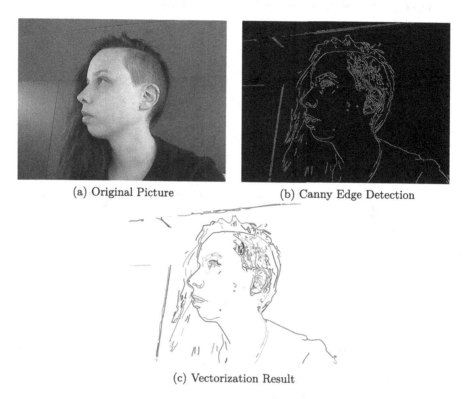

(a) Original Picture   (b) Canny Edge Detection

(c) Vectorization Result

Fig. 2. Vectorization process: on the original picture a canny edge detection is applied that is then vectorized by the algorithm described in Sect. 3. Each vectorized line is assigned a random color for better differentiation. (Color figure online)

Fig. 3. Graph representation of a single stroke, with the original stroke on the left. Then partitioned into n points of equal distance on the stroke and finally converted to a graph structure as described in Sect. 4. An equal distance of points was chosen to allow for the placement of a distinct amount of points. While a flexible distance determined by a fitting algorithm (for example the fitting used in StrokeCoder [16]) would likely result in a better approximation, the number of points used is harder to control.

The complete neural net structure can be found in Fig. 4:

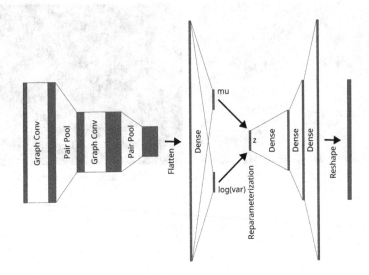

Fig. 4. Variational Graph Autoencoder Overview (Color figure online)

In a final step, the color space of the original image is reduced to only very few colors by choosing the most dominant colors and matching each pixel to this color palette. To achieve this, we use modified median cut quantization as described by Frederick at al [4]. Next, all pixels containing the darkest quantized color are chosen as possible area to fill with a pattern. In a final step, generated strokes from the trained autoencoder are placed on these pixels with the constraint that they must not touch any other already placed line. Utilizing the possibility to also generate slightly altered strokes helps the image to obtain a hand-drawn look. This process creates a random pattern that acts as shading in the portrait sketch, though not taking the stroke line placement of the original template sketch into account. We see many ways of possible improvements here, as described in future work.

5   Results and Discussion

With the use of variational autoencoders, the latent space is forced to represent a gaussian distribution whereas regular autoencoders create a manifold with unknown shape. This regularization comes with the benefit that latent space vectors can be sampled randomly and still represents to the learned model.

Figure 5 shows results from the trained vector line model: all lines in the top sample sketch have been used as training data. Their representation is learned well, as can be observed in the left line set: here the original lines are retrieved from the latent space and resemble the original shapes very well. It is now possible to add noise to the latent vectors to receive slight alterations of the original strokes. This works well, but also deformities can be observed (center set), especially visible on pointy triangles and circles. As mentioned above, it is possible to sample random latent vectors (right set). Here, most of the strokes are some mixed form of the original lines and do visually not resemble the original strokes well.

We decided to add slight noise to our latent vectors to benefit from the novel strokes but still keeping the original style recognizable.

Fig. 5. Latent space observations: a model is trained with all strokes in the sample sketch shown at the top. A first (left) set is generated by observing the decoded original strokes. The second set (middle) is produced by adding noise to the original strokes latent vectors. And finally a last set (right) is created by sampling random latent vectors.

Final results can be seen in Fig. 8, which are a subset of all 280 PatternPortraits drawn so far. Figure 6 shows four additional close ups, containing the final image retrieved from the picture shown in Fig. 2.

Besides using humans as depiction subjects, some further experiments were made with animals and inanimate objects. Though with an additional step of removing the subject's background manually, as the pictures were taken in visually very noisy environments. Exemplary results can be seen in Fig. 7.

While creating about 280 "PatternPortraits", we could observe recurring reactions from visitors that we summarize as qualitative results in the following section:

Fig. 6. Close up of four resulting "PatternPortrait" images. The bottom right image is the result from the process shown in Fig. 2.

(a) Poppy Seed Capsules   (b) Snail   (c) Succulent Plant

Fig. 7. "PatternPortrait" experiments with animals and plants as picture subjects.

Fig. 8. Subset of "PatternPortraits" taken at a festival on a bright and sunny day with white plain background.

- Either while plotting an image or when seeing a finished plot of another person: people were quick to identify and match the depicted face, often verbally discussing who is currently drawn or stating how well they think the person was captured in the final image.
- The pen plotter drew the final images by choosing the fastest way through all lines. This drawing order was often verbally described as non-human and even confusing.
- Most visitors very much enjoyed the abstract image style with all it's quirks and imperfections. Though also some visitors mentioned they did not feel represented well enough, especially when some of their facial featured weren't captured well by the edge detection or a pattern stroke was placed in a way where it interfered with their facial features. For example a large stroke is placed in the eyebrow area resulting in a visually heavily enlarged eyebrow.
- People enjoyed posing and experimenting in front of the camera and were overall curious to see the results. It occurred multiple times that a person created a "PatternPortrait" to then later come back with their friend group to create more. We interpret behavior as very positive and joyful attitude towards the project.

Besides visitor's reactions, we also would like to note some technical details:

- The canny edge detection can be run in real time, which es very helpful to give a preview to the visitors and overall for proper framing of the picture.
- The used webcam with low SD resolution (640 by 480 pixels) is enough to create "PatternPortraits". So no additional expensive camera equipment is needed.
- The canny edge detection kernel size and also the pattern stroke size and number are parameters that can be used to tunc the overall process time from taking the image to finished pen plot. We tuned the process to have an overall process time between 4 and 10 min, which we found a suitable waiting time for visitors while maintaining enough detail in the image.
- The vectorization process is rather slow, especially when visitors wear high-contrast clothing, which creates lots of extra lines to calculate. This could be optimized by either using a faster algorithm or even train a neural net to perform the task.
- The perceived gray value from the resulting patterns does not match the original image's perceived darkness. Also the pattern's gray value can shift dramatically depending on the strokes used and their currently random placement. Ways to optimize this behavior is discussed in the next section containing ideas for future work.

6   Conclusion and Future Work

In this paper we presented a process to create abstract portrait drawings from images that playfully utilize single freehand sketches as reference to create shading patterns. We showed a strategy to extract facial and body features and transform then into vector lines. Additionally we presented a graph neural net architecture that is capable of learning sketch stroke representations in their vector represented form and can generate variations of those strokes. Combining these two approaches results in interesting and joyful abstract drawings that were well received by about 280 participants.

While already receiving very positive feedback on the generated portraits, this project only resembles the first step into the creative usage of generative pattern models.

A possible next step is to not only learn the single line representations but also the overall arrangement of lines in the template sketch. This would allow to experiment with different (co-)creative tasks like automatic extension of patterns, outer shape-dependent pattern filling or the creating of co-creative drawing tools [17].

Continuing in the line of transferring pictures to abstract line drawings, it would be important to better control the gray value. This would not only heavily improve the overall image quality but would also be an interesting parameter to use in the above mentioned (co-)creative tasks.

A further improvement directly regarding "PatternPortrait" would be to include image segmentation to remove dark areas from the background. This would ensure that the depicted person is always the focus point and no unnecessary lines or patterns are created. Additionally, the face region could be excluded from the area of possible pattern placement to benefit the facial feature depiction and also prevent unnecessary pattern placement in case of a darker skin.

While in this project it would also have been possible to interpolate between sketch lines to retrieve derivative stroke lines, variational autoencoder and similar architectures encode more information than only the pure shape. Instead similar elements are clustered together. We hope to exploit this feature in future work for learning more complex patterns. Another interesting approach for future generative systems would be the implementation of diffusion models on vector-based sketch data [15].

References

1. Sketchy. http://www.jarkman.co.uk/catalog/robots/sketchy.htm. Accessed 01 Nov 2023
2. Canny, J.: A computational approach to edge detection. IEEE Trans. Pattern Anal. Mach. Intell. **6**, 679–698 (1986)
3. Chung, S.: Sketching symbiosis: towards the development of relational systems. In: Vear, C., Poltronieri, F. (eds.) The Language of Creative AI. Springer Series on Cultural Computing, pp. 259–276. Springer, Cham (2022). https://doi.org/10.1007/978-3-031-10960-7_15
4. Fredrick, S.M.: Quantization of color images using the modified median cut algorithm. Ph.D. thesis, Virginia Tech (1992)
5. Ha, D., Eck, D.: A neural representation of sketch drawings. arXiv preprint arXiv:1704.03477 (2017)

6. Hilaire, X., Tombre, K.: Robust and accurate vectorization of line drawings. IEEE Trans. Pattern Anal. Mach. Intell. **28**(6), 890–904 (2006). https://doi.org/10.1109/TPAMI.2006.127

7. Kingma, D.P., Welling, M.: Auto-encoding variational bayes. arXiv preprint arXiv:1312.6114 (2013)

8. Kipf, T.N., Welling, M.: Semi-supervised classification with graph convolutional networks. In: International Conference on Learning Representations (2017). https://openreview.net/forum?id=SJU4ayYgl

9. Maes, H.: What is a portrait? Br. J. Aesthetics **55**(3), 303–322 (2015). https://doi.org/10.1093/aesthj/ayv018

10. McCormack, J.: Niche constructing drawing robots. In: Correia, J., Ciesielski, V., Liapis, A. (eds.) EvoMUSART 2017. LNCS, vol. 10198, pp. 201–216. Springer, Cham (2017). https://doi.org/10.1007/978-3-319-55750-2_14

11. Noris, G., Hornung, A., Sumner, R.W., Simmons, M., Gross, M.: Topology-driven vectorization of clean line drawings. ACM Trans. Graph. **32**(1) (2013). https://doi.org/10.1145/2421636.2421640

12. Pearson, J.E.: Complex patterns in a simple system. Science **261**(5118), 189–192 (1993)

13. Schneider, P.J.: An algorithm for automatically fitting digitized curves. Graphics gems, pp. 612–626 (1990)

14. Secord, A.: Weighted voronoi stippling. In: Proceedings of the 2nd International Symposium on Non-photorealistic Animation and Rendering, pp. 37–43 (2002)

15. Wang, Q., Deng, H., Qi, Y., Li, D., Song, Y.Z.: SketchKnitter: vectorized sketch generation with diffusion models. In: The Eleventh International Conference on Learning Representations (2023). https://openreview.net/forum?id=4eJ43EN2g6l

16. Wieluch, S., Schwenker, F.: StrokeCoder: path-based image generation from single examples using transformers. arXiv preprint arXiv:2003.11958 (2020)

17. Wieluch, S., Schwenker, F.: Co-creative drawing with one-shot generative models. In: Romero, J., Martins, T., Rodríguez-Fernández, N. (eds.) EvoMUSART 2021. LNCS, vol. 12693, pp. 475–489. Springer, Cham (2021). https://doi.org/10.1007/978-3-030-72914-1_31

18. Xu, P., Hospedales, T.M., Yin, Q., Song, Y.Z., Xiang, T., Wang, L.: Deep learning for free-hand sketch: a survey. IEEE Trans. Pattern Anal. Mach. Intell. **45**(1), 285–312 (2022)

MAP-Elites with Transverse Assessment for Multimodal Problems in Creative Domains

Marvin Zammit$^{(\boxtimes)}$ (ID), Antonios Liapis (ID), and Georgios N. Yannakakis (ID)

Institute of Digital Games, University of Malta, Msida, Malta
{marvin.zammit,antonios.liapis,georgios.yannakakis}@um.edu.mt

Abstract. The recent advances in language-based generative models have paved the way for the orchestration of multiple generators of different artefact types (text, image, audio, etc.) into one system. Presently, many open-source pre-trained models combine text with other modalities, thus enabling shared vector embeddings to be compared across different generators. Within this context we propose a novel approach to handle multimodal creative tasks using Quality Diversity evolution. Our contribution is a variation of the MAP-Elites algorithm, MAP-Elites with Transverse Assessment (MEliTA), which is tailored for multimodal creative tasks and leverages deep learned models that assess coherence across modalities. MEliTA decouples the artefacts' modalities and promotes cross-pollination between elites. As a test bed for this algorithm, we generate text descriptions and cover images for a hypothetical video game and assign each artefact a unique modality-specific behavioural characteristic. Results indicate that MEliTA can improve text-to-image mappings within the solution space, compared to a baseline MAP-Elites algorithm that strictly treats each image-text pair as one solution. Our approach represents a significant step forward in multimodal bottom-up orchestration and lays the groundwork for more complex systems coordinating multimodal creative agents in the future.

Keywords: MAP-Elites · Quality Diversity · Image Generation · Text Generation · Text-to-image Generation · Digital Games

1 Introduction

Evolutionary search in creative domains such as visual, audio, or text generation has traditionally struggled to evaluate the artefacts it produces. This is mostly because there is no universal metric to assess the quality of media content [13,38]. Early approaches relied on ad-hoc metrics such as timing intervals in music generation [1] and compression-based indices for image generation [29], or tasked humans to evaluate the evolving population [20,27,41,45]. As more refined deep learning algorithms became available, models trained for a specific task have been employed more frequently as a fitness measure of evolved artefacts [21,40].

© The Author(s), under exclusive license to Springer Nature Switzerland AG 2024
C. Johnson et al. (Eds.): EvoMUSART 2024, LNCS 14633, pp. 401–417, 2024.
https://doi.org/10.1007/978-3-031-56992-0_26

For generative media, the most interesting development in the field of deep learning is the training and release of multimodal models. These models map multiple modalities to the same latent space, thereby enabling the direct comparison of different types of media. Contrastive Language-Image Pre-Training (CLIP) [35] was one such model which demonstrated excellent zero-shot image classification to any input set of semantic labels. Similar models map text with other modalities, such as audio [9], which in unison with CLIP (or similar models) may compare images to another modality via an intermediary text modality. Alternatively, models such as Meta's ImageBind [14] directly combine multiple input and output modalities into a single embedding space, which facilitates *multimodal generation* and assessment but also opens up new possibilities for cross-modal learning and transfer learning.

The advent of more nuanced metrics based on large-scale corpora (unimodal or multimodal) does not quite address the limitations of optimising a universal "quality" metric in creative domains. Such a singular drive may lead to a narrow view of human creativity which often builds on niches such as art movements, music genres, and literary paradigms. To address this, more recent research in evolutionary computation has focused on the diversity of the output instead of its quality [26], or combining the two in Quality Diversity (QD) algorithms [15,34]. QD algorithms promote diversity in the artefacts while maintaining some minimal criteria on quality [25,28] or keeping only the fittest individuals within each phenotypic niche [32]. The latter approach is followed in Novelty Search with Local Competition [26], which only compares neighbouring individuals to assess their (local) dominance. Another prominent QD algorithm is Multi-dimensional Archive of Phenotypic Elites (MAP-Elites) [32]. MAP-Elites partitions the solution space into a multi-dimensional grid (the *feature map*), where each axis represents varying properties within a specific behavioural characteristic (BC) or phenotypic trait of the solutions. Each cell stores the optimal individual (elite) according to the global fitness function, promoting only competition within the phenotypic niche. The most popular implementation of MAP-Elites operates in a steady-state fashion, selecting a parent among the elites (at random) and mutating it to produce an offspring. The offspring is then mapped to a cell of the feature map according to its BCs and may replace the elite in that cell if it has a higher fitness. As MAP-Elites illuminates a problem space, it is particularly apt for creative domains where it has already shown successes [2,3,8,12,15,49].

This paper applies the MAP-Elites algorithm to a multimodal creative domain, specifically generating text descriptions and cover images for hypothetical video games. To address this challenge, we propose an algorithmic improvement on QD search: MAP-Elites with Transverse Assessment (MEliTA). MEliTA introduces an inter-modal evaluation process that shares partial artefacts (e.g. image or text) among phenotypically similar elites in order to find more coherent pairings. This innovative approach enhances the creative co-evolutionary process, resulting in the discovery of fitter and more diverse outcomes.

2   MAP-Elites with Transverse Assessment

MAP-Elites with Transverse Assessment (MELiTA) is a variant of MAP-Elites designed to evolve multimodal artefacts to minimise incongruity between modalities. MELiTA builds on a number of assumptions, which in our use case revolve around two modalities but can scale to any number of modalities:

- each evolved individual (A) is a collections of N (separable) artefacts, each encompassing a single modality M_i, e.g. $A = \{a_{M1}, a_{M2}, \ldots\}$
- variation operators ($V_{M1}, V_{M2}, \ldots$) can be applied on each artefact type ($a_{M1}, a_{M2}, \ldots$) separately, potentially informed by other modalities but not modifying the other artefact types, e.g. $a'_{M1} = V_{M1}(a_{M1}, a_{M2}, \ldots)$ etc.
- there exists a function (f) that returns a value q indicating coherence between all modalities, e.g. $q(A) = f(a_{M1}, a_{M2}, \ldots)$
- there exist N functions ($g_{M1}, g_{M2}, \ldots$) each returning a value ($\beta_{M1}, \beta_{M2}, \ldots$) indicating properties of one artefact type ($a_{M1}, a_{M2}, \ldots$) separately, e.g. $\beta_{M1}(A) = g_{M1}(a_{M1})$, $\beta_{M2}(A) = g_{M2}(a_{M2})$, etc.

Following the above notations, MELiTA produces an N-dimensional archive of elites (N being the number of modalities in the artefacts), characterised by N behaviour characterisations ($\beta_{M1}, \beta_{M2}, \ldots$).

During the evolutionary cycle, an existing elite $E = \{e_{M1}, e_{M2}, \ldots\}$ is selected from the archive—the selection operator can be uniform selection or more sophisticated [10]. One artefact of this individual is chosen randomly (e.g. e_{M1}) and changed via the appropriate variation operator, creating in this example $e'_{M1} = V_{M1}(e_{M1}, e_{M2}, \ldots)$. The new artefact ($e'_{M1}$) is assigned a behaviour characterisation based on its modality (i.e. $g_{M1}(e'_{M1})$). A new individual E' is created by combining the new artefact with unchanged artefacts of the parent E, i.e. $E' = \{e'_{M1}, e_{M2}, \ldots\}$. In vanilla MAP-Elites (as applied in this paper), the individual E' would be compared with the elite (E'_{old}) with BCs $g_{M1}(e'_{M1}), g_{M2}(e_{M2}), \ldots$ in terms of q, replacing it if $q(E') > q(E'_{old})$ or occupying the cell if no elite exists for those BCs. In MELiTA, the artefact e'_{M1} is iteratively paired with the artefacts of other modalities of each elite $R = \{r_{M1}, r_{M2}, \ldots\}$ that occupies a cell with BC $g_{M1}(e'_{M1})$, producing a new candidate solution $R' = \{e'_{M1}, r_{M2}, \ldots\}$ and computing the coherence between modalities for the new individual ($q(R')$). The collection of candidate solutions $\boldsymbol{R'}$ along with E' are sorted by their coherence score q and form an ordered list of candidate solutions $\boldsymbol{L}$. In order, each member $\phi \in \boldsymbol{L}$ is checked against the occupying elite (ϕ_{old}) in a cell with the BCs of ϕ. If no such elite exists (the cell is empty), the individual ϕ occupies that cell and the process ends. If the current elite ϕ_{old} is worse than the new candidate ($q(\phi_{old}) < q(\phi)$) then ϕ replaces ϕ_{old} and the process ends; if the current elite is not worse, then the process continues with the next member in $\boldsymbol{L}$. This results in only one new individual being inserted into the archive (at most) per evolutionary cycle, and ensures that empty cells can also be filled in the archive (by individual E', if it is better than other alternatives).

The process of MELiTA will become clearer through the use case of Sect. 3.

3   Use Case: Generating Text and Visuals for Game Titles

As an exploratory use case, we select images and text as modalities for MEliTA, as both benefit from the availability of reliable AI generators. In this use case, the goal is to generate fictitious video games in the form of a *game title*, a *short description* of the game, and a *cover image*. Through this experiment, we wish to create a diverse set of coherent and appropriately game-like art and text blurbs that can inspire players and game developers alike.

For each modality, core considerations are how the artefact is generated (or changed via mutation) and how it is characterised for the purposes of the MAP-Elites feature map [32]. The sections below clarify how artefacts of each modality are generated and characterised, followed by a rundown of the MEliTA process.

3.1   Text Modality

The Generative Pre-trained Transformer 2 (GPT-2) [36] is an auto-regressive model based on the transformer architecture [48]. This model has undergone extensive pre-training through a substantial corpus of English text through a self-supervised learning approach [4]. While GPT-2 has since been eclipsed by more cogent models [6], particularly the more recent large-language models (LLMs) [33,46,47], it distinguishes itself with considerably faster inference, at the expense of reduced performance. For an iterative evolutionary algorithm, the substantial speed gain GPT-2 offers was considered a good trade-off.

Text Generation. To generate believable titles and descriptions for fictitious games, a GPT-2 language model was fine-tuned on a dataset composed of real titles and descriptions from the extensive catalogue of the Steam platform[1]. The data was curated, removing entries without English text, non-game entries (e.g. utilities, videos), and entries labelled with adult material or nudity. The resulting dataset contains approximately 72,000 pairs of game titles and descriptions.

For game titles, the pre-trained GPT-2 model of [36] was fine-tuned exclusively on video game titles of the above Steam dataset. Given the modest scale of this dataset, we used the most compact variant of the model (approximately 124 million parameters). The transformer was trained on the list of titles demarcated by distinctive beginning and end tokens: "`<|begin|>game title<|end|>`". First attempts exhibited over-fitting, and the model echoed existing titles. To mitigate this, the weights and biases of the last layer were reset prior to the fine-tuning process. This resulted in a more robust model capable of generating novel titles exhibiting minimal overlap with their original counterparts.

A second GPT-2 model was fine-tuned from the pre-trained model of [36], this time incorporating both title and description (see example in Table 1) in the format "`<|begin|>game title<|body|>description<|end|>`". A new game title can thus be generated through the first model using an input prompt of "`<|begin|>`". A description can be generated by priming the second model with this new title in the format "`<|begin|>game title<|body|>`".

---

[1] https://store.steampowered.com/.

Table 1. Text variation samples: Partial mutation chooses a space or punctuation at random, removes the text after it and uses the sequence up to that point to generate the remaining description (in red). Full mutation removes the description and only uses the game title (in bold) to generate the description (in red).

Original	**Hooey! You Got a Monster!?**, You've been selected for an experiment at the University of Chicago's Animal Research Center. A girl, who has lost her memory in this creepy and magical place will use your memories to discover why you got kidnapped by Professor Teller on his way there...
Partial mutation	**Hooey! You Got a Monster!?**, You've been selected for an experiment at the University of Chicago's Animal Research Center. A professor with no memories about his past finds you. An unforgettable story about two animals, their lives and what happened to them in this dream-like place.
Full mutation	**Hooey! You Got a Monster!?**, The best game of all time is back, and better than ever. This year we've reimagined the classic arcade platformer with over 100 levels to conquer. Gameplay has been revamped for even more fun, it's more difficult then ever but much harder in this new version!

Text Mutation. Two variation operators may be applied during mutation of text descriptions: one (partial mutation) retains coherence with the previous description while the other (full mutation) resets the description in order to avoid early convergence. For partial mutation, the first part of the description was retained, split along a selected space or punctuation mark around the middle of the text[2]. The rest of the description is removed. The game title together with the first description fragment are reintroduced as input to the second GPT-2 model to complete the description. For full mutation, the entire description is removed and only the game title is used as input to the second GPT-2 model to produce a new description from scratch. The chance of full mutation is 20%, otherwise partial mutation is applied. Text variation samples are shown in Table 1.

Text Characterisation. Several approaches to characterising text were explored for this use case, e.g. via the Gunning-Fog readability index [16]. We settled on topic modelling using topics extracted from the Steam dataset, as we want to match against dominant patterns for game descriptions, typical [38] of the domain we attempt to emulate. We used the Latent Dirichlet Allocation [5] algorithm (LDA) for topic modelling and trained the algorithm across different numbers of topics (from 4 to 30). Using the perplexity and complexity metrics as guides, we settled on an optimal setup of 16 topics.

To characterise each generated description, we subject it to the tuned LDA topic model, which results in a set of probabilities designating the likelihood of the description to be aligned with each of the predefined topics. In situations where the most probable topic assignment falls short of achieving a probability threshold of at least 40% above other topics, the description is deemed as *unclassified* and is not added to the feature map (i.e. it is ignored during evolution).

3.2   Image Modality

We leverage Stable Diffusion (SD) for the image generation tasks in this paper as it can produce high-quality images with low compute. SD [39] has openly

---

[2] We randomly select three spaces or punctuation marks within the text and keep the middle one. This makes it likely that the split will be in the middle of the description.

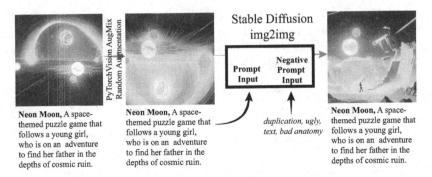

Fig. 1. Image variation sample: The parent's (unchanged) text modality is used as a prompt for image repair based on SD, alongside "standard" negative prompts.

accessible source code and pre-trained weights. Where diffusion models [44] are trained to reconstruct noisy versions of the desired output (e.g. image), SD models are trained on a noisy latent vector of the output, thus reducing inference time and improving robustness. Using a classifier-free guidance approach [19], both a conditional and an unconditional diffusion model are trained, and by comparing their responses during inference, a balance can be struck between image quality and faithful image adherence to the input prompt.

Image Generation. For this paper we utilise a text-to-image SD model to generate the cover art for the fictitious games. The input textual prompts contain both a hypothetical game title and a corresponding description, presented in the format "title, description" (see example in Table 1). Based on early experiments, we also add a set of negative prompts ("duplication, ugly, text, bad anatomy") to enhance the aesthetic quality of the generated outputs.

Image Mutation. As with text, we mutate the image on the phenotype level in two stages. First, we apply the AugMix augmentation function [18] from the TorchVision software library [30] to distort the original image. The distorted image is then paired with the original prompts (game title, description, negative prompts) as inputs to an image-to-image SD model (see Fig. 1). In order to strike a balance between image quality and computational efficiency, all tasks pertaining to image generation and mutation were executed over a sequence of 40 diffusion steps. Since the augmentation function may lead to more or less distorted images, resulting images after SD may match the parent image to a lesser or greater degree. Unlike text mutation, we have less control over the chances of a large phenotypic change, but consider such a change beneficial to avoid early convergence and a slow evolutionary process.

Image Characterisation. As with text classification we explored several BCs for images, but we settled on two straightforward metrics: complexity and colour-

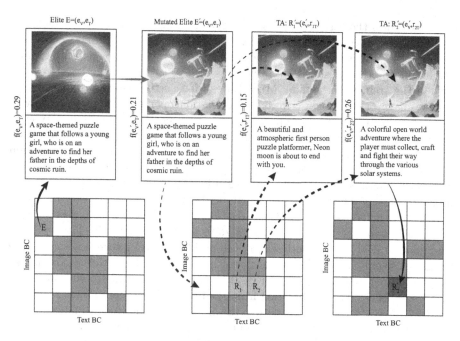

Fig. 2. The MEliTA process in a simplified feature map for this use case, with grey cells occupied by elites. From one selected elite E, the changed image (e'_V) produces three candidate solutions from elites E, R_1, R_2. Based on their CLIP score, the ordered list of candidates is $L = \{R'_2, E', R'_1\}$. Since $q(R'_2) > q(R_2)$ the candidate R'_2 (that merges the image from E' and text from R_2) replaces R_2. If $q(R'_2) \leq q(R_2)$ then E' would occupy the empty cell at (5, 0). Dotted lines denote temporary individuals that are lost after this parent selection.

fulness. *Image complexity* is calculated as the ratio of edge pixels found via the Holistically-Nested Edge Detection (HED) model [51] over the total pixel count. While previous approaches in evolutionary art relied on Sobel or Canny filters [29] for complexity estimation, HED edges offer a notable advantage in terms of accuracy and noise reduction. *Image colourfulness* is calculated via a quantitative measure of the perceived chromatic richness or saturation [17]. To combine the two image metrics into a concise BC, each numerical value (complexity, colourfulness) is categorised into four bins (very low, low, high, very high) and the image is classified in terms of the combinations of these bins. Therefore, the image BC comprises a total of 16 bins, matching the dimensions of the text BC.

3.3 MEliTA Applied to the Use Case

MEliTA in this bimodal use case performs parent selection via an Upper Confidence Bound (UCB) algorithm [23], which takes into account frequency of parent selection and can improve coverage of the final elites [42]. The feature map consists of a grid of 16×16 cells, with a text BC for membership in one of 16 topics found in the Steam database (see Sect. 3.1) and a visual BC for membership in one of 16 combinations of image complexity and colourfulness categories (see Sect. 3.2). This use-case uses the CLIP score between image and description (including the title as shown in Table 1) as fitness, to decide whether elites survive in the archive. In 'vanilla' MAP-Elites the selected parent produces either a new image or a new description via variation operators, then pairs it with the remaining artefact (description or image, respectively) in its genotype to produce a new offspring. However, Transverse Assessment operates as follows.

When a parent is selected in MEliTA, either a new image or a new description (chosen randomly) is produced from the parent's existing artefacts via variation operators described in Sects. 3.1 and 3.2. Figure 2 shows an example where the image (e_V) is modified (to e'_V), and will be described below. The new image e'_V is paired with the parent's existing text (e_T) and the CLIP score, text BC and visual BC coordinates of this new candidate solution E' are calculated (in this case, placing it at the currently empty cell at 5,0). For transverse assessment the image is also paired with the title and description of all elites with the same image BC of E' (the row with R_1, R_2 in Fig. 2), creating new candidate solutions; their CLIP score is calculated[3]. Starting from the fittest of these temporary solutions, if the CLIP score of an existing elite in the archive with these coordinates is lower than the new solution or if the cell is unoccupied, the candidate solution occupies that cell and the process stops (in this case replacing R_2). If no candidate solution is added to the archive, the process ends and a new parent selection is made.

4   Experimental Protocol

In order to evaluate the performance of MEliTA as a QD algorithm, we leverage the bimodal generation challenge described in Sect. 3 and aim to validate the high-level hypothesis that *MAP-Elites with Transverse Assessment leads to better quality and diversity in the generated artefacts than MAP-Elites without Transverse Assessment*. We formalise this into more concise hypotheses:

H1 The final archive of solutions is better and more diverse for MEliTA compared to MAP-Elites without transverse assessment.

H2 MEliTA discovers better and more diverse solutions faster than MAP-Elites without transverse assessment.

H3 The final archive of solutions for MEliTA can be perceived as more diverse by humans compared to MAP-Elites without transverse assessment.

H1 assesses the final product of evolution, while H2 assesses the performance of the evolutionary process over time. Both H1 and H2 assess diversity based on the ad-hoc BCs of the feature map, while H3 uses orthogonal diversity metrics which are aligned to human perceptions of diversity per modality.

---

[3] The BC coordinates for these candidate solutions do not need to be recalculated as they are combinations of text BCs and visual BCs that are already known.

Performance Metrics. To evaluate H1 and H2, we rely on traditional performance metrics for QD evolutionary search [32]. Specifically, for H1, we evaluate the final archive's mean fitness (among occupying elites), max fitness (i.e. the highest fitness among elites), coverage (i.e. ratio of occupied cells over all cells in the grid) and QD score (i.e. sum of all the elites' fitness). The first two metrics measure quality, coverage measures diversity (according to the chosen BCs) and QD score measures a combination of the two[4]. For H2, the same metrics are tracked over time (i.e. after every parent selection) and assessed as an area under the curve (AUC); high AUC values may mean that the final metrics have high scores or that reaching a high score was done earlier in the process. For H3, the final elites are assessed in terms of orthogonal measures of diversity to those used to populate the feature map. For visual diversity two measures are used: the Learned Perceptual Image Patch Similarity (LPIPS) measure is used as a distance metric trained on human annotations of visual distance [53], and the structural similarity (SSIM) measure commonly used for quantifying image degradation suffered through transmission or compression losses. For textual diversity in the descriptions, their embeddings generated by the SBERT encoder [37] are employed, since they capture the semantics of each sentence. Latent vectors of two individuals' descriptions are compared in terms of cosine similarity [11] to derive the *SBERT distance*. For each distance metric, we calculate for each elite the mean distance with all other elites, as well as the nearest-neighbour distance with the closest elite as a more reserved measure of visual/text overlap.

Test Cases. In order to verify these hypotheses, we follow the below protocol. We generated 100 game titles via the GPT-2 model described in Sect. 3.1 and select 7 titles which are varied in terms of theme and length. These titles are:

T1 "Neon Moon"
T2 "Lion King"
T3 "Hexgrave"
T4 "Fantasy Fables: The Legend of the Flying Sword"
T5 "The Princess of Thieves"
T6 "The Shadow Warrior 2: Shadows of the Past"
T7 "Hooey! You Got a Monster!?"

For each title, we perform 10 evolutionary runs per tested method (MEliTA and MAP-Elites). To provide a fair and controllable initial population for these methods, we use GPT-2 to produce 100 descriptions per title and for each description we produce 4 images via SD. In each evolutionary run, all text descriptions and a random image among the 4 candidates per description produce the initial 100 individuals which are then assigned to the feature map.

---

[4] Unlike [32], we do not normalise the values to the maximum found across runs and across methods. Instead, we present the non-normalised results (e.g. the ratio of occupied versus the maximum size of the feature map for coverage).

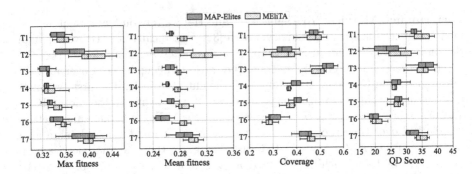

Fig. 3. Metrics of the archives after 2000 selections in MAP-Elites and MEliTA. Box plots summarise values from 10 runs per title.

5   Results

In order to validate the hypotheses of Sect. 4, MEliTA and MAP-Elites with no Transverse Assessment were ran with the same initial populations per game title (minor variation introduced through stochastic image selection as described in Sect. 4) for 2000 parent selections. Results throughout this section referring to the final archive are derived from the elites in the feature map after 2000 parent selections. Statistical significance between the results of 10 evolutionary runs of different methods is established via the non-parametric two-tail Wilcoxon Rank-Sum test, at a significance level $p < 0.05$.

5.1   Evaluating the Quality and Diversity in the Final Archive

Figure 3 shows the four performance metrics of the QD algorithm applied on the final archive (after 2000 parent selections). It is evident that overall MEliTA results in fitter individuals than MAP-Elites, with results from 6 of 7 game titles having a statistically higher mean fitness (all except T7) and 3 of 7 game titles having a statistically higher maximum fitness (T2, T5, T6). This indicates that the evolved solutions are overall more coherent between the visuals and the text, likely due to the fact that a generated text or image can be paired with another image or text from the archive that is a better fit than the image-text combination produced by MAP-Elites. This prioritisation of pairing artefacts from different elites together leads to a lower coverage of the feature space. While for many game titles this drop in coverage is slight, MAP-Elites has significantly higher coverage for 2 of 7 game titles (T3 and T5). Since MEliTA produces fewer but fitter elites compared to MAP-Elites, QD scores of the two methods tend to be comparable. The only significant difference in QD score is for T7 where MEliTA has a higher QD score than MAP-Elites. Based on this analysis, we can claim that MEliTA leads to better results at the cost, at times, of feature map coverage. Therefore, H1 is only partially validated—for quality but not diversity.

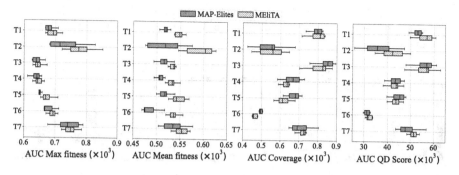

Fig. 4. Area under curve (AUC) of QD metrics over 2000 selections in MAP-Elites and MEliTA. Box plots summarise values from 10 runs per title.

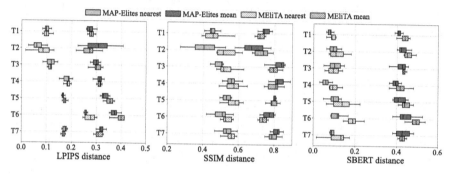

Fig. 5. Visual and textual distance metrics (mean and nearest-neighbour) among final elites of MEliTA and MAP-Elites without Transverse Assessment. Box plots summarise values from 10 runs per title.

5.2   Evaluating Quality and Diversity Throughout Evolution

Figure 4 shows the area under the curve (AUC) scores over 2000 parent selections for the different QD performance metrics. The findings from the AUC corroborate those of Sect. 5.1: mean and maximum fitness rises faster in MEliTA than in MAP-Elites (significantly so in 3 and 4 out of 7 game titles respectively). Coverage rises faster in MAP-Elites than MEliTA in most cases (significantly so in 2 out of 7 game titles), and AUC of the QD score is mostly comparable between the two methods; MEliTA has significantly higher AUC for the QD score only for T1. Based on these findings, we can claim that MEliTA can find fitter individuals quicker than MAP-Elites without Transverse Assessment. H2 is thus only partially validated—for quality but not diversity.

5.3   Evaluating Visual and Textual Diversity of Final Artefacts

To assess the perceivable differences in the two methods' archives, we use visual distances (via LPIPS and SSIM) and textual distance (via SBERT embeddings)

Table 2. Example output of MELiTA for the fictitious game title "The Shadow Warrior 2: Shadows of the Past" (T6). The description (after removal of the title) is left with its original errors.

Game 1	Game 2	Game 3	Game 4	Game 5
In this game, players have to use their sword and weapons wisely because there are many enemies who will try very hard for you. After many defeats in a long time, each one offers new challenges.	After a global catastrophe, you are left alone to face your past. The greatest evil - an alien race called Korda is planning on tearing Earth apart as they do every few centuries in this new adventure from Arkane Studios and Infamous Games.	Play as a shadow warrior, taking on different threats in this third person shooter that's more brutal than ever before.	In this game, players have to use their swords and bows in order not only survive but also fight with others.	After years' long struggle, Dark Lord Arthur's dark legacy is coming to light, and a new evil seems lurking at every corner. TheShadow Warrior2 : Shadows OfthePast.

of the final elites in each run of MELiTA and MAP-Elites after 2000 parent selections. Figure 5 shows the mean and nearest-neighbour distances for each of these metrics, which are largely orthogonal to the BCs used in MAP-Elites.

Figure 5 indicates that, as a whole, the final results of MAP-Elites are not more or less diverse than MELiTA. In terms of visuals, MELiTA has significantly higher mean and nearest-neighbour LPIPS distance than MAP-Elites in results of 2 game titles (T5, T6) and 1 game title (T5) respectively, and significantly higher nearest-neighbour SSIM distance than MAP-Elites in 2 game titles (T2, T3). While these findings are hardly consistent, they indicate that MELiTA tends to produce more diverse images than MAP-Elites. Clearer patterns are gleaned for textual diversity: MELiTA has significantly higher mean and nearest-neighbour SBERT distances in results of 3 game titles (T1, T2, T6) and 4 game titles (T1, T4, T5, T6) respectively. No game title has significant diversity improvements for MELiTA across all metrics. However, we can claim that H3 is validated: the final archives of MELiTA, even if smaller, are more visually and textually diverse than MAP-Elites without Transverse Assessment.

It is also interesting to observe differences in elites directly. We choose T6 ("The Shadow Warrior 2: Shadows of the Past") as it has high LPIPS and SBERT distances overall, and an indicative run that had high values for both. We apply k-medoid clustering using a Euclidean distance combining SBERT and LPIPS. The medoids for $k = 5$ are shown in Table 2. These samples' text descriptions vary, with some merely describing the gameplay (Games 1, 3 and 4) and some describing a science fiction narrative (Game 2) or a Camelot narrative (Game 5). Some text descriptions include "noise" (e.g. the end of the description in Game 5) while others overlap due to the partial mutation operation (e.g. Game 1 and Game 4). In terms of the images, most depict "action shots" which are likely not part of gameplay, although Game 5 does show a first-person shooter view (presumably in-game) complete with a gun within an otherwise fantasy setting. Depicted characters are mostly armoured and their faces covered, although it is unclear whether this is because of ninja tropes or because of the negative prompts on e.g. "bad anatomy". Characters are often seen wielding swords, although there are also modern guns shown in Games 3 and 5. While backgrounds mostly show a foggy forest setting, Game 2 deviates with warmer colours and fiery "wings" on the character at the image centre. Overall, while the images of these medoids do not depict as much diversity as one would expect (in terms of "ninja" characters and colour palettes), there is an overall consistency between the (sometimes generic) descriptions and the associated images. Perhaps the most concerning fact is that Game 2 includes names of actual game studios, likely due to the Steam dataset that GPT-2 was trained on; such text could be problematic due to intellectual property concerns if the use case would be made widely available.

6   Discussion

Through the use case for generating fictional game descriptions and cover images, we ascertained that MEliTA can produce fitter—if fewer—elites. We also recognise that coverage decreases since new elites are only added to the archive if there is no better alternative when pairing an offspring's changed modality with the modalities of an existing elite. This drop in coverage did not result in a drop in visual and textual diversity of produced elites when assessed on metrics decoupled from the BCs. Both in terms of mean distance and nearest-neighbour distance (which can counter the impact of a different number of elites), the elites of MEliTA were more visually and semantically diverse, as well as more coherent (higher fitness) than their counterparts produced without transverse assessment. The diversity metrics (SBERT, LPIPS, SSIM) represent the state-of-the-art for these purposes, but should be corroborated with human feedback. Future work could explore, via a user study with players and game developers, to which degree the generated multimodal artefacts are deemed diverse or inspirational.

It is worth noting that MEliTA produces far more candidate solutions (via transverse assessment) than the single offspring produced by MAP-Elites. This means that for MEliTA the number of fitness evaluations can be much higher than for MAP-Elites, although the number of BC evaluations is the same (only one individual with the parent's unchanged modality). For a 16×16 feature map, as in this use case, MEliTA may perform as many as 16 fitness evaluations per parent selection compared to 1 fitness evaluation for MAP-Elites. Calculating the CLIP score is not expensive, so the computational overhead of MEliTA is negligible; this may not be the case for simulation-based fitness evaluation [32].

The use case in this paper included a crude form of constrained optimisation [7], as the death penalty was applied on all unclassified text descriptions [31]. Additions to MEliTA could explore better ways of handling infeasible individuals, e.g. via a two-population approach as in [22,43]. Preliminary experiments using additional cells for unclassified individuals did not yield any substantial differences from MEliTA, but future work could combine constrained QD [43] with a minimal fitness threshold to distinguish feasible individuals.

Examples of the generated results in Table 2 also highlighted some limitations in the chosen variation operators. By operating on the phenotype (images or text) rather than on a latent representation, variation operators produce more controllable and less noisy output—compared to latent variable evolution [12,52]. Text variation specifically tends to cause overlap between individuals' descriptions due to partial mutation (see Sect. 3.1). Alternative text mutation operators could leverage models capable of generating text tokens in reverse [50], i.e. generating the first part of the descriptions given the end part, or use genotypic operators which alter the embeddings of the text [24]. Moreover, the GPT-2 trained model is admittedly dated in the current ecosystem of LLMs. While future work could explore a more modern text generator such as OpenAI's ChatGPT [33], the main challenge is its closed-source nature which hinders both fine-tuning its weights and keeping track of the provenance of its output.

While this use case explored a bimodal problem that was easy to visualise in a feature map with two axes, MEliTA could conceivably work with more artefact modalities and more BC dimensions—and those two do not need to necessarily match. Each artefact modality could easily have more than one BC; even in our case the two metrics for images (complexity and colourfulness) could have become different BC dimensions leading to a 3-dimensional feature map. Moreover, MEliTA could be applied to a more complex multi-modal creative problem such as evolving text, image, and audio for an interactive story application. In that case, an LLM can be used to generate the story and the locations over which it takes place, SD can be used to generate background images per location, and a text-to-music generator such as MusicGen [9] can generate a soundscape for each part of the story. In this case fitness can be assessed via a multimodal network such as ImageBind [14], by computing the vector similarity of each embedding.

7　Conclusion

This paper aimed to address the recent emergence of image-to-text and text-to-image generators through a QD perspective. To this end, we adapt MAP-Elites to operate on solutions that consist of artefacts of different modalities, and implement a transverse assessment method that allows such (partial) artefacts to be shared with other individuals. We show that MEliTA can outperform MAP-Elites, at the cost of fewer solutions. Extensions of this work should explore more state-of-the-art (but open-source) text generation algorithms while also extending the transverse assessment approach to incorporate quality constraints and integrate more types of artefacts and modalities.

Acknowledgements. This project has received funding from the European Union's Horizon 2020 programme under grant agreement No 951911.

References

1. Alfonseca, M., Cebrián, M., De la Puente, A.: A simple genetic algorithm for music generation by means of algorithmic information theory. In: Proceedings of the IEEE Congress on Evolutionary Computation, pp. 3035–3042 (2007). https://doi.org/10.1109/CEC.2007.4424858
2. Alvarez, A., Dahlskog, S., Font, J., Togelius, J.: Empowering quality diversity in dungeon design with interactive constrained MAP-elites. In: Proceedings of the IEEE Conference on Games (2019). https://doi.org/10.1109/CIG.2019.8848022
3. Alvarez, A., Font, J.: TropeTwist: trope-based narrative structure generation. In: Proceedings of the Foundations of Digital Games conference (2022). https://doi.org/10.1145/3555858.3563271
4. Balestriero, R., et al.: A cookbook of self-supervised learning. arXiv preprint arXiv:2304.12210 (2023). https://doi.org/10.48550/arXiv.2304.12210
5. Blei, D.M., Ng, A.Y., Jordan, M.I.: Latent Dirichlet allocation. J. Mach. Learn. Res. 3(Jan), 993–1022 (2003)
6. Brown, T., et al.: Language models are few-shot learners. In: Proceedings of the Neural Information Processing Systems Conference (2020)
7. Coello Coello, C.A.: Constraint-handling techniques used with evolutionary algorithms. In: Proceedings of the Genetic and Evolutionary Computation Conference (2010)
8. Colton, S.: Evolving neural style transfer blends. In: Romero, J., Martins, T., Rodríguez-Fernández, N. (eds.) EvoMUSART 2021. LNCS, vol. 12693, pp. 65–81. Springer, Cham (2021). https://doi.org/10.1007/978-3-030-72914-1_5
9. Copet, J., et al.: Simple and controllable music generation. arXiv preprint arXiv:2306.05284 (2023)
10. Cully, A., Demiris, Y.: Quality and diversity optimization: a unifying modular framework. IEEE Trans. Evol. Comput. 22(2), 245–259 (2017)
11. Dangeti, P.: Statistics for Machine Learning. Packt Publishing (2017)
12. Fontaine, M.C., Nikolaidis, S.: Differentiable quality diversity. In: Proceedings of the Neural Information Processing Systems Conference (2021)
13. Galanter, P.: Artificial intelligence and problems in generative art theory. In: Proceedings of the Conference on Electronic Visualisation & the Arts, pp. 112–118 (2019). https://doi.org/10.14236/ewic/EVA2019.22
14. Girdhar, R., et al.: ImageBind: one embedding space to bind them all. In: Proceedings of the IEEE/CVF Conference on Computer Vision and Pattern Recognition (2023)
15. Gravina, D., Khalifa, A., Liapis, A., Togelius, J., Yannakakis, G.N.: Procedural content generation through quality-diversity. In: Proceedings of the IEEE Conference on Games (2019)
16. Gunning, R.: The Technique of Clear Writing, pp. 36–37. McGraw-Hill Book Co. (1973)
17. Hasler, D., Suesstrunk, S.: Measuring colourfulness in natural images. In: Proceedings of the Conference on Electronic Imaging (2003). https://doi.org/10.1117/12.477378
18. Hendrycks, D., Mu, N., Cubuk, E.D., Zoph, B., Gilmer, J., Lakshminarayanan, B.: AugMix: a simple data processing method to improve robustness and uncertainty. In: Proceedings of the International Conference on Learning Representations (ICLR) (2020)

19. Ho, J., Salimans, T.: Classifier-free diffusion guidance. In: Proceedings of the NeurIPS Workshop on Deep Generative Models and Downstream Applications (2021)
20. Hoover, A.K., Szerlip, P.A., Stanley, K.O.: Interactively evolving harmonies through functional scaffolding. In: Proceedings of the Genetic and evolutionary Computation Conference (2011)
21. Johnson, C.G.: Stepwise evolutionary learning using deep learned guidance functions. In: Bramer, M., Petridis, M. (eds.) SGAI 2019. LNCS (LNAI), vol. 11927, pp. 50–62. Springer, Cham (2019). https://doi.org/10.1007/978-3-030-34885-4_4
22. Khalifa, A., Lee, S., Nealen, A., Togelius, J.: Talakat: bullet hell generation through constrained Map-Elites. In: Proceedings of the Genetic and Evolutionary Computation Conference (2018)
23. Kocsis, L., Szepesvári, C.: Bandit based Monte-Carlo planning. In: Fürnkranz, J., Scheffer, T., Spiliopoulou, M. (eds.) ECML 2006. LNCS (LNAI), vol. 4212, pp. 282–293. Springer, Heidelberg (2006). https://doi.org/10.1007/11871842_29
24. Lehman, J., Gordon, J., Jain, S., Ndousse, K., Yeh, C., Stanley, K.O.: Evolution through large models. In: Banzhaf, W., Machado, P., Zhang, M. (eds.) Handbook of Evolutionary Machine Learning. Genetic and Evolutionary Computation, pp. 331–366. Springer, Singapore (2023). https://doi.org/10.1007/978-981-99-3814-8_11
25. Lehman, J., Stanley, K.O.: Revising the evolutionary computation abstraction: minimal criteria novelty search. In: Proceedings of the Genetic and Evolutionary Computation Conference (2010)
26. Lehman, J., Stanley, K.O.: Evolving a diversity of virtual creatures through novelty search and local competition. In: Proceedings of the Genetic and Evolutionary Computation Conference (2011)
27. Liapis, A., Yannakakis, G.N., Togelius, J.: Adapting models of visual aesthetics for personalized content creation. IEEE Trans. Comput. Intell. AI Games 4(3), 213–228 (2012)
28. Liapis, A., Yannakakis, G.N., Togelius, J.: Constrained novelty search: a study on game content generation. Evol. Comput. 23(1), 101–129 (2015)
29. Machado, P., et al.: Computerized measures of visual complexity. Acta Physiol. (Oxf) 160, 43–57 (2015). https://doi.org/10.1016/j.actpsy.2015.06.005
30. Marcel, S., Rodriguez, Y.: Torchvision the machine-vision package of torch. In: Proceedings of the ACM International Conference on Multimedia (2010). https://doi.org/10.1145/1873951.1874254
31. Michalewicz, Z.: Do not kill unfeasible individuals. In: Proceedings of the 4th Intelligent Information Systems Workshop (1995)
32. Mouret, J.B., Clune, J.: Illuminating search spaces by mapping elites. arXiv preprint arXiv:1504.04909 (2015). https://doi.org/10.48550/arXiv.1504.04909
33. OpenAI: GPT-4 technical report. arXiv preprint arXiv:2303.08774 (2023). https://doi.org/10.48550/arXiv.2303.08774
34. Pugh, J.K., Soros, L.B., Stanley, K.O.: Quality diversity: a new frontier for evolutionary computation. Front. Robot. AI 3, 40 (2016)
35. Radford, A., et al.: Learning transferable visual models from natural language supervision. In: International Conference on Machine Learning, pp. 8748–8763. PMLR (2021)
36. Radford, A., et al.: Language models are unsupervised multitask learners (2019). https://openai.com/research/better-language-models. Accessed 11 Jan 2024
37. Reimers, N., Gurevych, I.: Sentence-BERT: sentence embeddings using Siamese BERT-networks. In: Proceedings of the Empirical Methods in Natural Language Processing Conference (2019)

38. Ritchie, G.: Some empirical criteria for attributing creativity to a computer program. Mind. Mach. **17**, 76–99 (2007)
39. Rombach, R., Blattmann, A., Lorenz, D., Esser, P., Ommer, B.: High-resolution image synthesis with latent diffusion models. In: Proceedings of the IEEE/CVF Conference on Computer Vision and Pattern Recognition (CVPR) (2022)
40. Roziere, B., et al.: EvolGAN: evolutionary generative adversarial networks. In: Proceedings of the Asian Conference on Computer Vision (2021)
41. Secretan, J., Beato, N., D'Ambrosio, D.B., Rodriguez, A., Campbell, A., Stanley, K.O.: Picbreeder: evolving pictures collaboratively online. In: Proceeding of the SIGCHI Conference on Human Factors in Computing Systems (2008)
42. Sfikas, K., Liapis, A., Yannakakis, G.N.: Monte Carlo elites: quality-diversity selection as a multi-armed bandit problem. In: Proceedings of the Genetic and Evolutionary Computation Conference (2021)
43. Sfikas, K., Liapis, A., Yannakakis, G.N.: A general-purpose expressive algorithm for room-based environments. In: Proceedings of the FDG Workshop on Procedural Content Generation (2022)
44. Sohl-Dickstein, J., Weiss, E., Maheswaranathan, N., Ganguli, S.: Deep unsupervised learning using nonequilibrium thermodynamics. In: Proceedings of the 32nd International Conference on Machine Learning (2015)
45. Takagi, H.: Interactive evolutionary computation: fusion of the capabilities of EC optimization and human evaluation. Proc. Inst. Electr. Electron. Eng. **89**(9), 1275–1296 (2001)
46. Touvron, H., et al.: LLaMA: open and efficient foundation language models. arXiv preprint arXiv:2302.13971 (2023). https://doi.org/10.48550/arXiv.2302.13971
47. Touvron, H., et al: LLaMA 2: open foundation and fine-tuned chat models. arXiv preprint arXiv:2307.09288 (2023). https://doi.org/10.48550/arXiv.2307.09288
48. Vaswani, A., et al.: Attention is all you need. In: Proceedings of the Neural Information Processing Systems Conference (2017)
49. Viana, B.M.F., Pereira, L.T., Toledo, C.F.M.: Illuminating the space of enemies through MAP-Elites. In: Proceedings of the IEEE Conference on Games (2022). https://doi.org/10.1109/CoG51982.2022.9893621
50. West, P., Lu, X., Holtzman, A., Bhagavatula, C., Hwang, J.D., Choi, Y.: Reflective decoding: beyond unidirectional generation with off-the-shelf language models. In: Proceedings of the 59th Annual Meeting of the Association for Computational Linguistics and the 11th International Joint Conference on Natural Language Processing (2021). https://doi.org/10.18653/v1/2021.acl-long.114
51. Xie, S., Tu, Z.: Holistically-nested edge detection. In: Proceedings of the IEEE International Conference on Computer Vision (ICCV) (2015). https://doi.org/10.1109/ICCV.2015.164
52. Zammit, M., Liapis, A., Yannakakis, G.N.: Seeding diversity into AI art. In: Proceedings of the International Conference on Computational Creativity (2022)
53. Zhang, R., Isola, P., Efros, A.A., Shechtman, E., Wang, O.: The unreasonable effectiveness of deep features as a perceptual metric. In: Proceedings of the IEEE/CVF Conference on Computer Vision and Pattern Recognition (2018). https://doi.org/10.1109/CVPR.2018.00068

Author Index

Printed in the United States
by Baker & Taylor Publisher Services